Climate Change and Carbon Markets

A Handbook of Emission Reduction Mechanisms

Edited by Farhana Yamin

from Routledge

First published by Earthscan in the UK and USA in 2005

Typeset by Pantek Arts Ltd., Maidstone, Kent
Cover design by Danny Gillespie

For a full list of publications please contact:
Earthscan
2 Park Square, Milton Park, Abingdon, Oxfordshire OX14 4RN
Simultaneously published in the USA and Canada by Earthscan
711 Third Avenue, New York, NY 10017

First issued in paperback 2014

Earthscan is an imprint of the Taylor & Francis Group, an informa business

A catalogue record for this book is available from the British Library

Library of Congress Cataloging-in-Publication Data

Climate change and carbon markets : a handbook of emission reduction mechanisms / edited by Farhana Yamin.
    p. cm.
  Includes bibliographical references and index.
  ISBN 1-84407-163-4
    1. Climatic changes–Government policy. 2. Emissions trading–Government policy. 3. Carbon dioxide mitigation–Government Policy. I. Yamin, Farhana.
QC981.8.C5C511346 2005
363.738'747–dc22
                                                2004022383
ISBN 13: 978-1-84407-163-0 (hbk)
ISBN 13: 978-1-138-00196-1 (pbk)

Contents

List of abbreviations   *ix*
List of figures, tables and boxes   *xiv*
Foreword   *xvii*
Preface and acknowledgements   *xxi*
About the contributors   *xxv*

Introduction   xxix
Farhana Yamin
Purpose   xxix
Structure of book   xxxii
The nature of the climate change problem   xxxvi
An overview of the UNFCCC   xxxvii
An overview of the Kyoto Protocol   xxxix
An overview of the Kyoto mechanisms   xli

Part I: The international rules on the Kyoto mechanisms   1

Farhana Yamin
I.1   Introduction   1
I.2   Activities implemented jointly   11
I.3   Cross-cutting mechanism issues   15
I.4   Participation/eligibility requirements   19
I.5   Emissions trading   26
I.6   Clean Development Mechanism   29
I.7   Joint Implementation (Article 6)   53
I.8   Compliance procedures and mechanisms under the Protocol   61
Notes   67
References   73

Part II: The EU Greenhouse Gas Emission Allowance Trading Scheme   75

Jürgen Lefevere
II.1   Introduction   75
II.2   The EU burden-sharing agreement   77

II.3 EU environmental policy: from command and control towards
    market-based mechanisms   81
II.4 The concept of emissions trading   86
II.5 The ethical dimension of emissions trading   92
II.6 The development of emissions allowance trading in the EU   95
II.7 Core elements of the ET Directive   101
II.8 The ET Directive and the Kyoto project-based mechanisms   126
II.9 Conclusion   139
Notes   139
References   148

Part III: Development and implementation of the Kyoto mechanisms worldwide   151

Chapter
1   Emissions trading under the Kyoto Protocol: how far from the ideal? 153
    Richard Baron and Michel Colombier

    1.1 Introduction and scope   153
    1.2 Emissions trading: focusing on economic efficiency   154
    1.3 Article 17 of the Kyoto Protocol: throwing governments into
        the cost-minimization game   157
    1.4 The EU Emission Allowance Trading Scheme: a step closer
        to the ideal?   160
    1.5 Conclusion   163
    Notes   164
    References   164

2   Trading through the flexibility mechanisms: quantifying the size of
    the Kyoto markets   166
    Odile Blanchard

    2.1 Methodology and assumptions   167
    2.2 Characteristics of the five cases   169
    2.3 Analysis of the market features of the five cases   171
    2.4 Conclusion   180
    Notes   181
    References   182

3   Implementation challenges: insights from the EU Emission
    Allowance Trading Scheme   183
    Fiona Mullins

    3.1 Introduction   183
    3.2 National Allocation Plans   184

3.3 Permitting procedures   196
3.4 Monitoring and verification   197
3.5 Registries   197
3.6 Conclusions   198
Notes   199

4 Joint Implementation and emissions trading in Central and Eastern
   Europe   200
   Jason Anderson and Rob Bradley

   4.1 Overview   200
   4.2 Interest in Joint Implementation and emissions trading in EITs 200
   4.3 CEE, international emissions trading and 'hot air'   203
   4.4 The preference for domestic action   205
   4.5 AIJ in CEE: early experience with projects   210
   4.6 Mechanism participation requirements and CEEs   213
   4.7 Early 'JI' experiences   215
   4.8 The future potential of JI   218
   4.9 European emission trading in Central and Eastern Europe 222
   Notes   229
   References   229

5 Implementing the Clean Development Mechanism and emissions
   trading beyond Europe   231
   Martijn Wilder

   5.1 Introduction   231
   5.2 Emissions trading (Article 17)   232
   5.3 Joint Implementation (Article 6)   244
   5.4 The Clean Development Mechanism (Article 12)   246
   5.5 International development agencies   259
   5.6 Other emissions trading activities   259
   5.7 Conclusion   261
   Notes   261

6 The Clean Development Mechanism: a tool for promoting long-term
   climate protection and sustainable development?   263
   Mark Kenber

   6.1 Introduction   263
   6.2 Assessing the CDM's contribution to sustainable development 265
   6.3 Tools to assess CDM project eligibility and sustainability 269
   6.4 Evolution of the CDM market   278
   6.5 Future issues and options   284

Notes   287
References   288

7   Determination of baselines and additionality for the CDM: a crucial
     element of credibility of the climate regime   289
     Axel Michaelowa

     7.1   Introduction   289
     7.2   Baseline determination   290
     7.3   Why baseline and additionality determination are not the same   296
     7.4   Conclusions   302
     Notes   303
     References   303

8   Creating the foundations for host country participation in the CDM:
     experiences and challenges in CDM capacity building   305
     Axel Michaelowa

     8.1   Introduction   305
     8.2   Capacity requirements to successfully implement the CDM   306
     8.3   Donor activities   310
     8.4   Challenges   317
     8.5   Conclusions   319
     Notes   320
     References   320

Part IV: Conclusion: Mechanisms, linkages and the direction of the future climate regime   321

Erik Haites

IV.1 Domestic GHG trading programmes   321
IV.2 Links among domestic GHG emissions trading programmes   328
IV.3 Direction of the future climate change regime   334
Notes   342
References   346

Appendices
1   Documents related to the EU emission allowance trading
     Scheme   353
2   EU Emission Allowance Trading Scheme Directive   355
3   EU Directive 2004/101/EC   371
4   EU Guidelines on Allocations of Allowances   383

Index   413

List of abbreviations

4AR	Fourth Assessment Report (IPCC – forthcoming 2007)
A6SC	Article 6 Supervisory Committee (for JI)
AAU	Assigned Amount Unit (under emissions trading)
ACEA	European Automobile Manufacturers Association
ADB	Asia Development Bank
AE	Applicant Entities
AGBM	Ad Hoc Group on the Berlin Mandate
AIJ	Activities Implemented Jointly
AIPs	Annex I Parties
ALGAS	Asia Least-Cost Greenhouse Gas Abatement Project (ADB)
AOSIS	Alliance of Small Island States
ASEAN	Association of South-East Asian Nations
ASPEN	Analyse des Systèmes de Permits d'Emissions Négociables
BAT	'best available techniques'
BAU	'business as usual'
BM	Bonn-Marrakesh Accords
BREF	BAT reference document
BTU	British thermal unit
CACAM	Central Asia, Caucasus, Albania and Moldova (negotiating coalition)
CAN	Climate Action Network
CBO	Congressional Budget Office (USA)
CCA	Climate Change Agreement
CCX	Chicago Climate Exchange
CDCF	Community Development Carbon Fund
CDET	Canadian Domestic Emissions Trading
CDM	Clean Development Mechanism
CDM-AP	CDM Accreditation Panel
CDM-AT	CDM Assessment Team
CEE	Central and Eastern Europe (UN regional group)
CEMD	Conservation and Environmental Management Division (MOSTE, Malaysia)
CER	Certified Emission Reduction
CERUPT	Certified Emission Reduction Unit Procurement Tender (The Netherlands)
CFC	chlorofluorocarbon (controlled under the Montreal Protocol)

CG	Central Group
CG-11	Central Group 11
CH_4	methane
CHP	combined heat and power
CO	carbon monoxide
CO_2	carbon dioxide
COP	Conference of the Parties (also under other MEAs, e.g. CBD)
COP/MOP	Conference of the Parties serving as the meeting of the Parties to the Kyoto Protocol
CPR	Commitment Period Reserve
DAC	Development Assistance Committee (OECD)
DC	developing country
DEFRA	UK Department for the Environment and Rural Affairs
DNA	Designated National Authority
DOE	Designated Operational Entity
EBCDM	Executive Board of the CDM
EBRD	European Bank for Reconstruction and Development
ECCP	European Climate Change Programme
EEA	European Economic Area/European Environment Agency
EEE	Eastern European Economies
EIA	Environmental Impact Assessment
EIG	Environmental Integrity Group
EIT	Economy in Transition (former Soviet Union and Eastern Europe)
ELV	emission limit value
ENB	Earth Negotiations Bulletin
ENGO	environmental non-governmental organization
EPER	European Pollutant Emission Register
EPS	Environmental Portfolio Standard (Arizona, USA)
ERPA	Emission Reduction Purchase Agreement
ERT	Expert Review Team
ERU	Emission Reduction Unit (under Article 6 projects – JI)
ERUPT	ERU Procurement Tender (The Netherlands)
ET	emissions trading
EU ETS	EU Emission Allowance Trading Scheme
EUA	EU allowances
FAO	Food and Agriculture Organization of the United Nations
FAR	First Assessment Report (IPCC – 1990)
FIELD	Foundation for International Environmental Law and Development
FSU	Former Soviet Union
G-77	Group of 77 (UN-wide negotiating coalition of developing countries)

GATT	General Agreement on Tariffs and Trade
GCOS	Global Climate Observing System (WMO programme)
GDP	Gross domestic product
GEF	Global Environment Facility
GERT	Greenhouse Gas Emission Reduction Trading Project (Canada)
GHG	greenhouse gas
GIS	Green Investment Scheme
GRULAC	Group of Latin America and the Caribbean (UN regional group)
GWP	Global Warming Potential
HCFC	hydrochlorofluorocarbon
HFC	hydrofluorocarbon
IACC	Inter-Agency Committee on Climate Change (Philippines)
ICAO	International Civil Aviation Organization
IDR	In-Depth Review (of an Annex I Party national communication)
IE	Independent Entity (under JI)
IEA	International Energy Agency
IET	International Emissions Trading (among Parties under Article 17)
IETA	International Emissions Trading Association
IGO	Intergovernmental organization
IMO	International Maritime Organization
INC	Intergovernmental Negotiating Committee for the UNFCCC (1990–1994)
IPCC	Intergovernmental Panel on Climate Change
IPPC	Integrated Pollution Prevention and Control
JAMA	Japanese Automobile Manufacturers Association
JDET	Japanese Domestic Emissions Trading
JI	Joint Implementation
JUSCANNZ	Japan, US, Canada, Australia, Norway, New Zealand
KAMA	Korean Automobile Manufacturers Association
KP	Kyoto Protocol
lCER	long-term CER
LDC	least developed country
LFE	large final emitter
LULUCF	land-use, land-use change and forestry
M&P	modalities and procedures
MEA	Multilateral Environmental Agreement
MoE	Ministry of Environment (Japan)
MOST	Ministry of Science and Technology (China)
MOSTE	Ministry of Science, Technology and Environment (Malaysia)

MoU	Memorandum of Understanding
MRET	Mandatory Renewable Energy Target (Australia)
N_2O	nitrous oxide
NAIPs	Non-Annex I Parties
NAP	National Allocation Plan
NCCC	National Committee on Climate Change (Indonesia)
NDRC	National Development and Reform Commission (China)
NEDO	New Energy Development Organization (Japan)
NGA	Negotiated Greenhouse Agreement (New Zealand)
NGGIP	National Greenhouse Gas Inventories Programme (IPCC)
NGO	non-governmental organization
NMVOCs	non-methane volatile organic compounds
NO_x	nitrogen oxides
NSS	National Strategy Studies (World Bank)
OAIC	Argentine Office of Joint Implementation (defunct)
OAMDL	Oficina Argentina del Mecanismo para un Desarollo Limpio
ODA	Official Development Assistance
ODS	ozone-depleting substances
OE	Operational Entity (Japan)
OECD	Organization for Economic Cooperation and Development
OPEC	Organization of Petroleum Exporting Countries
OPG	Ontario Power Generation (Canada)
PAMs	policies and measures
PCF	Prototype Carbon Fund (World Bank)
PDD	Project Design Document (CDM)
PERT	Pilot Emissions Reductions Trading Project (Canada)
PFC	perfluorocarbon
PIN	project idea note
POLES	Prospective Outlook on Long-term Energy Systems
PP	project participants
PPM	production processes or methods
QELROs	Quantified Emission Limitation and Reduction Objectives
REC	Renewable Energy Certificate (Australia)
REEA	Renewable Energy (Electricity) Act 2002 (Australia)
RES	Renewable Energy Source (Australia)
RGGI	Regional Greenhouse Gas Initiative (USA)
RMU	Removal Unit (under LULUCF)
SAR	Second Assessment Report (IPCC – 1996)
SB	subsidiary body
SBI	Subsidiary Body for Implementation
SBSTA	Subsidiary Body for Scientific and Technological Advice

SCAQMD	South Coast Air Quality Management District
SCC	small-scale projects (CDM)
SDPC	State Development Planning Commission (China)
SEPA	State Environmental Protection Administration (China)
SO_x	sulphur oxides
TAR	Third Assessment Report (IPCC – 2001)
tCER	temporary CER (under the CDM)
TEAP	Technology and Economic Assessment Panel (under Montreal Protocol)
TEC	Total Emissions Control (China)
TERI	The Energy Research Institute (India)
TFI	Task Force on Inventories (IPCC)
UN	United Nations
UNCED	UN Conference on Environment and Development
UNCTAD	UN Conference on Trade and Development
UNDP	UN Development Programme
UNEP	UN Environment Programme
UNFCCC	UN Framework Convention on Climate Change
UNIDO	UN Industrial Development Organization
UNITAR	UN Institute for Training and Research
URF	Uniform Reporting Format (under AIJ Pilot phase)
VER	Verified Emission Reduction
WEOG	Western Europe and Others Group (UN regional group)
WHO	World Health Organization
WMO	World Meteorological Organization
WSSD	World Summit on Sustainable Development
WTO	World Trade Organization
XA	Exchange Allowance (CCX)
XO	Exchange Offset (CCX)

List of figures, tables and boxes

Figures

I.1   CDM project activity cycle   32
2.1   The permit market equilibrium in the Initial Deal   173
2.2   Permit supply and demand in the Bonn-Marrakesh case   175
3.1   Levels of analysis for National Allocation Plans   185
4.1   Anticipated 2010 emissions for countries acceding to the EU in 2004

                                                                                 203
5.1   Approval process for CDM projects in Malaysia   251
6.1   Comparison of abatement costs, carbon credit prices and forecasts

                                                                                 283
7.1   Principle of the baseline   290
7.2   Status of methodology submissions   293
7.3   Baseline methodology submissions according to sector   294
7.4   Accepted baseline methodologies according to sector   294
7.5   Approaches used for electricity baselines   295
7.6   Different baselines for the same CDM project   297
7.7   CER price increase through additionality checks   298
7.8   Grubb's paradox   299
7.9   Hot Air supports Grubb's paradox   300
8.1   Tasks and actors within the CDM process   307
8.2   DNAs according to region   308

Tables

I.1   Kyoto mechanisms and related COP Decisions   3
I.2   Overview of the Kyoto flexibility mechanisms   10
I.3   Difference between Track 1 and Track 2 JI (Article 6)   58
II.1   Member State targets under the initial EU burden-sharing
       agreement (before the finalisation of the Kyoto negotiations)   79
II.2   Member State targets under the final EU burden-sharing agreement   80
II.3   Overview of key dates and documents relating to the development
       of the EU ETS Directive   95
II.4   Key dates for the implementation of the EU ETS Directive   102

2.1 Marginal abatement costs of a few countries 168
2.2 Assumptions of the five cases 170
2.3 Trade characteristics of the Initial Deal 172
2.4 Trade characteristics of the Missed Compromise 174
2.5 Supply and demand characteristics in the Bonn-Marrakesh case 176
2.6 Trade characteristics of the Committed Countries case 178
2.7 Trade characteristics of the European Union leadership case 179
4.1 A summary of potentially available AAUs for EITs during
     the commitment period 202
4.2 Kyoto targets, 1990 emissions, 2010 anticipated emissions 204
4.3 AIJ projects in EITs, by country 210
4.4 Projects and reductions estimated under AIJ 211
4.5 Project types in AIJ 211
4.6 Status of JI administratively in CEE countries 215
4.7 The scope for low-cost JI 218
4.8 Ranking of conditions for JI on three different scales 218
4.9 Status of Memorandum of Understanding between buyers and
     sellers of credits 219
6.1 Eight indicators of sustainable development 272
6.2 Gold Standard technologies 273
6.3 Key variables in the components of the sustainability matrix 275
6.4 Criteria and indicators in multi-attribute assessments of CDM
     projects 277
6.5 Breakdown of CDM project types 279
8.1 NSS completion 312
8.2 CDM awareness-building programmes 312
8.3 CDM institution-building programmes 314
8.4 CDM project development programmes 316
9.1 Hypothetical backstop intensity targets 323
9.2 Allocation of allowances to Firm Q for 2010 323
9.3 Compliance by Firm R in 2010 324

Boxes

I.1 Information and review provisions and stakeholders and
     observers in CDM project cycle 41
I.2 CDM additionality, baselines and crediting 44
I.3 Simplified modalities and procedures for small-scale
     CDM projects 47
I.4 Joint implementation: history and characteristics 54
II.1 Command and control legislation: the 1996 Integrated
      Pollution Prevention and Control Directive 81

II.2 Emissions trading: the ultimate market-based instrument? 83
II.3 Negotiated agreements in the EU 84
II.4 The EU's co-decision procedure 100
II.5 The German climate change agreements 105
II.6 The EU's comitology procedure 107
4.1 Evidence for caps and technology forcing 207
7.1 How to calculate grid emission factors 292

Foreword

Jos Delbeke

This book comes at a timely moment. Carbon markets are developing rapidly across the globe as a result of national and international climate policy initiatives. It is less than six months before the start of the European Union's Emission Allowance Trading Scheme and it is already clear that it will, in many respects, become a landmark in EU and international climate change policy.

The EU Emission Allowance Trading Scheme (EU ETS) demonstrates the EU's political commitment to the Kyoto Protocol and its innovative mechanisms. One should remember that the European Union for a long time was not very fond of market-based environment policy instruments. Even today, the major part of EU environmental legislation is based on 'command and control' through setting technical standards for individual companies, factories and consumers. The implementation of the EU ETS could become the first step to a deeper change of the EU's environment policy approach. In a world characterized by global economic integration and enhanced competition, using market-based mechanisms would drive environmental policy towards more cost-effectiveness while ensuring at the same time the achievement of its environmental objectives.

The EU ETS will be introduced three years in advance of the first Kyoto commitment period. For those that were involved in the legislative process, the Emissions Trading Framework Directive will certainly be remembered as one of the most ambitious pieces of EU legislation as regards its timetable and its scope.

Between reaching the final political agreement in July 2003 and the actual start on 1 January 2005, Member States and the Commission have only 18 months to transpose the Directive into national legislation, to implement all related secondary legislation such as the monitoring and reporting guidelines as well as the Registries Regulation, and to set up the electronic registry system and the EU's transaction log. This would not have been possible without the determination of many individuals in the Commission, the Member States, the European Parliament, the UNFCCC Secretariat and private sector as well as non-governmental organizations.

The fact that, during the first phase, only carbon dioxide and six sectors will be covered might give the false impression that the scheme is rather restrictive. The opposite is true: 12,000 to 14,000 industrial installations in 25 EU Member States will participate. More than 40 per cent of all CO_2 emissions in the Union will be covered. This will make it the largest emissions trading scheme worldwide.

One important reason for establishing the EU ETS is to engage private sector companies and the financial sector actively in the fight against climate change. No doubt, it will be the ingenuity and creativity of engineers, economists and financial specialists in these companies that will generate the new ideas and technologies that are necessary to reduce greenhouse gas emissions. Since Kyoto in 1997, those actively involved have grown from a handful of experts that negotiated the Kyoto Protocol and academics that fed them with ideas to hundreds of 'carbon professionals' that recently attended the first International Carbon Fair in Cologne. Many foreign companies that have operations in Europe are also becoming familiar with the idea of emissions trading.

Another key characteristic of the EU scheme is its openness to other schemes and to emission reduction projects abroad. This should reassure those that feared that the EU ETS is about 'Fortress Europe'. Instead, the EU is open to cooperate with others in the most constructive manner acknowledging that no country or political block in this world will be able to combat climate change on its own.

Norway, Switzerland, Canada and New Zealand have already expressed their keen interest to link up with the European scheme. Colleagues in Japan, Russia, Australia and the US are studying it with great interest. The aspiration for the EU ETS is to become the reference for a much larger international emissions trading scheme at the company level.

Something that many around the world have yet to realize is that the EU ETS already puts a value on CDM credits as of 1 January 2005. In the National Allocation Plans that have been scrutinized to date Member States pronounce already that they will buy more than 200 MT CO_2 over the first commitment period. This provides a clear signal to project developers around the world and will hopefully spur the transfer of technologies. It should inject a new dynamic into the debate of future technology development and transfer.

The EU ETS is also a good example of transatlantic cooperation. The Commission has learned a lot from the sulphur and NO_x trading schemes that are in place in the United States and excellent professional relationships have been built. After completing the legislative process in Europe and coming close to full implementation, an increased interest in many states in the US in emissions trading to combat climate change can be recognized.

On a final note, there is no doubt that, for the months and years to come, the EU will still be on the steep end of the learning curve as regards emissions trading. Nobody could realistically expect that the EU will get it 100 per cent right the first time. At the present moment we see sometimes fierce debates in the capitals of the

Member States on the allocation of emissions rights which will overnight create assets worth billions of euros. There is also anxiety in parts of business that, in future, emissions of carbon and other greenhouse gases would be priced and this is only the beginning of moving into a carbon constrained world. For others, the number of allowances that is allocated to industry is far too high. Time will tell. 'Learning by doing' is an important in-built feature of the Emissions Trading Directive.

For all these reasons, Farhana Yamin's handbook explaining emerging emissions reduction mechanisms in Europe and elsewhere comes at an opportune moment. She has been able to gather a very competent, highly professional group of authors. All of them have been deeply involved in the process over the last years, whether in the international and multilateral negotiations or in the debate in the European Union. Their contributions provide an excellent overview of emissions reduction mechanisms such as emissions trading and the detailed technical issues associated with their implementation across the globe. It will hopefully reach out and widen the audience to bring those on board that had not been involved in the matter in the past. I am confident, therefore, that this handbook will contribute to widening the understanding of emissions trading internationally over the coming years and how we came to where we are today.

Jos Delbeke
Director
DG Environment
European Commission
Brussels

Preface and acknowledgements

This book has taken more than a decade to write. Literally. My interest in the Kyoto mechanisms started on a plane journey coming back from the Earth Summit held in Rio de Janeiro in 1992. With a long flight ahead I began to browse through the countless documents I had collected but managed to finish only one: *Combating Global Warming: A Study on a Global System of Tradeable Carbon Emission Entitlements* published by the United Nations Conference on Trade and Development (UNCTAD, 1992). As a young lawyer dedicated to the progressive development of international law, and one searching for ways to advance the cause of vulnerable members of the Alliance of Small Island States (AOSIS) I had begun advising, utilizing permits and markets to protect the global climate seemed a progressive and intriguing idea – if a little far fetched and risky. Writing in our journal a year later, and with the benefit of discussions with other FIELD lawyers working on climate change, Philippe Sands, James Cameron and Jake Werksman, I expressed concerns about whether forms of global trading schemes could be agreed internationally, citing by way of example the difficulties the 12 members of the European Community were then having agreeing a limited scheme covering sulphur and nitrogen oxides among themselves. I would certainly not have believed that a decade later the international community would put over two hundred pages of rules operationalizing an array of global trading mechanisms on the international statute book, nor that the European Union would be on the verge of commencing the largest greenhouse gas emissions trading scheme in the world. This book is the end point of the amazing intellectual and professional journey that I have been on since that flight from Rio. A journey I have had the good fortune of sharing with many of the contributors to this book who have been working on the mechanisms issues, in many cases on joint projects and papers.

The chief purpose of this book is to describe the rules, institutions and procedures agreed internationally and at the European Union level to implement the three trading-based mechanisms included in the 1997 Kyoto Protocol: joint implementation, the Clean Development Mechanism and international emission trading, and to map out how current implementation and future evolution of these and related mechanisms is unfolding in jurisdictions outside the EU. Devising mechanisms rules has involved a huge amount of shared learning by the international community with many heated arguments along the way. In many ways I feel privileged that my differing professional roles have enabled me to see these debates from many perspectives: as

AOSIS advisor in the climate negotiations who along with other developing country negotiators remains a little cautious about the mechanisms, as consultant to the EU Commission advising the EU on the legal/policy framework to implement the mechanisms domestically and finally as an academic submitting myself to the cross-examination of students who always seem to split 50:50 into mechanisms enthusiasts and critics.

Involvement in these differing capacities also convinced me there was a need for a book that brought together mechanisms rules developed in different legal spheres. It was becoming increasingly difficult to keep track of the primary materials and sources of further reading. Somewhat ironically, the rejection of the Kyoto Protocol by the Bush Administration provided further impetus because it brought home the fact that linking the mechanisms under the Kyoto Protocol with current and future developments in non-Parties to the Protocol would require a better understanding of how domestic trading schemes could be designed to be Kyoto-independent as well as Kyoto-consistent.

The collection of contributions in this book is designed to support the implementation and evolution of the climate change regime by providing an accessible account of the rules, procedures and institutions governing access and use of the Kyoto flexibility mechanisms, and of catalysing discussions about the contribution these mechanisms might play in the design of the future climate regime.

My biggest thanks go to my fellow authors: Jason Anderson, Rob Bradley, Michel Colombier, Richard Baron, Odile Blanchard, Erik Haites, Mark Kenber, Jürgen Lefevere, Axel Michaelowa, Fiona Mullins and Martijn Wilder. All of them are busy mechanisms specialists advising a wide range of governments and intergovernmental and non-governmental organizations and all responded to my call to put pen to paper to explain the mechanisms rules they are working on right now to others less familiar. An extra special thanks goes to Jürgen Lefevere not only for his contribution in this book but also for being a wonderful colleague during the 1998–2002 period when the intensity of mechanisms discussions, internationally and within the EU, left us with little time to think about anything else.

My deep appreciation also goes to the main funder of this book, the EU Commission, DG Environment, for providing the financial incentives needed to write it, and also for stewarding the European Emission Allowance Trading Scheme through the EU legislative labyrinth in record time so that we would have something exciting near home to write about! Well deserved thanks are due to the past and present members of the Climate Change Unit with deep respect and appreciation of the role played by Jos Delbeke, Peter Vis, Damien Meadows, Matthieu Wemaere, Peter Zapfel, Yvonne Slingenberger, Artur Runge-Metzger and Matti Vianii, to name but a few, and also to other climate change experts who have over the years provided good advice and always found time to answer difficult questions, namely Michael Grubb, Fanny Misfeldt, Jan Corfee-Morlot, Kevin Baumert, Nancy Kete, Murray Ward, Christina Zumkeller, Bill Hare, Jennifer Morgan and Malte Meinshausen. I must also express my thanks to the many developing country delegates I have had the privilege of working with over the years, including Christiana Figueres, Xuedu Lu, Gao Feng, Raul Estrada, V. J. Sharma, Chow Kok Kee, John Kilani, Patricia Itturegui, Jose Miguez, Everton Vargas, Glyvan Miera Filho and, of course, the AOSIS leadership, Neroni Slade, John Ashe, Espen Ronnenberg and Philip Weech, who have helped me appreciate the complexities of turning policy into practice.

Former and present FIELD staff also deserve mention for providing research and administrative support for the work on the mechanisms and other aspects of the climate regime, and for assisting me to complete this project even when I moved in 2003 to the Institute for Development Studies at the University of Sussex where Oliver Burch has smoothed the final completion of the manuscript. My deepest thanks go as always to my husband, Mike Yule, for patiently putting up with mechanisms-related absences and long monologues about Kyoto and the momentum behind the carbon markets. I am not quite sure if we are there yet but we are a long, long way from where we started.

Farhana Yamin,
Fellow, Institute of Development Studies
University of Sussex, UK
30 March 2004

About the contributors

Farhana Yamin is a Fellow in Environment at the Institute of Development Studies, University of Sussex, England, where she undertakes research, postgraduate teaching and consultancy work on global environmental issues. Before joining IDS, Ms Yamin was Director of the Foundation for International Environmental Law and Development (FIELD) and Programme Director of its Climate Change and Energy Program from 1992 to 2001. From 1992 to 2002, Ms Yamin also represented AOSIS as a member of the Samoan delegation and was a senior negotiator and policy advisor on the Kyoto mechanisms. She has been a lead author for the IPCC for the SAR, TAR and 4AR and has coordinated several multi-partner research and policy collaborations for a number of international organizations, including most recently leading the consortia that worked with the EU Commission from 1999 to 2001 to define the legal/policy framework for the EU Emissions Allowance Trading Directive. With Joanna Depledge, Ms Yamin has recently co-authored *The International Climate Change Regimes: A Guide to Rules, Institutions and Procedures* which will be published by Cambridge University Press in 2004.

Jason Anderson works at the Institute for European Environmental Policy. He was a policy analyst at Climate Action Network Europe, the European node of CAN International where he coordinated NGO work on CDM and JI. He was also responsible for CAN Europe's work on energy market liberalisation, carbon capture and storage and F gases. He has previously worked for Enersol Associates in Honduras promoting the use of solar energy and at the Solar Energy Office of the United States Department of Energy. He has degrees from Harvard and Berkeley.

Richard Baron has worked as a climate change policy analyst at Iddri (institute du développement durable et des relations internationales). He works on climate change policy and international emissions trading at the International Energy Agency. He also worked on the economic modelling of GHG reductions at CIRED (France) and PNNL (USA).

Odile Blanchard is an Associate Professor in Economics at the University of Grenoble, France, and a researcher at LEPII (Production and International Integration Economics Laboratory), in the department of Energy and Environmental Policies (EPE, formerly IEPE). In 2001–2002, Dr Blanchard spent a year as a visiting fellow at the Climate and Energy Program of the

World Resources Institute (WRI), USA. Her research focuses on the economic analysis of climate negotiations and options to mitigate future climate change, as well as the development of analytical tools to support climate-related decisions. She is also undertaking a climate-friendly project at Grenoble university to define a strategy to reduce the university's greenhouse gas emissions.

Rob Bradley is an energy policy specialist at Climate Action Network Europe where he was an Associate at the World Resources Institute. Previously he was responsible for European NGO input into the European Union's Emission Trading System for the past five years. He has also been an analyst and NGO representative in the UNFCCC negotiations during that time. During the last nine years he has consulted for European governments, the European Parliament, the European Commission and private corporations on energy and environment issues. He has degrees from University College London and the University of East Anglia.

Michel Colombier, Deputy Director of IDDRI, is an energy policy specialist. He held positions at the French Ministry of Economy, the agency for the environment and energy efficiency (Ademe) and was a French delegate at the Kyoto Protocol negotiations. As a private consultant he has developed energy efficiency projects in developing and transition countries.

Erik F. Haites, of Margaree Consultants Inc., has contributed to the design of proposed emissions trading programmes in Canada, the UK, the European Union and the United States. He has assisted the UNFCCC Secretariat on issues related to the Kyoto mechanisms since 1998 and served as Head of the Technical Support Unit for Working Group III of the Intergovernmental Panel on Climate Change for the preparation of its Second Assessment Report.

Jürgen Lefevere is an administrator at the Climate Change and Energy Unit of the Environment Directorate General (DG ENV) of the European Commission in Brussels, Belgium. At the time of writing Part II, Jürgen Lefevere was Programme Director of the Climate Change Programme of FIELD, London. During 1999–2003 he was involved in a number of studies for the European Commission, including studies relating to the Green Paper on Emissions Trading (March 2000), the proposal for a Directive establishing a scheme for greenhouse gas emission allowance trading within the Community (October 2001), the proposal for a Directive to link the ET Directive with the project-based mechanisms (July 2003) and guidelines for the monitoring and reporting of greenhouse gas emissions.

Axel Michaelowa has been doing research on transboundary market mechanisms for the mitigation of greenhouse gas emissions since 1994. He is head of the Programme International Climate Policy at the Hamburg Institute of International Economics in Germany, associate editor of the journal *Climate Policy* and member of the UNFCCC roster of experts on CDM methodologies.

He is consulting for the German Technical Cooperation on the design of its CDM capacity-building programmes in several developing countries of Asia and Africa.

Fiona Mullins is an independent consultant specialising in climate change, with 13 years of experience in this field. She provides advice to governments and industry on a wide range of climate change issues. Fiona's recent work has focused on   analysis of the Kyoto mechanisms, including greenhouse gas mitigation policies, the design of emissions trading systems, emissions baseline issues and institutional capacity. Previously Fiona worked at the OECD where she carried out analytical work on climate change policies, including emissions trading. Fiona also worked for the New Zealand government assessing measures to respond to climate change.

Martijn Wilder is a Partner of Baker & McKenzie where he is responsible for the global climate change and emissions trading practice. Martijn advises national and state governments, major corporations and multilateral organizations. In particular Martijn is an adviser to the World Bank's Carbon Finance Unit, has been advising the European Commission, UNEP and leading financial institutions on climate change issues and CDM transactions, and is heavily involved in capacity building and developing climate change and emissions trading laws for a number of developing countries. Martijn is also the Adjunct Lecturer in International Environmental Law at the University of NSW. He has published widely in all areas of environmental and public international law. With honours degrees in both Economics and Law, Martijn has an LLM from the University of Cambridge which he attended as a Cambridge Commonwealth Trust Scholar.

Introduction

Farhana Yamin

Purpose

Climate change is widely acknowledged to be the most important environmental problem facing humankind. Because the atmosphere knows no boundaries and the world's economies are linked through trade and capital flows, international cooperation to curb greenhouse gases is essential. The institutional framework for such cooperation is provided by the 1992 United Nations Framework Convention on Climate Change (UNFCCC) which came into force ten years ago and by its supplementary 1997 Kyoto Protocol. The Protocol is one of the most complex treaties ever negotiated. At its core lie legally binding targets for the world's wealthier countries to reduce greenhouse gas emissions over 2008–2012. Countries with Kyoto targets can achieve these targets through domestic efforts. They can also reduce greenhouse gases (GHG) in other countries at lower cost than at home by making use of the three flexible mechanisms set out in the Protocol: joint implementation (JI), the Clean Development Mechanism (CDM) and international emissions trading (IET). Because the geographic location of GHG emissions is environmentally irrelevant, cost-effectiveness considerations speak for allowing countries to take credit for overseas actions that curb GHG emissions at source or enhance the removal of GHGs by sinks.

Although there are other political, economic, social and ethical reasons for preferring domestic action, one of the most remarkable aspects of the Protocol is how implementation of its flexible mechanisms is proceeding rapidly around the world, ahead of its formal entry into force. One of the Kyoto mechanisms, the CDM, is already being implemented in over 50 developing countries under the legal authority of the Conference of the Parties (COP) of the UNFCCC which is acting as the interim institutional body of the Protocol. More significantly still is the establishment of a scheme for GHG emission allowance trading within the 25 Member States of the European Union (EU) which will formally commence on 1 January 2005. The EU Emission Allowance Trading Scheme (EU ETS) will cover some 15,000 entities with a market potential estimated at euro 10 billion a year. The EU scheme will link with the existing CDM market and the nascent JI market focused on Central and Eastern European countries. Additionally, notwithstanding rejection of the Protocol by the Bush administration, domestic trading and offset schemes are also being devised in other parts of the world, including the US and Australia, that are Kyoto consistent and in many cases actually anticipate future linkages with the Kyoto mechanisms. These developments signify that interest in implementing the Kyoto mechanisms will increase worldwide over time – among Parties and non-Parties to the Protocol, particularly when the Protocol formally enters into force as is now expected in 2005 with the decision by President Putin, backed by The Russian Duma, to ratify the Protocol.

This book is intended to support the implementation and evolution of the climate change regime by providing an accessible account of the rules, procedures and institutions governing access to and use of the Kyoto flexibility mechanisms. These rules span three different legal planes:

- international rules as set out in the Kyoto Protocol and elaborated through decisions taken by the COP of the UNFCCC;
- regional rules, principally those agreed by the European Union governing the EU ETS; and
- national schemes, agreed at the federal or state level, establishing trading or offset provisions for national entities.

The carbon markets established by these schemes will interact in complex ways. Cost-minimization will be the core motivation for trades among and across schemes, particularly at the entity level, but legal and institutional provisions, and broader political and strategic considerations, will also play a significant role in determining whether, where, when and how trades take place. One of the strength of this book is that it brings together Kyoto mechanisms developments from the international, EU and national jurisdictions in a single place providing a clear understanding of the linkages between these levels. By doing so, it aims to help the growing community of policy-makers in developed and developing countries understand the rules governing emerging carbon markets and how they will function and evolve in the future. It will therefore be useful for companies, market intermediaries and service providers such as brokers, traders, auditing and certification entities, consultants, law firms and stakeholder organizations already active in the Kyoto markets.

The development of international and EU mechanisms related rules has proceeded at breakneck speed feeding off, and in turn sustaining, a highly specialist cadre of mechanisms policy advisors and researchers. Every such community develops its own lexicon and share of jargon, and the mechanisms community is no different. By cutting down as much jargon as possible and by explaining the rules related to the mechanisms currently found in many disparate sources in one place, this book tries to appeal to those who may have been put off by the specialized and highly technical nature of the mechanisms discussions to date.

The engagement of a wider group of policy-makers in the Kyoto mechanisms is essential for many reasons. First, implementation of the Kyoto mechanisms must impinge on a wide swathe of economic activity if business-as-usual GHG trends in the energy, transport, agriculture and forestry sectors are to be deflected downwards. Those active in these sectors can only make use of the Kyoto mechanisms if they first understand them. A failure to engage these actors could result in increased GHG emissions which in turn will adversely affect the achievement of internationally agreed economic, social and developmental objectives such as the UN Millennium Development Goals. Second, as the scale of mitigation action increases over the coming decades – as it must if the 60–80 per cent reductions pronounced necessary by the Intergovernmental Panel on Climate Change are to achieved – the role of the mechanisms in the climate regime will likely increase. A sound understanding of the strengths, potential and limitations of the Kyoto mechanisms for affecting global GHG emissions is therefore necessary.

As the material in this book makes clear, the evolution of the flexibility mechanisms has gone hand in hand with the evolution of commitments under the climate change regime. This suggests there is a positive linkage between the two. As they begin to address the nature of additional commitments, policy-makers will need to gauge the efficacy of the Kyoto mechanisms vis-à-vis other policy instruments, such as taxes and more traditional forms of command and control regulation or, more likely, how to enact an appropriate mix of such instruments to match their particular national circumstances. Current levels of interest in the mechanisms will increase significantly because uncertainties relating to the Protocol's entry into force have been resolved by President Putin's decision to back ratification. This book is therefore of particular relevance for those interested in the design of the future climate regime, in particular those interested in the negotiation of new commitments, including for the post-2012 period, which would increase interest in carbon markets and draw in developing countries as well as non-Parties to the Protocol such as the US which could be expected to engage more deeply in GHG reductions in the coming years.

Structure of the book

This introductory chapter provides readers who are not familiar with the climate change regime a brief overview of the climate change problem and a capsule guide to the key provisions of the UNFCCC and the Kyoto Protocol, including through the adoption of the Marrakesh Accords and subsequent COP decisions.[1] It also provides a brief historical sketch of the evolution of the mechanisms and their main features. A glossary of frequently used acronyms and abbreviations is included at the back of the book.

Part I of the book explains in detail the provisions in the UNFCCC and the Kyoto Protocol, governing access to and use of the Kyoto Protocol flexible mechanisms, including all the COP decisions adopted up until December 2003. By way of background, it provides an overview of the main negotiating positions taken by key players in the climate regime, such as the EU, developing countries and non-EU OECD countries, over the last decade. More detailed accounts of the negotiations on individual mechanisms are provided in Part III by Anderson and Bradley (JI), Mark Kenber (CDM) and Richard Baron (IET).

After the historical overview, Part I then sets out the international mechanisms-related rules agreed pursuant to the UNFCCC and the Protocol. Section 2 of Part I provides an explanation of activities implemented jointly (AIJ) projects under the UNFCCC as these were an early precursor to the Kyoto mechanisms. Section 3 then addresses cross-cutting issues which affect all three of the Kyoto mechanisms, namely: adoption and review of mechanisms, modalities, equity issues, supplementarity, fungibility and stakeholder involvement. The cross-cutting participation requirements that must be met by Parties to the Protocol with Kyoto targets to be eligible to enact mechanisms transactions require submission of GHG inventories and supplementary information under Articles 5, 7 and 8 of the Protocol and its expert review. These participation and eligibility requirements are set out in Section 4. Each of the three mechanisms is then explained in Sections 5, 6 and 7 starting with international emissions trading under Article 17 of the Protocol before turning to the CDM under Article 12 and JI under Article 6. The concluding Section 8 provides an overview of the way in which compliance procedures will assess adherence with the Protocol's binding targets and the mechanisms' eligibility conditions as both are fundamental for securing the environmental integrity of the mechanisms.

Part II of the book shifts to the European stage. It describes the evolution and implementation of the EU ETS, including details of the proposed 'Linking Directive' which will allow the EU scheme to link to the Kyoto mechanisms. It also deals with issues relating to joint fulfilment under Article 4 of the Kyoto Protocol as currently only the EU has been able to make use of these burden-sharing provisions in the Protocol. Although elements of EU environmental policy have been moving towards the adoption of market-based mechanism such as taxes, charges and trading, many of the EU's international prescriptions for the climate regime have favoured domestic action and more direct forms of regulation, such as policies and measures (PAMs). The adoption of the EU ETS is thus a bold venture for the EU and one which in the absence of US participation in the Kyoto, puts the EU in the role of vanguard. The rationale for the EU ETS and its development through a series of stakeholder consultations are not well known outside the EU. The insights and experience from this process may prove instructive for policy-makers elsewhere and are therefore explained in detail in Part II. The theoretical, sometime voluminous, literature on the merits of the Kyoto mechanisms assumes regulators start with a blank page on which to write climate policies. Accordingly, they neglect discussion of how these mechanisms should be incorporated into a pre-existing framework of environmental regulation. By way of background, Part II sets out how the EU ETS has had to fit into existing EU legislation which in some cases had to be modified to accommodate the EU ETS. The core of Part II then describes the key elements of the EU ETS and key policy issues raised by the anticipated linkages with the Kyoto mechanisms.

Part III contains eight chapters which collectively examine the evolution of trading-based GHG mechanisms in a number of countries. The underlying objective of these contributions is to shed light on key issues relevant to understanding the contribution the Kyoto mechanisms might make in future climate policy.

The opening chapter by Richard Baron and Michel Colombier explains the theoretical assumptions about the efficacy of trading-based approaches that guided the inclusion of all three mechanisms in Kyoto before evaluating how these assumptions fare when judged against reality. Although implementation on a significant scale has yet to be reached, Baron and Colombier's assessment suggests a more modest contribution from the Kyoto mechanisms than justified by the early hype.

Chapter 2 by Odile Blanchard presents research results quantifying the size of the Kyoto carbon markets. The size of these markets and the price at which carbon is traded varies according to the supply of Kyoto units and the demand made of the market by participating governments. Blanchard compares a range of scenarios, including and excluding the US and Russia as well as scenarios in which pro-Kyoto countries give Kyoto legal effect in the face of US and Russian non-ratification, as well as if Europe takes the leadership role on its own. Her analysis provides important new information about who trades what with whom under these scenarios. Interestingly, the economic models suggest that committed countries that go ahead with Kyoto, even if the US stays out of the regime, are not faced by significantly additional costs from the expectations at Kyoto itself and at the same time still achieve environmental gains.

Fiona Mullins focuses in Chapter 3 on a range of administrative and technical implementation issues relating to the EU ETS, in particular on the critical issue of allocation of allowances, including how identification of installations, permitting procedures, sources of relevant information and new sources of entry and closure, might be handled under the EU scheme. These 'core' issues are likely to be of interest to mechanisms experts worldwide. Her chapter outlines the different approaches to allocation within the EU, focusing on the UK and Germany as they are the largest GHG emitters in the EU and with the most installations. Her chapter draws out early insights about the way in which institutions and procedures are taking the lead on National Allocations Plans (NAPs) which Member States have to provide to the EU Commission by 31 March 2004.

Chapter 4 by Jason Anderson and Rob Bradley looks at opportunities and prospects for joint implementation particularly in Central and Eastern Europe (CEE), including countries that have economies in transition (EIT) status under the Protocol. They explain the negotiations on JI, including political issues, and explain in detail the economic and moral basis for preferring domestic actions. Their chapter then quantifies the size of available 'hot air' credits in specific countries and the political economy driving interest in the Kyoto markets. As many of the countries of the CEE and Russia have been involved in AIJ projects, Chapter 5 provides details of the various AIJ projects and the lessons learnt, for example issues relating to problems with approval processes to those concerning baselines, including the participation of CEE countries in the ERUPT and World Bank PCF. It also discusses which of these countries is actually likely to meet the mechanisms' eligibility requirements documenting the limited capacity in 13 CEE countries.

Chapters 5–8 address challenges arising out of the CDM and JI outside Europe by developed and developing countries. In Chapter 5, Martijn Wilder describes prospects and progress made to date by Annex I Parties in establishing emissions trading and JI/CDM programmes, focusing on developments in Australia, the US, New Zealand, Canada and Japan. For non-Annex I Parties, he looks at developments in some of the large developing countries, namely China, India, Indonesia, Argentina, Brazil, Thailand and the Philippines. Although policy uncertainties about the Protocol's entry into force are dampening efforts to get the CDM going, all of the countries discussed here are clarifying their domestic regulatory and administrative structures to enable them to play a bigger role in the CDM. The development of a pilot emissions trading system in China dealing with sulphur and nitrous oxides is also reported as it has the potential for reducing GHGs as well as in future expanding to include them more explicitly.

Chapter 6 examines the evolution of the CDM and asks searching questions about its role in delivering sustainable development and climate benefits.

These questions go to the heart of the level of international and regulatory effort that should be made internationally and nationally to develop the CDM. Although policy literature on the CDM stresses win-win benefits on both counts, in practice, projects that score highly on sustainable development benefits may not score highly with investors concerned with maximizing their Certified Emissions Reductions (CERs). The entry into force of the Kyoto Protocol is vital for securing the future of the CDM.

Chapter 7 by Axel Michaelowa discusses baselines and additionality issues as ultimately these will define the environmental credibility of the CDM and the amount of investment it is likely to attract in coming years. The chapter focuses on the policy implications of the recent work on baselines and methodologies undertaken by the Executive Board of the CDM and sets out a clear basis for the consideration of approaches to baselines and additionality that will secure both greater environmental and developmental gains from the CDM.

Chapter 8 looks at the critical issue of capacity building for the CDM. As Michaelowa points out, the CDM will not fall out of the sky and start generating CERs. Interest in developing projects in developing countries and the establishment of host countries' institutions to propose them internationally will require capacity building. Current efforts are in their early stages and, as the analysis in this chapter shows, have been hampered by the proliferation of uncoordinated donor initiatives. This chapter suggests how these problems might be overcome in order to produce better partnerships between developed and developing countries based on what will be key to the successful implementation of the CDM.

The concluding chapter of the book by Erik Haites describes the status of proposals concerning the evolution of emissions trading in Canada, Japan, Norway, Switzerland and the United States. But its core purpose is to look beyond the horizon and to think about their implications for the future direction of the climate regime. It sets out the rationale for linking domestic trading programmes in the long term as well as concrete ways in which such linkages might be operationalized. The chapter therefore focuses on examining the evolution of the Kyoto mechanisms in the context of negotiations on mitigation commitments under the future climate regime – a subject which will be of critical importance in bringing not only the United States back on board the climate regime, but also deepening engagement in climate mitigation by developing countries.

The Appendix to the book sets out a list of key documents relating to the EU ETS. The text of the following documents is included here for ease of reference and to facilitate access to the EU ETS by policy-makers outside the EU:

- Directive 2003/87/EC of the European Parliament and of the Council of 13 October 2003 establishing a scheme for greenhouse gas emission allowance trading within the Community and amending Council Directive 96/61/EC.

- Directive 2004/_/EC of the European Parliament and of the Council of 13 September 2004 amending the Directive 2003/87/EC establishing a scheme for greenhouse gas emission allowance trading within the Community, in respect of the Kyoto Protocol's project mechanisms of 23 July 2003, COM(2003)403 – 'Linking Directive Proposal'.
- Communication from the Commission on guidance to assist Member States in the implementation of the criteria listed in Annex III to Directive 2003/87/EC establishing a scheme for greenhouse gas emission allowance trading within the Community and amending Council Directive 96/61/EC, and on the circumstances under which *force majeure* is demonstrated of 7 January 2004, COM(2003)830.

The nature of the climate change problem

The scientific assessment of the causes and impacts of climate change has been undertaken by the Intergovernmental Panel on Climate Change (IPCC). Established in 1988, the IPCC has produced three major assessment reports that have helped policy-makers understand that the Earth's climate system is the result of complex and dynamic interactions between the Earth's atmosphere, biosphere and oceans which human activities are beginning to throw out of balance. The main GHG is carbon dioxide (CO_2). Other gases include methane, nitrous oxides and man-made chemicals: hydrofluorocarbons (HFCs), perfluorocarbons (PFCs), sulphur hexafluoride (SF_6). Emissions of these GHGs have risen considerably due to fossil fuel burning, deforestation, livestock farming and other human activities. All countries contribute to GHG emissions: industrialized countries are responsible for the greatest share of past and current emissions but increased contributions from developing countries are projected to match industrialized countries' current levels somewhere around 2020.[2]

The most recent IPCC report, the 2001 Third Assessment Report (TAR), concluded that if current trends continue the concentration of GHGs in the atmosphere will double by the end of the century. The IPCC has warned of the potentially serious effects of climate variability caused by increased concentrations and concluded that 'business-as-usual' (BAU) scenarios predict a rate of increase in global mean temperatures greater than that seen over the past 10,000 years.[3] Resultant climate impacts include sea-level rise, changes in agricultural yields, forest cover and water resources and an increase in extreme events, such as storms, cyclones, landslides and floods. These impacts will have adverse consequences, including serious health consequences, particularly for developing countries and poorer communities, who generally lack the financial and institutional resources necessary for coping or adapting to climate change.

Although the IPCC has declined to define what a safe level of GHG concentration in the atmosphere should be, it has concluded that stabilization of CO_2 concentrations at *any* level requires eventual reduction of global CO_2 emissions to a small fraction of the current emission level. The IPCC has also reviewed the literature on the costs and benefits of mitigation and adaptation. The economic benefits of avoiding climate change remain poorly quantified because they accrue in the future and are widely dispersed. Although they vary considerably depending on modelling assumptions, the costs of mitigating climate change, on the other hand, are incurred in the short term and on particular actors. These considerations tend to skew discussions on the benefits of taking action to avoid climate change even though the majority of assessments conclude that there are many zero- or low-cost options for reducing GHGs. They also enhance the economic viability of low-cost, but essentially temporary, sink options, which again skews action towards short-term policy responses which tend to shift the risk that such measures will prove insufficient for future generations to bear.

An overview of the UNFCCC

The Convention establishes an objective, guiding principles, commitments and institutional provisions to help ground the international response to climate change. Article 2 of the Convention establishes an ultimate objective for the Parties: stabilization of GHG concentrations in the atmosphere at a level that would prevent dangerous anthropogenic interference with the climate system. This goal emphasizes mitigation while recognizing some degree of adaptation will be necessary.

The UNFCCC was negotiated between 1990 and 1992, against the backdrop of preparations for the UN Conference on Environment and Development (UNCED). Held in Rio de Janeiro, UNCED was convened to promote the integration of environmental protection in economic and social development and resulted in the adoption of five instruments to protect the environment and promote sustainable development.[4] UNFCCC negotiations were strongly influenced by the North/South dynamics of UNCED. Variants of the Rio Declaration principles are set out in Article 3 of the Convention. The principle of 'common but differentiated commitments', a central component of the compact agreed by developed and developing countries at Rio, is set out in Article 3.1 and is critical for understanding the structure of the Convention. Reflecting their lesser historical contribution to global environmental degradation and more limited present-day resources, the principle of differentiated responsibilities translates into commitments by developing countries to protect the global environment that are less onerous than those taken by developed countries.

Furthermore, the implementation of developing country commitments is conditional on the provision of technology and financial resources, mainly Official Development Assistance (ODA), from developed countries.

Nearly all the commitments set out in the UNFCCC are differentiated: more detailed commitments have been taken on by a total of 41 developed countries that are listed in Annex I of the Convention and hence known as Annex I Parties.[5] In terms of mitigation commitments, Annex I Parties are required to take the lead in modifying long-term GHG emission trends by enacting policies and measures.[6] The Convention also includes a quantified aim for Annex I Parties: to stabilize their CO_2 emissions and other GHGs at 1990 levels by the year 2000. To monitor progress, Annex I Parties have to submit *annual* GHG inventories and implementation reports, called national communications, usually every three years, to the UNFCCC supreme body, the COP.[7] This information is subject to expert scrutiny in a process called in-depth review. The information contained in the communications is then compiled and synthesized so that the COP can consider whether to take further action.

The term non-Annex I Parties refers by default to 130 developing countries that negotiate as a bloc called the G-77. It also includes other countries that do not see themselves or are regarded by others as developing countries, such as Mexico and Korea, and countries from Central Asia, such as Kazakhstan. The mitigation commitments of non-Annex I Parties requires them to prepare inventories (but not on an annual basis) and national programmes addressing climate change but without specifying that these must contain policies and measures that reduce GHGs or enhance sequestration through sinks, such as by cutting levels of deforestation.[8]

The Convention also contains financial, technological and adaptation assistance provisions mandating resource flows from the wealthier Annex I Parties that are listed in Annex II of the Convention in favour of developing countries.[9] Annex II Parties have commitments to providing developing countries with financial and technological assistance to meet the *full costs* of preparing GHG inventories/national communications and the *incremental costs* of implementing their other Convention commitments.[10] All Parties must factor adaptation into their responses to climate change.[11] Annex II Parties must also assist developing countries that are vulnerable to climate change impacts meet the costs of adaptation.[12]

Because Parties have very diverse circumstances, the Convention recognizes the need for these to be accorded consideration by the COP. Accordingly, the Convention contains provisions requiring the COP to give special consideration to countries that are undergoing the process to a market economy (known as EITs),[13] to the poverty and social developmental priorities of developing countries,[14] to specific geographic and economic circumstances of particular groups of developing countries,[15] to the needs of least developed countries,[16] and to the heavy reliance on the production and consumption of fossil fuels by certain developed countries.[17] A range of general obligations related to enhancing international cooperation on research and systematic observations and education, training and public awareness are also set out in the Convention.[18]

The Convention establishes institutional machinery to oversee the implementation of these commitments and to ensure that further action is taken by Parties to respond to the latest scientific and technical information, including the negotiation of new commitments adopted in the form of amendments and protocols to the Convention.[19] The main Convention institutions are:

- Conference of the Parties (COP);[20]
- Secretariat;[21]
- Subsidiary Body for Implementation (SBI);[22]
- Subsidiary Body for Scientific and Technological Advice (SBSTA);[23] and
- Financial Mechanism operated by the Global Environment Facility (GEF).[24]

The COP is the main policy-making body. It meets annually and provides the chief forum for international discussions about climate change. Nearly two hundred decisions have been adopted by the nine COPs held to date addressing matters related to the implementation and evolution of commitments and linkages with other multilateral environmental agreements (MEAs) and the World Trade Organization.

The IPCC is an independent scientific network with a separate legal existence. It supports the climate regime by liaising with SBSTA and by providing advice and technical support through its main assessment reports and work on methodological and technical issues, such as the preparation of GHG inventories.

An overview of the Kyoto Protocol

The first session of COP held in Berlin in 1995 agreed that the mitigation commitments of Annex I Parties were inadequate and set in motion negotiations that led to the adoption by COP-3 of the 1997 Kyoto Protocol.[25] Although a number of developed countries, principally the US and Australia, wanted to include additional commitments for developing countries in the Protocol, all Parties eventually agreed the Protocol should focus on advancing the implementation of developing countries' existing commitments. The Protocol thus preserves the differentiated structure of the Convention's commitments. It also shares the Convention's objectives and guiding principles, and deploys, with some necessary legal modifications, its institutional machinery. The modifications create a legally distinct institutional body for the Protocol: referred to as the Conference of the Parties serving as the meeting of the Parties to the Protocol (COP/MOP). The remaining modifications ensure that only Parties to the Protocol can make decisions relating to the Protocol.[26]

The rationale for the negotiation of the Protocol was to strengthen the mitigation commitments of Annex I Parties. This is achieved through the establishment of legally binding targets for Annex I Parties.[27] These targets are differentiated and in the form of absolute national emissions caps to be achieved from 2008 to 2012 with a specific requirement on Annex I Parties to have made demonstrable progress by 2005.[28] The targets cover a basket of six main GHGs from defined sectors and sources listed in Annex A of the Protocol. In addition, Annex I Parties are allowed to count net sequestration from certain land use, land use change and forestry activities (LULUCF) towards compliance with their Kyoto targets.[29] The Protocol also establishes a collective target for Annex I Parties amounting to 5 per cent below 1990 levels in the commitment period 2008–2012.

Annex I Parties' mitigation commitments are further strengthened by a requirement that they implement and/or further elaborate policies and measures such as those listed in Article 2 of the Protocol. Annex I Parties may achieve their Article 3 targets through such PAMs. They can achieve compliance with Article 3 by making use of three flexible mechanisms: JI under Article 6, CDM under Article 12 and IET under Article 17 which are explained in detail in the remainder of this book. Parties can also group together to redistribute the efforts they need to make to reach their targets using the joint fulfilment provisions set out in Article 4 which are currently only utilized by the EU.

The inclusion of binding targets and the Kyoto mechanisms necessitated detailed, consistent and transparent reporting of GHG inventory data and mechanisms-related transactions. Supplementary reporting and review requirements for Annex I Parties are set out in Articles 5, 7 and 8 of the Protocol. Compliance with many of these is part of the mechanisms, eligibility requirements for Annex I Parties which will be assessed by the Compliance Committee in accordance with the procedures and mechanisms relating to compliance under the Kyoto Protocol adopted by COP-7.[30] The reporting, review and compliance provisions were deemed essential for ensuring the environmental integrity of the Kyoto targets.

One feature of the Kyoto Protocol which distinguishes it from the Convention is the higher profile given to issues concerning the economic impacts of response measures on developing countries.[31] These provisions were included because many oil-producing developing countries wanted minimization of the potential adverse economic impacts they might experience as a result of mitigation policies being implemented by Annex I Parties. Use of the mechanisms is likely to reduce these potential impacts.

An overview of the Kyoto mechanisms

The three mechanisms established by the Kyoto Protocol allow different kinds of partnerships between Parties and results in different units that can be counted equally towards compliance with Annex I Parties' Article 3 Kyoto commitments. This book refers to these collectively as 'Kyoto units' and uses the term 'credits' to refer more generally to the notion of permits and allowances.

Two of the Kyoto mechanisms cover Annex I Parties with binding quantitative commitments (JI and emissions trading (ET)) and the third (CDM) covers projects involving non-Annex I Parties without quantitative commitments and Annex I Parties. JI allows Annex I Parties to jointly undertake projects that reduce emissions or enhance sinks in an Annex I Party and results in emission reduction units (ERUs). The CDM allows non-Annex I and Annex I Parties to jointly undertake emission reduction in non-Annex I Parties that contribute to sustainable development and result in CERs. Because permanence is a specific issue for afforestation/ reforestation CDM projects, COP-9 agreed that such projects would lead to the creation of either temporary CERs (tCERs) or long-term CERs (lCERs). For simplicity, however, the term CERs will be used in the book to cover tCERs and lCERs. As host countries have no quantitative targets, methodological and institutional procedures underpinning the creation of CERs are particularly important for screening out projects that would have happened anyway. The concept of 'additionality' is critical for the CDM and its effective application is overseen multilaterally by the Executive Board of the CDM (EB/CDM).

JI, on the other hand, takes place between Parties with quantitative targets. Consequently, the need for international oversight is, in theory, correspondingly reduced. The Marrakesh Accords create two 'tracks' for ensuring the environmental integrity of JI projects. Track 1 applies when a host Annex I Party is in conformity with its reporting and review requirements and is thus allowed to effectively self-regulate JI projects and to create and transfer ERUs. Where the host Annex I Party fails to meet its reporting and review eligibility conditions for participating in JI under Track 1, it may still participate in JI projects under Track 2. This requires international oversight of JI activities provided by the Article 6 Supervisory Committee (A6SC) whose functions are similar to those of the EB/CDM. A Party that meets the reporting and review requirements may still allow/require JI projects to use Track 2 if it so wishes as this might be seen to make the projects more credible.

Emissions trading provides for an inventory-based system of transfers and acquisitions that is only open to Annex I Parties with Article 3 commitments. ET allows for the transfer and acquisition of assigned amount units (AAUs) as established pursuant to Articles 3.7 and 3.8. Sequestration achieved by LULUCF under Articles 3.3 and 3.4 will result in the creation of removal units (RMUs) which, subject to rules set out in the Marrakesh Accords, can also be transferred and acquired among Annex I Parties through ET as can ERUs and CERs acquired through JI and the CDM. Thus ET allows Parties to transfer and acquire the full range of Kyoto units. The environmental integrity of ET is particularly critical to the integrity of the entire Protocol and is underpinned by requiring all Annex I Parties to comply with strict reporting and review conditions as well as the requirement to maintain a 'commitment period reserve' to limit overselling. Adherence to these requirements is overseen by the Compliance Committee established pursuant to the Protocol.

Notes

1. For a more detailed explanation of the climate regime see Yamin and Depledge (2004). The Marrakesh Accords refers to a package of COP decisions relating to the Convention and the Protocol adopted at COP-7, held in Marrakesh in 2001. The Accords set out many details necessary to implement the Protocol, including detailed rules and procedures governing the Kyoto mechanisms.
2. This does not take past emissions into account.
3. IPCC (2001).
4. These other instruments are Agenda 21, the Rio Declaration, the Forests Principles and the Convention on Biological Diversity. On the negotiations of the Convention see Bodansky (1993).
5. Primarily the OECD countries plus the former planned economies of Central and Eastern Europe.
6. UNFCCC, Article 4.2(a) and (b).
7. UNFCCC, Article 12.
8. UNFCCC, Article 4.1.
9. Annex II covers the OECD, minus Turkey and Korea and minus the former planned economy countries.
10. UNFCCC, Article 4.3.
11. UNFCCC, Article 4.1.
12. UNFCCC, Article 4.4.
13. UNFCCC, Article 4.6.
14. UNFCCC, Article 4.7.
15. UNFCCC, Article 4.8.
16. UNFCCC, Article 4.9.
17. UNFCCC, Article 4.10.
18. UNFCCC, Articles 5 and 6.
19. For more detailed explanation of the institutional aspects of the climate regime see Chapters 13 and 14 in Yamin and Depledge (2004).
20. UNFCCC, Article 7.
21. UNFCCC, Article 8.
22. UNFCCC, Article 9.
23. UNFCCC, Article 10.
24. UNFCCC, Article 11.
25. For a detailed history of the Kyoto Protocol (KP) see Grubb et al (1999) and Oberthur and Ott (1999).
26. UNFCCC, Article 17.5, KP, Article 13.2.

27. KP, Article 3.1 and Annex B.
28. KP, Article 3.2.
29. KP, Articles 3.3 and 3.4.
30. Decision 24/CP.7, COP-7 Report.
31. KP, Articles 2.3 and 3.14.

References

Bodansky, D. (1993) 'The United Nations Framework Convention on Climate Change: A commentary', *Yale Journal of International Law*, vol 18, no 2, pp 451–558.

Grubb, M., Vrolijk, C. and Brack, D. (1999) *The Kyoto Protocol. A Guide and Assessment*, London, Earthscan/RIIA.

IPCC (2001a) *Climate Change 2001: The Scientific Basis. Contribution of Working Group I to the Third Assessment Report of the Intergovernmental Panel on Climate Change*, eds J. T. Houghton et al. Cambridge: Cambridge University Press.

IPCC (2001b) *Climate Change 2001: Impacts, Adaptation and Vulnerability. Contribution of Working Group II to the Third Assessment Report of the Intergovernmental Panel on Climate Change*, eds J. McCarthy et al. Cambridge: Cambridge University Press.

IPCC (2001c) *Climate Change 2001: Mitigation. Contribution of Working Group III to the Third Assessment Report of the Intergovernmental Panel on Climate Change*, eds B. Metz et al. Cambridge: Cambridge University Press.

Oberthur, S. and Ott, H. (1999) *The Kyoto Protocol: International Climate Policy for the 21st Century*, Berlin, Springer Verlag.

Yamin, F. and Depledge, J. (2004) *The International Climate Change Regime: A Guide to Rules, Institutions and Procedures*, Cambridge, Cambridge University Press.

Part I

The international rules on the Kyoto mechanisms

Farhana Yamin

I.1 Introduction

This section explains the international rules, institutions and procedures governing access to and use of the Kyoto mechanisms agreed pursuant to the UNFCCC and the Kyoto Protocol. The bare-bone provisions in these two treaties have been extensively elaborated by the UNFCCC governing institutional body, the Conference of the Parties (COP), which has met nine times since the Convention entered into force in 1994. The institutional body with authority for the Protocol is formally called the Conference of the Parties serving as the meeting of the Parties (COP/MOP) and this will come into operation when the Protocol takes legal effect. Until that time the annual session of the COP will continue to operate as the interim institutional body for the Protocol.

As discussed below, negotiations on the mechanisms proved controversial in the run-up to Kyoto for a variety of reasons. The mechanisms-related textual provisions in Articles 6, 12 and 17 of the Protocol had to be substantially elaborated before Annex I Parties felt sufficiently comfortable about submitting the Protocol for domestic ratification proceedings. The international rules to elaborate the additional mechanisms rules and modalities took just under six years. The bulk of significant mechanisms-related issues were finally resolved by COP-6 Part II held in July 2001 with the adoption of the Bonn Agreements, with detailed textual decisions being set out in the Marrakesh Accords adopted by COP-7. Matters related to the inclusion of sinks in the CDM were resolved more recently at COP-9, held in Milan, December 2003. Many of the international rules related to the Protocol take the form of recommended draft decisions adopted by the COP which will be forwarded by the COP to the first meeting of the COP/MOP for formal adoption. Although they are not yet in full legal force, the COP decisions containing the recommended decisions COP/MOP-1 are expected to be confirmed in full by the COP/MOP without further negotiation as they are part of the finely balanced package of decisions adopted pursuant to the Bonn Agreements and the Marrakesh Accords.

The rules, institutions and procedures of governing access and use of the Kyoto flexibility mechanisms are thus now to be found in two separate treaties and more than a dozen COP decisions amounting to over two hundred pages of text. Rules relating to registration and validation of CDM projects are being further elaborated by the Executive Board of the CDM (EB/CDM). These disparate, intricately related outputs and procedures make international mechanisms-related rules difficult to track down. The purpose of this section of the book is to provide a guide to the rules, institutions and procedures of the Kyoto mechanisms which brings these disparate sources together in an accessible manner.[1] For ease of reference the main decisions agreed by the COP relating to the mechanisms are listed in Table I.1. For those not familiar with the Kyoto mechanisms, Table I.2 provides an overview of the key distinguishing features of each of the three Kyoto mechanisms: Article JI, CDM and ET.

The remainder of Part I is structured as follows. This introductory section sets out the context and historical background to the negotiations on the Kyoto mechanisms, including a brief overview of the negotiating stance groups of Parties have taken on the mechanisms, how their views have evolved and the international challenges to be faced by the mechanisms community in the future. **Section I.2** describes the rules relating to activities implemented jointly (AIJ) under the FCCC as these projects are precursors to the Kyoto mechanisms and many could be converted into JI or CDM projects. **Section I.3** then describes the cross-cutting related issues which affect all three of the Kyoto mechanisms, namely: adoption and review of mechanism modalities, equity issues, supplementarity, fungibility, stakeholder involvement and participation requirements. **Section I.4** then describes participation and eligibility requirements for the Kyoto mechanisms, including requirements relating to submission of GHG inventories and supplementary information under Articles 5, 7 and 8 of the Protocol. **Sections I.5, I.6 and I.7** then describe the rules for each of the three mechanisms starting with international emissions trading under Article 17 of the Protocol before turning to the CDM under Article 12 and JI under Article 6. The linkages with reporting and review provisions are explained in section I.4 with how compliance with them will be assessed set out in **section I.8**.

Table I.1 *Kyoto mechanisms and related COP Decisions*

Issue	FCCC/KP Article	COP Decision*
Accounting of assigned amounts	KP: 7.4; 3.7; 3.10–12	8/CP.4; **15/CP.7; 19/CP.7; 24/CP.8**
Activities implemented jointly (AIJ)	FCCC: 4.2(d) 14/CP.8; 20/CP.8	5/CP.1; 8/CP.2; 10/CP.3; 6/CP.4; 13/CP.5; **8/CP.7;**
Carry-over of Kyoto units	KP: 3.13	**19/CP.7**
CDM	KP: 12, 3.12 **17/CP.7** Guidance to EB: 21/CP.8, 18/CP.9 LULUCF projects: **19/CP.9**	7/CP.4; 14/CP.5; **15/CP.7;**
Compliance	KP: 18	8/CP.4; 15/CP.5; **24/CP.7**
Emissions trading	KP: 17, 3.10–11 **18/CP.7**	7/CP.4; 14/CP.5; **15/CP.7;**
Joint implementation	KP: 6, 3.10–11 **15/CP.7; 16/CP.7**	1/CP.3; 7/CP.4; 14/CP.5;
Land use, land-use change and forestry	KP: 3.3, 3.4, 3.7 Croatia: 22/CP.9	1/CP.3; 8/CP.4; 9/CP.4; 16/CP.5; **11/CP.7; 12/CP.7**
Methodologies under FCCC	FCCC: 4.1(a), 4.2(c)	4/CP.1; 2/CP.3
Methodologies under KP	KP: 5 **21/CP.7; 20/CP.9**	2/CP.3; 8/CP.4; **20/CP.7;**
Reporting under Kyoto Protocol	KP: 7 (Annex I Parties only) **22/CP.7**; 22/CP.8	8/CP.4 (Annex I) Guidelines (Annex I):
Review of emission inventories under FCCC (Annex I): Procedures/guidelines	FCCC: 12	2/CP.1; 6/CP.3; 6/CP.5; 19/CP.8; 12/CP.9
Review of national communications under FCCC (Annex I): Procedures/guidelines	FCCC: 12	2/CP.1; 6/CP.3;
Review process under KP (Annex I): Guidelines	KP: 8	8/CP.4; **23/CP.7; 22/CP.8; 23/CP.8; 21/CP.9**

Note: Bold type indicates the most significant decisions.
Source: Adapted from Yamin and Depledge (2004).

Evolution of the Kyoto mechanisms

Domestic experience of market-based instruments and emissions trading as well as underlying differences in regulatory culture and administrative capacity have influenced how countries have approached the negotiations on the Kyoto mechanisms. This section provides the context and historical background for understanding how Parties approached the inclusion of the Kyoto mechanisms. Three broad phases can be discerned as follows and are described below:

- 1991–1997: learning in a hostile environment;
- 1997–2001: acceptance and cautious engagement;
- 2001 onwards: implementation and experimentation.

Learning in a hostile environment (1991–1997)

In broad terms, many of the non-EU countries, particularly the US, had positive experience of using market-based instruments in the environmental context in the late 1980s and early 1990s. These countries were already using market-based instruments such as trading, taxes and charges in environmental policy, in place of, or as a supplement to, traditional command and control techniques. As a result many regulators, industry and policy researchers had come to regard command and control regulation as overly technology-prescriptive and innovation-unfriendly.

The US leadership within the group of countries called JUSCANNZ was important in converting others to the advantages of using mechanisms in combating climate change.[2] The US had already legislated for two kinds of trading schemes in the context of the 1990 amendments to the Clean Air Act. This legislation created a domestic 'cap-and-trade' scheme which involved setting a limit on total emissions, distributing permits equal to allowable emissions and requiring entities to hold sufficient permits to cover their emissions during a given compliance period as well as a 'baseline and credit' scheme whereby participants received 'credits' for emissions reductions achieved against a hypothetical baseline. A common feature of both schemes is that a financial asset, called a permit or an allowance, is created by governments which is usually given (rather than auctioned) to polluters in a cap-and-trade scheme or issued in the case of baseline and credit schemes upon demonstration that emissions have been reduced. Because they are financially valuable, the allocation of permits generates political 'buy-in' and lessens the 'pain' of pollution control – an aspect which is important in the context of the 'pork-barrel' nature of US politics. Finally, the experience of the Clinton Administration in the early 1990s with the failed introduction of a carbon/energy tax (subsequently the British thermal unit (BTU) tax) convinced US policy-makers that any US domestic policy to reduce GHGs would have to involve emissions trading. And as US contributions to GHG emissions accounts for nearly 20 per cent of the global total, emissions trading would be a sensible tool to limit the costs of global compliance.

Although some academics were pointing out the merits of the US sulphur trading programme as an innovative precedent for climate negotiators even during the UNFCCC negotiations, these early suggestions were not well enough understood to allow Parties to consider their global application in the UNFCCC (Grubb et al, 1999). The Convention did, however, include a reference to allowing groups of Annex I Parties to achieve their aim of returning their emissions of CO_2 and other GHGs to 1990 levels by 2000 jointly with other Parties (Bodansky, 1993; Yamin, 1993). The Convention's reference to 'joint implementation' in Article 4.2(a) and 4.2(d) was ambiguous, however, because it could be taken to refer to cap-and-trade and baseline and credit forms of trading among Annex I Parties only as well as a multitude of other forms of international cooperation. This made it difficult to operationalize criteria for joint implementation at COP-1, held in Berlin in 1995. Although some developing countries such as Costa Rica and other Central and Latin American states were by then advocates for project-based forms of trading, most developing countries were hostile to these because Annex I Parties had not committed themselves to binding legal targets at Berlin. Additionally, the creation of permits and credits at the international level raised distributional issues concerning their allocation among Parties with highly differentiated legal commitments as well as fundamental moral, equity and environmental considerations.[3] Finally, because JUSCANNZ were also pressing hard for developing countries to take on additional commitments, as well as being the leading advocates for mechanisms, developing countries were mistrustful that trading would be used by these richer countries to buy their way out of taking domestic action while shifting the actual burden of pollution control to the South.

After COP-1, negotiations on Annex I Parties' commitments commenced in the Ad Hoc Group on the Berlin Mandate (AGBM). JUSCANNZ continued to press for emissions trading and project-based mechanisms and sporadically to call for developing country commitments. The resulting atmosphere of hostility and mistrust underlined the doubts developing countries had about the prominence given to the mechanisms by JUSCANNZ. Although differences of view began emerging within the EU, it is true to say that the regulatory culture of the EU Member States as a whole was less embracing of trading instruments, although a number of countries were moving towards the use of environmental taxes and various forms of charging.[4] Developing countries, on the other hand, remained sceptical or hostile for the reasons outlined above. In addition, when AGBM negotiations commenced nearly a decade ago, the majority of developing countries were struggling to introduce first-generation forms of environmental regulations to tackle urgent environmental problems like water pollution and local air quality. This left them with little time to acquire more knowledge about new techniques dealing with climate change mitigation – a problem they regarded the North should be dealing with first and, given its long-term nature, one that did not justify the use of their scarce regulatory resources with experimental approaches.[5] The academic and policy literature on the mechanisms did not help matters as much of it was fervently either pro-taxes or pro-emissions trading with little discussion of how different instruments could be mixed, matched and overlaid in the context of existing regulations to suit countries' very different legal, administrative and regulatory structures.

The signal that the negotiations on the Protocol would have to take emissions trading seriously came from the US at COP-2 in 1996 when it announced that it wanted to negotiate an instrument that would contain legally binding targets and some form of emissions trading (Depledge, 2000). This was followed up by a US non-paper submitted in December 1996 which formally linked this position with the inclusion of emissions trading.[6] The shift in US position polarized divisions within the OECD as the EU continued to focus on mandatory policies and measures (PAMs) while adding its voice to developing country concerns that domestic action should be emphasized.

As negotiations continued into 1997, it became clear that many EITs would be negotiating targets that had the effect of giving them generous allowances which they could trade with others.[7] The formation of the Umbrella group – combining JUSCANNZ with Russia and Ukraine – raised serious questions about the detrimental affect of 'hot air' on the environmental effectiveness of the emerging regime.[8] The notion that the mechanisms would be used by rich countries to avoid making politically unpopular domestic reductions further stoked EU and developing country opposition. Finally, legal, institutional and procedural concerns also began to be raised by the research community as to whether domestic trading schemes, like the sulphur programme, could be applied to the international context in the rather naive manner being suggested by some economists (Yamin, 1993). This is because unlike domestic trading schemes which rely on high penalties and strict enforcement to deter non-compliance, monitoring, tracking and verifying GHG emissions in an international context is far more complex. Accurate and timely self-reporting by governments in the absence of an international authority enforcing compliance does not have a good track record. These concerns were compounded by the fact that under the Convention many Annex I Parties had failed to keep pace with their reporting commitments, giving credence to concerns that trading mechanisms that required governments to keep tabs on thousands of emissions sources would prove too taxing for many governments and might give rise to fraudulent transactions which would have the effect of undermining the environmental integrity of the regime.

Although developing countries continued to oppose cap-and-trade forms of emissions trading, and in fact did not participate at Kyoto in the drafting groups dealing with JI and IET, many of them became more and more interested in project-based mechanisms – an interest supported by the expansion of AIJ projects which helped institutional learning. In the last six months of 1997, Brazil took the lead in defining the essential features of the CDM. For reasons discussed elsewhere in this book, the CDM is particularly well suited to advancing quantified reductions of GHG in large, industrializing developing countries who have rapidly expanding economies that are inherently difficult to manage, involving as they do very large amounts of domestic and overseas capital investments with many vulnerable to external economic shocks, as evidenced by the financial crisis that devastated many Asian and Latin American economies in the late 1990s.[9] Although the underlying hostility of developing countries to trading almost led to a breakdown of negotiations on the inclusion of Article 17, COP-3 concluded with the inclusion of the three mechanisms and with clarification in Article 4 of how the joint fulfilment of the EU would work in practice.

Acceptance and cautious engagement (1997–2001)

The acceptance of the end deal in Kyoto marked a fundamental shift in the mechanisms negotiations. International fractiousness gave way to attention, some might say obsession, with detail as everyone realized the 'devil is in the details'. This realization was combined with a mild degree of intellectual fundamentalism about which mechanism was 'best' and ought to receive procedural and political priority, a question that was answered differently as blocs of Parties began to cluster round their favourite mechanism. Given their limited negotiating resources and economic interests, developing countries insisted, of course, that the CDM have priority. JUSCANNZ remained focused on emissions trading, regarding project-based mechanisms, with higher transactions costs, as an inferior policy form. Mindful of the difficulties of putting together credible inventory data that would allow them to participate in IET, the smaller EITs fought hard to ensure JI would not be squeezed by IET and CDM as these two mechanisms began to command the lion's share of policy attention and negotiating time.

While the negotiating thrust of JUSCANNZ and developing countries was fairly clear during this period, the EU took on the difficult task of trying to keep everyone happy while trying to maintain its environmental credentials. Although its stance on prioritizing domestic action went down well with environmental NGOs, it was hard to reconcile with the interests of its negotiating partners – and over time with growing domestic interest in emissions trading both at the Member State and Commission level.[10] Many developing countries were interested in ensuring the CDM was not squeezed out of the market as it would be if the EU's proposed across-the-board supplementarity restrictions on the mechanisms were to be applied – a point now being borne out with the introduction of the EU ETS.[11] The fear that the market for JI might turn out to be very small also worried the EITs who were by then entering into accession negotiations with the EU. All the while, the international agenda was clogged with numerous conferences, workshops and intersessional meetings discussing mechanism 'design' issues, signalling an intense period of learning on the part of negotiators, GHG emitting industries, stakeholders and an array of market intermediaries who all realized that the Kyoto mechanisms could be good news for business, or at least better than the other alternatives.

Growing international momentum behind the Kyoto mechanisms faced two major setbacks in this period. The first was the agreement reached on the inclusion of sinks under Articles 3.3 and 3.4 of the Protocol at COP-6 Part II and COP-7. As discussed by Blanchard in this book, for most Parties, the availability of sinks is the biggest factor impacting compliance costs and this has a critical bearing on the degree to which each country needs to take further GHG reduction action – domestic or overseas (Grubb and Yamin, 2001). The final deal struck in the Bonn Agreement and Marrakesh Accords means Annex I Parties will have less need to use the mechanisms: unless there is domestic pressure forcing them otherwise, many of them are likely to turn first to low- or zero-cost sinks options. The inclusion of sinks in the CDM, agreed in principle at COP-6 and in detail at COP-9, makes this pool of low-cost CERs even bigger. The second setback was the announcement by the Bush Administration in March 2001 that they would not ratify the Protocol. As the largest GHG producer and potentially the biggest buyer of Kyoto units, the US announcement threw the climate regime into a tailspin which dampened expectations about the viability and size of the emerging carbon markets.

Implementation and experimentation (2001 onwards)

The announcement by the Bush Administration that the Protocol was 'fatally flawed' did not, however, kill the Protocol. Rather it had the opposite effect: motivated by multilateralism and a genuine concern for the global environment, it simply galvanized the remainder of the international community to safeguard the Protocol – an instrument which had taken almost a decade to negotiate. The failure by the Bush Administration to come up with a politically viable alternative in time for COP-6 Part II in July 2001 was soon overshadowed by the Administration's preoccupation with responses to the events of 11 September 2001 which took place a few weeks prior to COP-7 held in Marrakesh.

From 2001 onwards, EU leadership on the mechanisms was a significant bonus for the climate community as a whole. The announcement by the EU that it would implement a domestic emissions trading scheme as of 2005, covering at a minimum 25 countries, ensured that interest in Kyoto and its mechanisms remained high. The initiatives of several governments in launching tenders for project-based credits, such as the Dutch ERUPT and CERUPT programmes and the establishment by the World Bank of its Prototype Carbon Fund (PCF), also signalled widespread support for continuation of the Kyoto mechanisms.[12]

The adoption of the Marrakesh Accords by COP-7 containing nearly a hundred pages of text operationalizing all three mechanisms, including complex interlinkages with the crucial issues of compliance, meant the focus shifted again to another tier of attention, as technical issues concerning registry design and baseline methodologies now began to preoccupy the mechanisms community.[13] The election of members of the Executive Board of the CDM at COP-7, which many feared might become a talking shop, proved misplaced. The EB/CDM commenced work smoothly evidencing a clear desire by all Parties to put the Kyoto mechanisms into swift operation. And it has done so notwithstanding the disruptive effects caused by the Bush Administration's insistence that the US not contribute financially to supporting CDM activities – a stance which is difficult to understand given that the CDM is contributing directly to GHG reductions by the very developing countries the Administration criticized as not doing enough when the US backtracked on Kyoto.

Evidence that the Bush Administration's disruptive tactics and non-engagement with the climate problem is out of step not only with scientific opinion and the international community but also with the thrust of climate policy in the remainder of the US is becoming more and more apparent.[14] As reported by Haites and Wilder in this volume, over 45 bills, resolutions and amendments addressing climate change were introduced in the US Congress in 2003–2004. Many state initiatives evidencing interest in the Kyoto mechanisms in the US are already being implemented. And an important federal initiative which would have established an emissions trading scheme covering 85 per cent of the US GHG emissions in 1990 was only narrowly defeated, with the Bush Administration playing a leading role in its demise.[15]

As the contribution of Wilder in this volume makes clear, implementation efforts are also advancing at the domestic level in other countries as they begin the practical task of designating national authorities and endowing them with resources and legal authority to undertake the functions outlined for them in the Marrakesh Accords. Notwithstanding significant policy uncertainties, the private sector on the whole has played a positive role with many leading players fronting high learning and transactions costs in putting together risky CDM projects, knowing full well that these will be subject to intense scrutiny from regulators and environmental NGOs alike. The EU's leadership on the mechanisms has continued with the adoption of a proposal to link the EU ETS with the Kyoto mechanisms being put forward in late 2003.[16] Although this will boost the JI and CDM markets, the link has also brought to light the difficulties the EU faces trying to advance the operation of the Kyoto markets prior to the entry into force of the Protocol – a scheme even as large as the EU ETS cannot absorb the potential supply of Kyoto units from non-EU countries. Although understandable on environmental grounds, the proposed exclusion of sinks projects from the EU ETS under the Linking Directive proposal might dampen expectations of some pro-CDM sinks countries about the inclusions of sinks in the CDM as agreed by COP-9. Concerns raised by Japan and other non-EU pro-Kyoto countries that they may be disadvantaged in their access to JI credits by the operation of the EU ETS evidence, for example, there are underlying political tensions about how centre-stage the EU scheme (which is a domestic scheme) should play. These tensions are likely to decrease once Kyoto enters into force and JI and IET come to the fore.

Looking to the future, it is clear that interest in the Kyoto project-based mechanisms, particularly the CDM, will dwindle over the next few years if policy-makers fail to ensure a market value for emissions reductions in the post-2012 period. Investors will not invest significantly in the CDM unless there is certainty that CERs will be valuable and can be banked against future targets. This requires international negotiations to address Annex I Parties' mitigation commitments for the post-2012 period. Many countries are sensitive about discussing the 2012 period because this will bring the question of US engagement to the fore as well as providing an opportunity for developing countries to contribute to the evolution of the Kyoto mechanisms and their role in the design of the future climate regime.

Table I.2 *Overview of the Kyoto flexibility mechanisms*

	Project-related mechanisms		Non-project mechanism
Name	Article 6/Joint Implementation	Clean Development Mechanism	Emissions Trading
Parties (subject to participation/ eligibility criteria)	Annex I – Annex I	Non-Annex I – Annex-I	Annex I – Annex I
Authorized Legal Entities (dependent on Party eligibility criteria	Yes	Yes	Yes
Kyoto unit	Emission reduction units (ERUs)	Certified emission reductions (CERs) Temporary CER (tCER) and long-term CER	Assigned amount units (AAUs)

		(ICER) from afforestation and reforestation projects	
Unit fungibility	Yes	Yes	Yes
Unit use restrictions	Refrain from using ERUs from nuclear facilities	CERs from afforestation and deforestation not to exceed 1% of Party's assigned amount Annex I are to refrain from using CERs from nuclear facilities	No restrictions
Unit carry over	Yes – 2.5% of a Party's assigned amount	Yes – 2.5% of a Party's assigned amount	Yes – without restriction
Unit availability	From 2008 to 2012	From 2000	From 2008 to 2012
Coverage of activities	All Kyoto eligible sources and LULUCF activities	All Kyoto eligible sources with priority to small scale Sinks limited to afforestation/ reforestation	Not applicable
Responsible institutions	Accredited Independent Entities, Article 6 Supervisory Committee, COP/MOP	Designated operational entities (DOEs), Executive Board, COP and COP/MOP	National Registries, Transaction Log, COP/MOP
Adminstratrative support	Secretariat	Secretariat	Secretariat
Administrative costs	To be borne by Participants	To be borne by Project Participant and DOEs	No specific provisions

Source: Adapted from Wollansky and Freidrich (2003).

I.2 Activities implemented jointly

This section describes the rules relating to activities implemented jointly under the pilot phase which commenced as a result of Decision 5/CP.1 adopted in Berlin. Article 4.2(a) and (b) of the Convention mandates Annex I Parties to adopt policies and measures to limit their GHG emissions and to modify their long-term emissions trend by implementing PAMs. This includes achieving the

Convention's 'quantified aim' to return emissions of CO_2 and other GHGs to their 1990 levels by the year 2000 (Yamin and Depledge, 2004, ch. 6).

Article 4.2(a) states such Parties 'may implement such policies and measures jointly with other Parties' but this provision is subject to criteria the COP might adopt under Article 4.2(d) regarding joint implementation. Cooperative measures to reduce or sequester GHGs can be undertaken between different governments, businesses and NGOs without sanction from the COP. Although many readings are possible, one intent of Article 4.2(d) was to provide an internationally accepted way for Parties to earn credits for undertaking or financing joint measures. Because the Convention obliges Annex II Parties to provide funding and technological assistance to developing country Parties to help the latter meet their obligations, identifying which activities should be considered by Annex II Parties as fulfilling their mandatory Convention commitments and which should earn credits proved problematic and continues to prove challenging in the context of consideration of how the issues of 'additionality' of AIJ projects should be assessed – an exercise which has many parallels with the assessment of additionality issues for CDM projects.[17]

AIJ pilot phase

COP-1 could not reach agreement on the fundamental issue of in what circumstances Annex I Parties should claim credits that could count towards compliance with their UNFCCC mitigation commitments. Because many developing countries were also opposed to the grant of credits, COP-1 agreed that 'no credits shall accrue to any Party' as a result of AIJ activities during the pilot phase. This phase was to be comprehensively reviewed and was scheduled to end not later than the end of 1999. The timeframe was chosen to allow the pilot phase to support sufficient 'learning by doing' as well as to enable completion of Protocol negotiations addressing political issues about the overall size and burden-sharing agreements relating to Annex I Parties' emissions targets. As discussed above, many developing countries and the EU were concerned about ensuring there was an appropriate balance between domestic and overseas mitigation abatement efforts by Annex I Parties.

Since COP-1, the COP has regularly reviewed the way in which information about AIJ projects should be provided (for example, to help understanding of how additionality may be assessed on a practical level) and to review the geographic spread of projects. COP-5, held in 1999, reviewed the pilot phase as a whole and decided to continue it beyond 1999, encouraging Parties that had not yet had experience with projects to take up such opportunities. Decisions to continue the pilot phase have also been adopted by COP-7 and COP-8. Until COP-8 Parties participating in AIJ projects were requested to provide AIJ information to the Secretariat to enable it to prepare an annual synthesis document considered by each COP. To simplify and reduce such reporting, COP-8 decided that in future the synthesis document should be prepared on a biennial basis, rather than on an annual basis as originally provided in Decision 5/CP.1.

Uniform reporting format

Decision 5/CP.1 called on the UNFCCC Subsidiary Bodies (SBs) to develop a framework for Parties to report in a transparent and credible manner on 'the possible global benefits and the national economic, social and environmental impacts as well as any practical experience gained or technical difficulties encountered.' Such reporting was critical for determining how baselines that met the financial and environmental 'additionality' criteria set by COP-1 (discussed below) were to be assessed. Because participation in AIJ is voluntary, such reporting was, and remains, distinct from information provided in Parties' national communications. COP-3 adopted the first 'uniform reporting format' (URF) which Parties could use to report, on a voluntary basis, on their AIJ projects.[18] A revised URF was adopted by COP-8 which incorporates experience with the first URF and takes into account the kind of information that would in future be needed to assess whether AIJ projects can meet eligibility criteria for CDM and JI.[19]

Substantive AIJ criteria

Although COP-1 could not reach agreement on credits, Decision 5/CP.1 was groundbreaking in setting out criteria for AIJ, many of which fed or were directly incorporated into CDM and JI modalities. Experience of developing a URF for reporting of AIJ projects assisted negotiations on reporting and methodological issues related to the CDM and JI projects. The substantive criteria for undertaking AIJ projects remain as set out in Decision 5/CP.1 and provide as follows:

- AIJ in no way modifies the commitments of each Party under the Convention and AIJ projects are supplemental to and only one subsidiary means of achieving the Convention's objective.
- AIJ can proceed among Annex I Parties and with non-Annex I Parties that so request.
- Participation in AIJ for all Parties is voluntary and requires prior acceptance, approval or endorsement by the governments of Parties concerned.
- AIJ projects should meet the 'environmental additionality' criterion which is policy shorthand for saying they 'should bring about real, measurable and long-term environmental benefits related to the mitigation of climate change that would not have occurred in the absence of such activities.'

- AIJ projects should meet a 'financial additionality' criterion which states that 'financing of AIJ shall be additional to the financial obligations of [Annex II Parties] within the framework of the financial mechanism as well as to current official development assistance (ODA) flows.'

The environmental and financial additionality criteria for AIJ projects have been incorporated in JI and the CDM modalities and are explained in greater detail below. Decision 5/CP.1 does not limit the scope of AIJ. Thus all types of projects that reduce or sequester emissions can, in principle, be AIJ projects. There are currently over 150 AIJ projects formally communicated to the UNFCCC Secretariat with a large share being in EITs.[20] Regular updates are provided on the number, geographic location and project type of AIJ projects are reported in *JI Quarterly*.[21]

AIJ projects and the Kyoto mechanisms

Although the modalities and procedures for the CDM and JI do not mention AIJ explicitly, COP-7 agreed the following. For the CDM:

> *a project activity starting as of the year 2000, and prior to the adoption of this decision, shall be eligible for validation and registration as a clean development mechanism project if submitted for registration before 31 December 2005. If registered, the crediting period for such project activities may start prior to the date of registration but no earlier than 1st January 2000.*[22]

COP-9 clarified that this wording inadvertently excluded the possibility of projects starting between the date of adoption of Decision 17/CP.7 and the date of the first registration of a CDM project activity earning CERs and accordingly decided that such CDM project activities may use a crediting period starting before the date of registration if the project activity is submitted for registration before 31 December 2005.[23]

JI projects starting as of the year 2000 may be eligible as Article 6 projects if they meet the requirements of the guidelines for the implementation of Article 6 of the Kyoto Protocol as agreed at COP-7. ERUs shall only be issued for a crediting period starting after the beginning of the year 2008 (as before then there is no assigned amount from which they can be issued).[24]

These provisions mean that there is no automatic conversion of AIJ projects to CDM or JI projects as some Parties had wished but that each project must fulfil the criteria set out for CDM and JI projects agreed at COP-7. The crediting start date of 2000 for CDM activities operationalizes the provisions of Article 12.10 of the Protocol which states that CERs obtained during the period 2000 to 2008 can be used by Annex I Parties for compliance with Article 3 commitments. There is no such equivalent start date for JI activities under Article 6. Thus AIJ projects carried out among Annex I Parties that qualify as JI projects will only generate ERUs from 2008 onwards because no Party will have an assigned amount before 2008.

I.3 Cross-cutting mechanism issues

The Kyoto mechanisms share a number of cross-cutting features. Many of these are set out in Decision 15/CP.7 which deals with the principles, scope and nature of all three mechanisms. Others with a larger technical component are also included in rules relating to the accounting of assigned amounts set out in Decision 19/CP.7, registry-related rules adopted by COP-8 and information-related reporting and review requirements pursuant to Articles 5, 7 and 8 of the Protocol.

Adoption and review of mechanism modalities

Given that the COP and the COP/MOP have their own distinctive legal authority, there was a shared desire to provide a smooth legal pathway from the adoption of mechanism modalities by the COP to their eventual endorsement by the COP/MOP. The transition is achieved by embedding the modalities for the mechanisms in the form of annexes that are attached to draft decisions that the COP recommends the COP/MOP adopt. In the case of the CDM, the COP has agreed to assume the responsibilities of COP/MOP until the entry into force of the Protocol. This means that the CDM is legally in operation in advance of the Protocol's entry into force.

The review of mechanism modalities had to balance the need to provide legal and regulatory certainty to Parties and legal entities that wanted to use the mechanisms on the one hand with the desire to make improvements based on the inevitable 'learning by doing' that will take place as Parties gain practical experience of these innovative mechanisms. Additionally, because all three mechanisms are related to the achievement of binding Article 3 Kyoto commitments, the timing and legal nature of revisions to mechanism modalities was also an issue. The modalities for all three mechanisms contain virtually identical review provisions which provide that:

- any future revisions of the modalities, rules and procedures for each mechanism shall be decided in accordance with the rules of procedures of the COP/MOP as applied;
- the first such review will be carried out no later than one year after the end of the first commitment period (i.e. 2013), based on recommendations

by the SBI, drawing on technical advice from SBSTA, as needed; and

* further reviews shall be carried out periodically thereafter.[25]

The JI and CDM modalities make clear that any changes resulting from overall reviews shall not have retrospective effect for projects that have already commenced.

Although these rules provide legal certainty for investors that the mechanisms will not change, this does not address the issue of ensuring that investors have incentives to actually invest in mechanism transactions. As set out in the evolution of the mechanisms discussed above, investors will need certainty that Kyoto units will have value post-2012 if significant levels of mechanism activity are to be sustained. Thus the issue of negotiating second commitment period targets, which underpin the financial value of Kyoto units, will have to be addressed in the next few years rather than being left to 2013.

Equity issues

In response to moral and equity concerns outlined in section I.1 dealing with the evolution of the mechanisms, the Marrakesh Accords state that the 'Kyoto Protocol has not created or bestowed any right, title or entitlement to emissions of any kind on Parties included in Annex I.'[26] The US Clean Air Act contains a similar provision which was intended to ensure that the federal government could still take decisions about permits where this was deemed necessary to protect the public interest without having to worry about paying off polluters for possible infringement of their legal rights (Yamin, 1999). Although the reference in Decision 15/CP.7 is to 'emissions' rather than to the actual units created by the Protocol, inclusion of this provision signals that Parties do not regard holdings of Kyoto units as property rights. Rather they see them as simply as unitized and divisible embodiments of promises accepted by sovereign states in the context of a multilateral agreement which for that reason can be revoked, revised and altered through further negotiation (Werksman, 1999b).

Concerns that use of the Kyoto mechanisms might entrench as well as exacerbate existing emissions inequalities by encouraging Annex I Parties to seek cheap reductions abroad led COP-7 to agree that AIPs 'shall implement domestic action in accordance with national circumstances and with a view to reducing emissions in a manner conducive to narrowing per capita differences between developed and developing country Parties while working towards achievement of the ultimate objective of the Convention.'[27] This provision will be taken into account in the review of demonstrable progress under Article 3.2 of the Protocol. The Secretariat has been mandated to prepare a report on the implications of the per capita paragraph every time the review process under

Article 8 of the Kyoto Protocol relating to national communications and supplementary information from AIPs is completed.[28]

Supplementarity

Prioritizing domestic action has moral as well as environmental effectiveness dimensions. The extent to which either trumped-up cost-effectiveness considerations and justified quantitative constraints on the use of the mechanisms was one of the most divisive elements of post-Kyoto negotiations. The Marrakesh Accords provide that 'use of the mechanisms shall be supplemental to domestic actions and domestic action shall thus constitute a significant element of the effort made' by each Annex I Party in meeting its Article 3.1 commitments. The word 'significant' carries no quantitative connotations and was chosen in preference to words such as 'principal' and 'primary' which were deleted by the US as they implied a quantitative priority in favour of domestic action.

A qualitative assessment of whether Annex I Parties will meet the supplementarity condition was agreed and this requires Annex I Parties to submit information about their use of the mechanisms and domestic action as part of the information that must be submitted in accordance with Article 7 which will be reviewed under Article 8. An additional report on how each Annex I Party is making 'demonstrable progress' under Article 3.2 of the Protocol will also form part of the information relevant to considering supplementarity. For clarity, questions of implementation raised by the qualitative assessment of supplementarity are to be explicitly mentioned as matters that will be addressed by the Facilitative Branch and cannot be addressed by the Enforcement Branch of the Compliance Committee.[29]

It should be noted that while they are not supplementarity limits, for the first commitment period, the total additions to a Party's assigned amount resulting from eligible LULUCF projects under the CDM shall not exceed one per cent of base year emissions of that Party, times five.[30] To the extent that the rules limiting the banking of CERs and ERUs, described below, constrain the use of the Kyoto mechanisms, they could create more incentives for domestic action.

Fungibility

The term fungibility embraces a range of issues relating to the nature of the initial assigned amount allocated to each Annex I Party pursuant to the Protocol and the interchangeability of Kyoto units with each other and their relationship with the initial assigned amount.

Articles 3.10 and 3.11 allow Parties to add and subtract from their assigned amount ERUs generated through JI under Article 6 and AAUs under

ET under Article 17. Article 3.12 on the other hand allows CERs to be acquired and added to a Party's assigned amount but does not state that Annex I Parties can transfer CERs to other Annex I Parties. The omission of the word 'transfer' from Article 3.12 was used by some developing countries to oppose CERs being traded among Annex I Parties after their initial acquisition. This also led to some of them opposing fungibility.

The Marrakesh Accords provide for full fungibility of ERUs, CERs, AAUs and RMUs.[31] Units generated by CDM LULUCF activities, known as tCERs and lCERs, are equal to other Kyoto units in terms of compliance. Irrespective of how they are created, Kyoto units can be exchanged on a one-to-one basis with each other as Decision 19/CP.7 defines each of these units to equal one metric tonne of carbon dioxide, calculated using agreed global warming potentials (GWPs). Thus for the purposes of compliance, these Kyoto units are equal.

Another aspect of fungibility is the differential ability to bank Kyoto units for the next commitment period. Decision 19/CP.4 limits banking of CERs and ERUs to 2.5 per cent of a Party's assigned amount and states that RMUs cannot be banked at all.[32] Thus there are restrictions on the banking of all Kyoto units except AAUs. Because there are no rules agreed otherwise, Annex I Parties could use enough other units for compliance purposes to meet the respective restrictions and then use AAUs for the balance. Thus the restrictions on banking could be circumvented, limiting the potential impact on the fungibility of the Kyoto units.

So far as the legal nature of assigned amounts and additions to them is concerned, a number of developing countries wanted to clarify the legal nature of Annex I Parties' assigned amounts and had thus pressed for the inclusion of language stating that the Protocol does not create any rights, title or entitlement as explained above. Additionally, Parties agreed that, once recorded in the compilation and database established by the Secretariat as part of the accounting modalities under Article 7.4, the assigned amount of 'each Party shall remain fixed for the commitment period.'[33] The fixed nature of the assigned amount was also intended to address developing country concerns that the additions and subtractions of Kyoto units to the assigned amount would somehow weaken or alter the nature of Annex I Parties' legal commitments. Decision 15/CP.7, paragraph 6, provides that such additions and subtractions will take place 'without altering the quantified emission limitation and reduction commitments inscribed in Annex B to the Kyoto Protocol.'

Decision 19/CP.7 on accounting modalities provides the answer because it states that Kyoto units are not 'added to' a Party's assigned amount until it designates those units to be used for purposes of meeting its commitment – which will be done at the end of the commitment period (which for compliance purposes includes an additional period for fulfilling commitments). Prior to this point in time all Kyoto units are simply held in a national registry. Thus

CERs can be freely transferred like the other Kyoto units and are as liquid as the other Kyoto units prior to being designated for compliance use. Then CERs can no longer be subtracted but other units could be in case the Party specified more units than necessary to achieve compliance with its commitment. By defining the point of 'addition' in this way, Decision 19/CP.7 aims to meet the concerns of developing countries but without practical impact on the transferability or liquidity of CERs.

Stakeholder involvement

Modalities for JI and the CDM contain a common definition of 'stakeholder' which means 'the public, including individuals, groups or communities affected, or likely to be affected' by the JI or CDM project.[34] To support the participation of stakeholders in JI and CDM projects, the modalities for JI and CDM provide that certain types of information must be made publicly accessible. Additionally, such information is necessary because there are various points in the JI and CDM project cycle where stakeholders may intervene to ensure that decisions – whether by national authorities or international bodies – are in conformity with the modalities set out in the various COP decisions. These points of intervention, the types of information that must be available and the timing of their availability are vital for ensuring that stakeholders and others perform 'watchdog' functions for the two project-based mechanisms which if successful could result in thousands of projects worldwide.[35]

Although the rules for emissions trading do not refer to stakeholders as such, rules on the establishment of national registries which will record Kyoto unit transactions, provide that non-confidential information in national registries must be publicly accessible through a user interface available via the Internet that allows 'interested persons to query and view' information held in national registries.[36] NGOs and stakeholders, particularly businesses engaged in the mechanisms, will therefore play a vital function in spotting the frequency, types and implications of discrepancies that might arise, including whether the transaction log established by the UNFCCC Secretariat is itself functioning correctly. Modalities for constructing national registries to enhance their public accessibility have been further elaborated since Marrakesh and are to be considered further, including how issues relating to confidential data should be handled.[37]

I.4 Participation/eligibility requirements

Another common feature of the mechanisms is that Articles 6, 12 and 17 emphasize that Annex I Parties 'may' use the mechanisms to fulfil their Article 3.1 commitments. Although the decision to participate is entirely voluntary, once made each Party has to fulfil certain participation requirements. All the mechanisms set out certain legal and administrative provisions to ensure that Parties retain sovereign responsibility over mechanism-related transactions taking place under their jurisdiction.

All three mechanisms allow Parties to authorize private actors to participate in the mechanisms.[38] In each case the mechanism modalities specify that the Party which authorizes legal entities shall remain responsible for the fulfilment of its obligations under the Protocol and shall ensure that such participation is consistent with the mechanism's modalities.[39] Legal entities may only transfer and acquire Kyoto units, however, where a Party itself meets the participation requirements set out in the mechanisms. From an economic perspective, the participation of entities in the mechanisms, under the responsibilities of Parties, was considered important because while mechanisms open only to governments are more efficient than no mechanisms at all, mechanisms limited to governments are not as efficient as mechanisms open to a wider range of actors. This is because governmental trading equates national marginal costs but does not equalize marginal abatement costs across sources within each country and it is the latter that lead to more significant reductions in compliance costs.

All the mechanisms define minimum environmental integrity-related standards that must be met by Parties that wish to participate in the mechanisms. The requirements are referred to here as 'eligibility conditions' to distinguish them from the broader legal and administrative participation requirements set out above. For Annex I Parties, these requirements will be assessed by the Enforcement Branch of the Compliance Committee, discussed in more detail in section I.8 below.

Protocol ratification

Whether participating as host or investor, an Annex I Party can only transfer and acquire ERUs under JI and all kinds of Kyoto units under Article 17 emissions trading if it is a Party to the Kyoto Protocol. This is simply because only Parties to the Protocol that have Article 3 commitments will have an 'assigned amount'. If countries that are not Parties to the Protocol, such as the US and Australia, establish national emissions trading schemes, the permits or allowances created by such national schemes will not form part of the Kyoto system.[40] Such non-Parties to the Protocol can unilaterally decide to allow the use of Kyoto units with their domestic obligations.

The CDM modalities specify that until the Protocol enters into force, all Parties to the Convention can participate in CDM projects.[41] After the entry into force of the Protocol, non-Annex I Parties may participate in the CDM only if they are Parties to the Protocol.[42] The CDM eligibility requirements for

Annex I Parties are not as clear and refer to eligibility at the time when such Parties will *use* CERs for compliance. Thus while the CDM modalities are really focused on how Parties to the Protocol can engage in CDM activities, they neither excluded the possibility that non-Parties to the Protocol could participate in CDM project activities nor spelled out how they may do so (Wilkins, 2002). To provide greater clarity, the CDM Executive Board at its eighth meeting clarified that with regard to validation requirements to be checked by a designated operational entity, before entry into force of the Kyoto Protocol, all Parties to the Convention may participate in CDM project activities. Subsequently, in accordance with provisions of paragraphs 37(a) and 40(a) of the CDM modalities and procedures, the registration of a proposed CDM project activity can, however, only take place once approval letters are obtained from Parties to the Convention that have ratified the Kyoto Protocol.

Designating national authorities

Both project-based mechanisms require Parties to designate national authorities to provide oversight of JI and CDM projects. Any Party involved in an Article 6 project shall inform the Secretariat of its designated 'focal point' while the CDM modalities require all Parties to designate a 'national authority for the CDM'.[43] The functions of these bodies are explained below. The requirement to establish a Designated National Authority (DNA) for the CDM were insisted upon by developing countries as they wanted to ensure that governmental bodies would be kept abreast of CDM projects and initiatives. One of the early experiences of the AIJ pilot phase had been the negotiation of projects by Annex I countries/companies directly with non-Annex I entities with the governments often not in the loop until a very late stage.

By February 2003, DNAs had been established in around 50 developing countries and in 10 Annex I Parties with many other Parties nearing the completion of domestic processes for doing so.[44]

Establishing assigned amount and pre-commitment period report

No Annex I Party can undertake JI and ET transactions, or use CERs towards compliance, unless its assigned amount pursuant to Articles 3.7 and 3.8 has been calculated and recorded in accordance with the annex to Decision 19/CP.7 on modalities for the accounting of assigned amounts. To meet this condition, each Annex I Party has to submit a pre-commitment period report which contains all the information needed to calculate its assigned amount as well as other kinds of information necessary to demonstrate that the Party is able to monitor, track and record mechanism-related transactions, such as having a national system and national registry in place (see below). This report is to be submitted prior to 1 January 2007 or one year after the entry into force of the Protocol for the Party, whichever is the latest. It is subject to a thorough review by the expert review teams (ERTs).

National system

No Annex I Party can undertake JI and ET transactions, or use CERs, unless it has in place by no later than 1 January 2007 a national system for the estimation of GHG emissions and removals pursuant to Article 5.1 of the Kyoto Protocol.

The term national system refers to the institutional, legal and procedural arrangements put in place by an Annex I Party to ensure that it can adequately estimate, report and archive GHG inventory data. Guidelines for the establishment of national systems and review processes which aim to ensure these are robust enough to meet mechanism requirements have been agreed under the Marrakesh Accords.[45] These guidelines require Annex I Parties to adhere to the 1996 IPCC Revised Guidelines for National Greenhouse Gas Inventories which incorporate common methodologies and reporting formats devised by the IPCC and subsequently endorsed by the COP for Parties to use when calculating emissions data and compiling their annual GHG inventories.[46] The guidelines for national systems require Annex I Parties to designate a national authority which is charged with maintaining a national system, including the adoption of quality assurance/control procedures to ensure GHG inventories are to the standard set by Kyoto. A description of the national system must be included in the pre-commitment period report which serves to identify whether Parties have the administrative capacity to track mechanism transactions and to monitor their GHG emissions.

As many EITs lack capacity to provide annual inventories following IPCC guidelines, the national system component of the mechanism participation requirements is likely to present many of them with problems unless additional capacity-building measures are taken. Anxiety that they could find themselves ineligible to participate in the Kyoto mechanisms because of their limited capacity to implement national systems to the standard set by the Protocol was a strong motivation for EITs to negotiate Track 2 JI, discussed below.

National registry

The term national registry refers to procedures and mechanisms set up by an Annex I Party to ensure the accurate tracking of Kyoto mechanism transactions and LULUCF accounting under Articles 3.3 and 3.4. National registries will be part of a broader system of accounting which will also include a separate CDM registry and an independent transaction log maintained by the

UNFCCC Secretariat. Registries, and these related components, are necessary not only for trading under the Kyoto mechanisms but also for linking the Kyoto mechanisms with national schemes that are Kyoto independent. To assist Parties establish registries that are compatible, the COP has developed technical guidelines which were adopted by COP-8 with SBSTA-19 agreeing conclusions on the need for the UNFCCC Secretariat to focus on developing the independent transaction log and for Annex II Parties to fund this work.[47] The chapter by Hobley in this volume describes the functions, design and implementation to date of registries under the Kyoto Protocol and the EU ETS in more detail.

So far as the Protocol participation requirements are concerned, whether or not they choose to participate in the mechanisms, all Annex I Parties must designate an organization that will serve as the 'administrator' of the national registry that must be established and maintained by all Annex I Parties as part of the accounting modalities necessary for tracking their assigned amounts.[48] Because registries are such a vital part of the administrative infrastructure for tracking transactions under the mechanisms, their effective functioning will be thoroughly reviewed and tested by expert review teams (ERTs) as part of the pre-commitment period report and review procedures. Confirmation that they are working effectively will be part of the review of the pre-commitment period report which is necessary for Annex I Parties to establish their assigned amounts (without this they have no assigned amount to trade).

Annual inventories and adjustments

Submission of accurate annual inventories, prepared in accordance with guidance adopted pursuant to Article 5.2 and Article 7.1, and submitted and reviewed annually and according to schedule, together with the supplementary information described below, is the backbone of the mechanisms eligibility requirements for Annex I Parties.[49]

The threshold conditions of failing to meet Article 5.2 and Article 7.1 eligibility conditions are set out in Decision 22/CP.7 which provides that any Annex I Party shall fail the mechanisms, methodological and reporting eligibility requirements for JI, the CDM and ET as follows:[50]

- if it has failed to submit an annual inventory of anthropogenic emissions by sources and removals by sinks, including the national inventory report and the common reporting format within six weeks of the submission date established by the COP;
- if it has failed to include an estimate for an Annex A source category (as defined in the IPCC Good Practice Guidance and Uncertainty Management in National GHG Inventories) that individually accounted for 7 per cent or more of the Party's aggregate emissions, defined as aggregate submitted emissions of the gases and sources listed in Annex A to the

Protocol, in the most recent of the Party's reviewed inventories in which the source was estimated;

- if any single year during the commitment period the aggregate adjusted GHG emissions of the Party concerned exceed the aggregate submitted emissions, defined as aggregate submitted emissions of the gases from the sources listed in Annex A to the Kyoto Protocol, by more than 7 per cent;
- if at any time during the commitment period the sum of the numerical values of the percentages calculated in relation to the single year eligibility requirement (stated above) for all years of the commitment period for which the review has been conducted exceeds 20; and
- if an adjustment for any key source category (as defined in IPCC Good Practice Guidance) of the Party concerned that accounted for 2 per cent or more the Party's aggregate emissions of the gases from the sources listed in Annex A was calculated during the inventory review in three subsequent years, unless the Party concerned has requested assistance from the Facilitative Branch of the Compliance Committee in addressing this problem, prior to the beginning of the first commitment period and assistance is being provided.[51]

The application of adjustments will be a critical element for many Annex I Parties, especially some EITs who continue to have capacity-related difficulties in submitting complete inventories. If current inventory standards are anything to go by, the application of adjustments to complete missing or inadequately justified emissions data is likely to be a not infrequent event. Where an adjustment is proposed by an ERT and is disputed by an Annex I Party, the matter is to be determined by the Enforcement Branch of the Compliance Committee. COP-9 adopted technical guidance in the form of recommendations for COP/MOP on methodologies for the application of adjustments by ERTs to ensure the circumstances in which these are proposed and applied are understood and acceptable to Annex I Parties in the hope this will limit the number of cases bought before the Enforcement Committee concerning disputed adjustments.[52] Decision 20/CP.9 requested the Secretariat to establish a process to enable ERTs to gain experience with adjustments in the inventory review process in 2003–2005 using real inventory data subject to the consent of the Party concerned.

The issue of guidance on how adjustments will be applied in respect of sink sources will be considered at COP-10, taking account of methodological work by the IPCC (completed for COP-9 in accordance with a request from COP-7) to develop Good Practice Guidance for LULUCF.[53] It is important to note that the mechanisms' eligibility provisions make clear that for the first commitment period, Parties will have to submit annual inventory data on sinks as part of their mechanisms eligibility requirements. The *quality* of sinks data, however, will be immaterial because the quality assessment for determining eligibility is limited to the parts of the inventory pertaining to emissions of GHGs from sources/sectors listed in Annex A to the Protocol. The lack of a quality assessment of sinks data was a significant issue at Marrakesh because a number of Annex I Parties, particularly EITs, still find it difficult to provide accurate data on sinks.[54] Parties that had wanted to link improvement of sinks data to mechanism eligibility had to be satisfied by the fact that Parties that fail to provide accurate data on sinks will only be allowed to issue RMUs in respect of sinks on which they have reported adequately – a provision which provides some degree of environmental integrity but which creates no incentives for general improvement of sinks data as a whole.[55]

Supplemental information

Submission of supplementary information on assigned amounts pursuant to Article 7.1 by Annex I Parties and adherence to guidelines for the accounting of assigned amounts adopted pursuant to Article 7.4 are eligibility requirements for all three mechanisms. The supplemental information to be included in national communications is set out in the KP reporting guidelines set out in Decision 22/CP.7 and 22/CP.8, and Decision 19/CP.7.

Questions of implementation relating to information about how an Annex I Party is striving to implement its Article 3.14 commitments (relating to the minimization of adverse impacts of response measures) are not an eligibility condition for the mechanisms.[56]

Commitment period reserve

The requirement to establish and maintain a commitment period reserve (CPR) is set out in Decision 18/CP.7 on emissions trading. Although it is not formally expressed as an eligibility requirement for all mechanisms, provisions relating to it must be adhered to if a Party is to engage in mechanism transactions in conformity with Decision 18/CP.7. The CPR is discussed further below under IET pursuant to Article 17.

Acceptance of compliance procedures

At Marrakesh, Parties agreed language that 'environmental integrity is to be achieved through sound modalities, rules and guidelines for the mechanisms, sound and strong principles and rules governing land use, land use change and forestry activities and a strong compliance regime.'[57] But unresolved differences about the binding nature and form of the compliance procedures that should be adopted by the COP/MOP left a number of JUSCANNZ countries unwilling to agree acceptance of the compliance procedures as an eligibility condition for use of the mechanisms.

The final wording agreed at Marrakesh provides that:

> *The eligibility to participate in the mechanisms by a Party included in Annex I shall be dependent on its compliance with methodological and reporting requirements under Articles 5.1 and 5.2 and Articles 7.1 and 7.4 of the Protocol. Oversight of this provision will be provided by the enforcement branch of the compliance committee, in accordance with the procedures and mechanisms relating to compliance as contained in decision 24/CP.7, assuming approval of such procedures and mechanisms by the [COP/MOP] in decision form in addition to any amendment entailing legally binding consequences, noting that it is the prerogative of the [COP/MOP] to decide on the legal form of the procedures and mechanisms to compliance.*

It is important to emphasize that this wording was not intended to undermine or prejudice the oversight of agreed eligibility conditions by the Enforcement Branch which all Parties agree is necessary.

Eligibility assessment, consequences and reinstatement

Oversight of the mechanisms' eligibility conditions is to be provided by the Enforcement Branch of the Compliance Committee in accordance with Decision 24/CP.7. Enforcement Branch procedures and timetables for how mechanism eligibility will be assessed are explained briefly in section I.8 below.

I.5 Emissions trading

Article 17 requires the COP (not the COP/MOP) 'to define the relevant principles, modalities, rules and guidelines, in particular for verification, reporting and accountability for emissions trading' among Annex B Parties with the proviso that such trading shall be supplemental to domestic action for the purposes of meeting the quantified commitments under Article 3. Because more substantive provisions on trading threatened adoption of the Protocol, the three short sentences of Article 17 represent all that could be agreed in Kyoto. Four years of negotiations later, the IET modalities agreed at Marrakesh barely run to two pages – evidence enough that if foundational principles can be agreed, the conceptual simplicity of emissions trading is an alluring feature, but one which requires binding targets, robust reporting and a strong national and international infrastructure to monitor, track, verify and compel compliance to make good its promise. The following sections provide an overview of IET under the Protocol and sets out the salient features of the trading modalities agreed at Marrakesh.

Overview

The key feature of Article 17 trading is that it is confined to Annex B Parties, i.e. countries with binding targets under the Protocol.[58] These Parties will have access to units of 'assigned amount' and, subject to the modalities agreed at Marrakesh, and not otherwise, they will be able to transfer and acquire these units and the full range of Kyoto units from each other to fulfil their Article 3.1 commitments.[59] The purpose of IET – to meet Article 3 commitments – is important to bear in mind because it means IET is a *means* to achieve a given environmental constraint. Article 17 trading is therefore a classic example of a cap-and-trade scheme of the kind first proposed by the USA in the Protocol negotiations in 1996.

As with the other mechanisms, the Marrakesh Accords provide that IET has to be undertaken in accordance with the modalities agreed by the COP which will be endorsed by the COP/MOP. This is significant because at Kyoto a number of   JUSCANNZ countries had argued that the provisions of the Protocol gave them a 'right to trade' without further reference to the COP. On the other hand, the need to develop further trading modalities quickly was one reason why Article 17 refers to the COP, and not the COP/MOP, as the body charged with defining further trading rules, the rationale being that the COP could provide institutional authority for interim trading which Parties thought might proceed even prior to the entry into force of the Protocol in a way that is now envisaged for the CDM. A clearer appreciation of the need to separate Convention from Protocol related matters has meant the reference to the COP in Article 17 is now largely of historic interest and the IET modalities confirm that it is the COP/MOP that will make further decisions relating to Article 17.

Principles and supplementarity

The reference in Article 17 to defining the relevant 'principles' for IET was a code word referring to a range of moral and equity issues and environmental concerns that had been touched on during the Protocol negotiations but which developing countries felt had not been adequately considered.[60] Decision 15/CP.7 sets out the principles, nature and scope of all three mechanisms and addresses issues relating to equity, fungibility and supplementarity which are discussed above.

Participation requirements

The reference in Article 17 to 'modalities, rules and guidelines, in particular for verification, reporting and accountability for emissions trading' relates to concerns that emissions trading could serve the purposes of helping Parties meet their Article 3 commitments and not adversely affect the environmental integrity of the Protocol. These concerns are addressed through the participation and eligibility requirements, explained above, and through the commitment period reserve, explained below.

Commitment period reserve

Because selling Kyoto units through emissions trading is likely to prove profitable, Article 17 creates the possibility that an Annex I Party could find itself in non-compliance with its Article 3 commitments through calculated or inadvertent overselling, particularly where there is weak international enforcement of compliance with international commitments (Yamin et al, 2001). Originally proposed by AOSIS, the concept of a commitment period reserve (CPR) emerged as a compromise to address large-scale selling by specifying the minimum quantity of Kyoto units a country must have in its national registry at any time and thus limiting the scope of non-compliance. One of the challenges was to set the CRP requirements so that they would protect against non-compliance yet not be so restrictive as to limit the liquidity of the market. The Bonn Agreement provided that 'each [Annex I] Party shall maintain ... a commitment period reserve which should not fall below 90 per cent of [its] assigned amount ... or 100 per cent of five times its most recently reviewed inventory, whichever is the lowest.'[61] The reserve can be made up of any Kyoto units valid for that commitment period. The limits adopted would prevent large-scale non-compliance.

At Marrakesh insistence by some Parties that the word 'should' did not make *maintenance* of the CPR at 90 per cent levels mandatory resulted in the CRP-related text being shifted from the eligibility section of Decision 18/CP.7 into a later portion of the text. Because lack of a mandatory level would have provided no safeguard against overselling, and thus undermined the sense of the compromise agreed in Bonn, negotiators agreed to leave the Bonn wording intact but agreed that 'a Party shall not make a transfer which would result in these holdings [of ERUs, CERs, AAUs and/or RMUs] being below the required level of the commitment period reserve.'[62]

Decision 19/CP.7 requires an Annex I Party to calculate the level of its CPR as part of the process of establishing its assigned amount. The transaction procedures set out in this Decision also make it difficult for any Annex I Party to breach the minimum level of the CPR without attracting immediate attention because the transaction log to be maintained by the Secretariat will verify whether all transactions of Kyoto units are in conformity with the required rules.[63] If any transaction is found not to be in order, because for example it breaches the CPR limits, the log will notify the national registry concerned which is legally obliged to stop the transaction. The acquiring registry will also be notified. Any units already transferred in breach of CPR limits will be deemed invalid for compliance purposes until the level of the CPR is re-established, with Parties being given 30 days to return to the required level. All discrepancies must be forwarded to the Secretariat as part of the review process for the Party or Parties concerned under Article 8. Thus any Party that has dropped its CPR below required levels will have to address questions relating to this in the context of the review of supplementary information under Articles 7.1 and 7.4 which is conducted annually by ERTs.[64] Any Party that allows its CPR to fall below required limits without taking prompt corrective measures risks finding itself before the Enforcement Branch and in the meantime not being able to transfer Kyoto units (other than those issued and transferred under Track 2 JI explained below).

Restraints and linkages

Because participation in Article 17 is voluntary, an Annex I Party is free to impose restrictions on who it will trade with under Article 17 and with respect to what. For example, a Party may refuse to accept certain ERUs and CERs towards its commitments and is free to decide the amount of CERs its legal entities can acquire for use towards compliance with domestic obligations it may establish.[65] No trade law implications would appear to flow from these types of restraints because transfers of Kyoto units represent exchanges of sovereign commitments which are not covered by international trade law disciplines as these are concerned with trade restraints relating to traditional types of goods and services (Werksman, 1999a, 1999b) and carbon permits cannot be considered as goods and services.

Parties that can participate in Article 17 emissions trading may do so without establishing a national emissions trading scheme. They may grant authorization to governmental bodies or authorize legal entities to participate even in the absence of a national trading scheme. Many Parties will, however, establish national trading schemes as one kind of domestic policy measure. Where such national trading schemes exist, the relationship between Kyoto units and the permit or allowance of the domestic scheme will need to be addressed.[66]

I.6 Clean development mechanism

Overview

Subject to meeting the participation requirements outlined above, the CDM modalities developed pursuant to Article 12 of the Protocol at Marrakesh allow non-Annex I and Annex I Parties and their authorized entities to jointly undertake emission reduction and afforestation/reforestation projects, in non-Annex I Parties that contribute to sustainable development and result in certified emissions reductions (CERs).[67] Article 12 of the Protocol therefore creates an innovative international 'baseline-and-credit' trading scheme with unprecedented levels of   coverage in terms of activities and types of partnerships. By creating assets with market value, CERs, the CDM is intended to help channel private sector investment towards climate-friendly projects that might not otherwise have taken place.

Lack of non-Annex I Parties' quantitative mitigation commitments in the CDM context creates incentives for those involved in CDM projects to inflate the amount of CERs claimed, through, for example, manipulation of counterfactual 'baseline' scenarios. Processes leading to CER issuance thus require multilateral oversight. This oversight is provided by the Executive Board of the CDM (EB) which itself draws on independent organizations to assess conformity of CDM project activities with internationally agreed modalities. An additional innovative aspect of the CDM is that a share of the proceeds from certified project activities (set at 2 per cent of the CERs), is automatically deposited in the CDM Registry, maintained by the UNFCCC Secretariat, to fund adaptation in developing countries vulnerable to climate change and, in due course, to cover CDM-associated administrative expenses.

Finally, apart from generating CERs, one strategic rationale for the CDM's inclusion in the Protocol was to provide a quantified means for non-Annex I Parties to contribute to mitigation commitments and to get a better understanding of trading mechanisms, but without such Parties having to take on legally binding mitigation targets. Participation by developing countries in the CDM is thus part of their broader efforts to contribute to climate mitigation in a manner which provides for 'learning by doing' while respecting their sustainable development priorities.

CDM project cycle

Article 12 of the Protocol lacked operational details specifying various actors' roles in CDM project activities. Many project cycles with different implications for actors were possible and there were many competing visions and preferences. These included the 'classic' model where an Annex I Party or its legal entities invests in projects in partnership with a developing country Party (bilateral approach), a unilateral approach where a non-Annex I Party undertakes CDM activities without an Annex I Party counterpart (unilateral CDM) and where an international financial institution or other intermediary puts together a portfolio of CDM activities on behalf of others (multilateral or portfolio approach). Like Article 12 itself, the project cycle set out in the Marrakesh CDM modalities can be tailored to fit all these approaches provided that the project concerned follows the five stages of the CDM project cycle and conforms to all the substantive requirements therein. Only projects that do so will result in the issuance of CERs by the EB into the CDM Registry.

Before any particular project can commence its journey through the project cycle, the following preliminary steps are necessary:

- designation of a national authority, called the Designated National Authority (DNAs) to provide written approval of the voluntary participation of each Party involved in the proposed project and to confirm the project's sustainable development credentials;
- designation of one or more Applicant Entities (AE) to carry out key functions in the project cycle by EB on a provisional basis. Once confirmed by COP/MOP these are known as designated operational entities (DOEs). DOEs must be hired on a contractual basis by project participants to perform specific functions;
- written clarification by Project Participants (PP) of their respective roles, including, crucially, how CERs arising from the project are to be distributed and the communication modalities necessary for PPs to liaise with the EB/CMD and the Secretariat. PPs can be Parties or private and/or public entities authorized by a Party to participate in the project under the responsibility of the Party; and[68]
- establishment of a CDM Registry by the EB to ensure the accurate accounting of the issuance, holding, transfer and acquisition of CERs by Parties not included in Annex I.[69]

The project cycle is set out in Figure I.1 which indicates the lead institutional actor with responsibility for the steps needed to progress from one stage to the next stage of the CDM project cycle. This project cycle applies to all CDM activities except for CDM small-scale project activities for which a more streamlined cycle has been agreed (see below).[70] The five steps of the CDM project cycle are:

Step 1: Design

A PP should design and submit information about a proposed project using a specific format. Appendix B of Decision 17/CP.7 outlines the key categories of information that 'shall' be included by PPs in the format of a document called the Project Design Document (PDD). This element of the PDD has been further refined by the Executive Board on the basis of Appendix B of the CDM modalities.[71] Submission of the PDD is necessary to commence the process for validation.[72]

Step 2: Validation and registration

Validation is the process of independent evaluation of a project activity by a DOE on the basis of the PDD to assess whether the proposed activity conforms to the CDM modalities.[73] Registration is the formal acceptance by the EB of a validated project as a CDM project activity. Registration is the prerequisite for the verification, certification and issuance of CERs related to that project activity. Validation ensures that a proposed project has the approval of the host Party. Registration ensures that all CDM projects fall under the purview of the EB/CDM and are thus subject to international scrutiny.

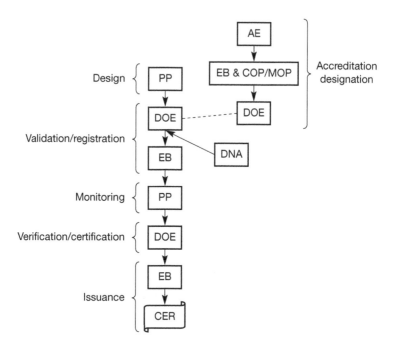

Figure I.1 *CDM project activity cycle*

Source: UNFCCC Secretariat CDM website

A 'pre-validation' stage is also now envisaged where a project is based on a new baseline and/or monitoring methodology. In such pioneering cases, the new baseline methodology shall be submitted by the DEO to the Executive Board for review, prior to a validation and submission for registration of this project activity, with the draft PDD, including a description of the project and identification of PPs.[74] This procedure is likely to be invoked more frequently in the early stages of the CDM. Where methodologies previously approved by the EB are being used, these methodologies must have been made publicly available along with any relevant guidance by the Board.

Step 3: Monitoring

Monitoring refers to the identification, collection and archiving of information necessary to design and implement a monitoring plan as required by CDM modalities. Implementation of the monitoring plan by the PP is a condition for the verification, certification and issuance of CERs.[75]

Step 4: Verification and certification

Verification refers to the periodic independent review and *ex post* determination by a designated operational entity (DOE) of the monitored reductions in anthropogenic GHG emissions by sources that have occurred as a result of a registered CDM project activity during the verification period.[76] Certification is the written assurance by the DOE that, during a specified time period, a project activity achieved the reductions in anthropogenic emissions as verified.

Step 5: Issuance of CERs

Certification results in a certification report by the DOE which forms the basis of a request by the DOE to the EB for the issuance of CERs.[77] Issuance refers to the instruction by the EB to the CDM Registry Administrator to issue (i.e. create) a specified number of CERs into the pending account of the EB. The responsibility to forward CERs to the registry accounts of the PPs rests with the CDM Registry which must also forward the CERs to cover the share of the proceeds for administrative expenses and adaptation to the appropriate accounts. Thus the EB has no role in the final allocation of CERs to PPs as this is a purely technical act undertaken by the CDM Registry Administrator.

CDM institutions and procedures

Article 12.4 provides that the 'CDM shall be subject to the authority and guidance of the COP/MOP and be supervised by an executive board of the CDM.' This section outlines the roles and responsibilities of various institutions and actors in the overall oversight of the CDM focusing on how the institutions, rules and procedures of the CDM have given effect to Article 12.7 which requires the COP/MOP 'to elaborate modalities and procedures with the objective of ensuring transparency, efficiency and accountability through independent auditing and verification of project activities.' Accordingly, apart from the actors involved in the project cycle set out above, it describes how stakeholders and UNFCCC accredited observers are involved in the CDM.

Prompt start of CDM

Until the entry into force of the Kyoto Protocol, the COP will play a significant preparatory role in Protocol issues. The 'prompt start' of the CDM envisaged in Article 12.10 has necessitated that in relation to the CDM, the COP plays a decision-making, rather than a preparatory, role. Decision 17/CP.7 at paragraphs 2 and 4 accordingly provides that the COP 'shall assume the responsibilities of the COP/MOP' as set out in the CDM modalities and that the EB and any DOEs 'shall operate in the same manner' as specified in the CDM modalities, i.e. as if the Protocol had entered into force. All the decisions taken by the COP, the EB and DOEs will be confirmed and given full retrospective effect by the COP/MOP upon the Protocol's entry into force, including issuance of CERs. For clarity the remainder of this section refers to the role of the COP/MOP but readers should bear in mind that presently the COP is legally already acting as the COP/MOP in matters relating to the CDM.

COP/MOP

The COP/MOP has ultimate authority over the CDM as a whole. Other than fundamental procedural matters of political relevance, such as amendment or elaboration of the rules of procedure of the EB, over which the COP/MOP retains decision-making powers, the CDM modalities confine the role of the COP/MOP essentially to examining broader strategic issues relating to the CDM, such as the geographical spread of projects and of DOEs and arranging funding of CDM project activities for those in need of assistance (discussed in the section on funding issues below). Because only the COP/MOP has the necessary legal authority, it also takes formal decisions involving external bodies, such as the designation of DOE.

So far as the relationship between the COP/MOP and the EB is concerned, the COP/MOP shall give 'guidance' to the EB on recommendations made to the COP/MOP by the EB as well as on the basis of annual reports submitted from the EB.[78] This formulation means that the COP/MOP is not intended to be involved in the day-to-day administration of the CDM or to re-open matters which the CDM modalities define as functions to be exercised by the EB. This division of labour was not readily accepted by a number of developing countries, particularly OPEC countries, as they wanted the EB either to formulate recommendations which the COP/MOP should decide and/or allow matters decided by the EB to be appealed to the COP/MOP. These views did not prevail because the lack of majority-voting decision-making rules for the COP/MOP would have meant PPs could not be certain any agreement would ever be reached by the COP/MOP, or even if it did, would have resulted in lengthy delays.

Executive board

Apart from the mention in Article 12.4 stating that subject to guidance from the COP/MOP, the Executive Board should supervise the CDM, Article 12 mentions only one other function for the EB: the provision of guidance on the involvement of private and/or public entities. Decision 17/CP.7 elaborates the functions, composition and rules of procedure for the EB. The creation of a limited membership body with substantive decision-making functions *and* majority voting was a breakthrough in the Bonn Agreement as until then Parties had been unable to agree composition and voting issues in relation to other bodies.

Because the EB is the first of the Marrakesh Accord bodies to commence operation, its rules of procedure and working practices are under greater scrutiny and will, in all probability, develop further with more experience. The iterative nature of the additional rules adopted by the Board is reflected in guidance provided by COP-8 which adopted the Board's rules of procedure but encouraged the Board 'to keep its rules of procedure under review, and if necessary, make recommendations … of any amendments or additions aimed at safeguarding its efficient, cost-effective and transparent functioning.'[79]

Functions
The Board's main functions are to formally accept validated projects as CDM projects, issue CERs and accredit operational entities provisionally, pending their formal designation by the COP/MOP. These functions, however, are carried out on the basis of rules and guidelines approved by the COP/MOP. The Marrakesh Accords include a number of procedural rules relating to the operation of the Board which were supplemented through a full set of rules of procedure elaborated by the Board and adopted (with some amendments) by COP-8.[80]

The operational functions of the EB involves close and direct liaison not only with Parties and intergovernmental organizations (IGOs) but also with businesses, project developers, non-governmental organizations (NGOs) and other market intermediaries involved in implementing CDM projects. To be effective, the EB has therefore had to adopt a more 'business-like' approach than that commonly seen in other IGOs. The environmental integrity of the CDM, and potentially of the Protocol, could come to depend on the quality of decisions made by the Board. Finally, because its work involves financial investments worth millions of dollars, it must be seen to be acting in an impartial manner. For this reason special requirements have been built in to safeguard commercially confidential information that members of the EB come across during their duties as well as after their term has expired.[81]

Membership
The Board is composed of ten members from Parties to the Kyoto Protocol, including one from each regional group plus one from the small island developing states (mirroring the COP Bureau formula), along with two representatives of Annex I Parties and two of non-Annex I Parties. In addition, the Marrakesh Accords specify that the formula will be applied taking into account the current practice in the COP Bureau, an implicit reference to the representation of oil exporting developing countries through one of the groups. Given that three of the five regional groups are almost exclusively composed of developing countries, this formula ensures a greater representative of developing countries in the EB. In an innovative move aimed at accommodating the high demand for representation on the EB, each full member is accompanied by an alternate from the same group. Alternates enjoy most of the rights of members, but not the right to vote. The addition of alternates has proved to be an especially useful resource to

draw upon because the Board has established a number of specialist panels and liaison points and the intensity of meetings (six per year) means not all members have been able to attend all meetings. References to 'members' below therefore also apply to 'alternates', unless otherwise specified.

Members are nominated by their groups and elected by the COP/MOP. As with other elected posts, Parties will be required to give active consideration to the nomination of women for the EB to improve the gender balance of FCCC and KP elected institutions.[82] The EB elects its own Chair and Vice-Chair at its first meeting each year, with one being from an Annex I Party and one from a non-Annex I Party and the positions rotating annually between the two groups.[83] The Chair's functions inscribed in the EB's rules of procedure are very similar to those set out for the COP President in the COP's Rules of Procedure, described above. The Vice-Chair's main function is to replace the Chair, should s/he be absent.

Members of the EB are elected for a period of two years and are eligible to serve for a maximum of two terms.[84] They must 'possess appropriate technical and/or policy expertise' (although the nature of such expertise is not specified). Importantly they must act in their personal capacities, a departure from other bodies under the Convention where members of bodies act as government representatives. To ensure financial integrity and independence, EB members must not have any 'pecuniary or financial interest' in any CDM project or designated operational entity. Additionally, EB members are forbidden from disclosing any confidential or proprietary information relating to their work, including after their term of office has expired. All EB members must take a written oath of service confirming their adherence to the above-mentioned stipulations before assuming their duties.[85] In addition, the Board may suspend any member, and recommend to the COP/MOP that his/her service be terminated, if that member is found to be in breach of the above provisions or fails to attend two consecutive EB meetings without proper justification.[86] However, the member concerned is given the right to a hearing and the matter must be put to a vote within the EB.

Board meetings

The EB must meet at least three times a year, where possible in conjunction with sessions of the regime bodies. It has actually done so 12 times since establishment – six per year (see below). The cost of participation by developing country members and their alternates as well as other Parties eligible under the FCCC are to be covered by the budget of the EB.[87] Details of these costs are provided in the EB second report to COP-9. Desire to reduce costs has lead to innovate ways for the Board to meet 'virtually' and use the Internet to facilitate decision-making (see below on voting).

The Chair, assisted by the secretariat, drafts the provisional agenda for each meeting, which is agreed by the EB at its previous meeting. Members may, however, propose additional items up to four weeks before the start of the meeting, with the provisional agenda then circulated to members by the secretariat at least three weeks before the meeting's opening. All documents for the EB are first sent to members two weeks before each meeting, and then posted on the secretariat website, subject to confidentiality provisions. Before a meeting of the EB can start, a quorum must be present, consisting of at least two-thirds of the members (not alternates), including a majority of Annex I and non-Annex I Party members.[88] The working language of the Board is English but all decisions of the EB must be made publicly available in all six UN languages.[89]

The EB began its work as soon as the modalities and procedures for the CDM were decided upon through the Marrakesh Accords, with an organizational meeting held immediately after COP-7. Since then it has met a total of 12 times. Its work programme and decisions are set out in its two annual reports submitted by EB to COP-8 and COP-9.[90] On the basis of these reports, COP-8 approved the rules of procedure of the EB and the simplified modalities and procedures for small-scale CDM projects (SCC) which are to be found at Annex I and Annex II of Decision 21/CP.8. This Decision also provided further guidance on the EB's work with additional guidance being provided by COP-9.[91]

Observers and stakeholders

According to both the Marrakesh Accords and the CDM rules of procedure, meetings of the EB are 'open to attendance, as observers, by all Parties and … UNFCCC accredited observers and stakeholders, except where otherwise decided.' In this regard, the CDM rules of procedure specify that Parties to the Convention that are not Parties to the Protocol may exercise the same rights as all other observers.[92]

To enhance the efficiency of the EB's work, the 'open to attendance by observers' rule has been interpreted in a novel, highly restrictive manner that runs contrary to the usual meaning given to the term 'attendance' which refers to physical presence at a meeting. In the CDM case, Parties and IGOs, along with representatives nominated by each NGO constituency up to a maximum of 50 persons, have not been permitted to enter into the room where the EB is meeting but are allowed to observe the proceedings of the Board by sitting in a nearby room where the meeting is broadcast live. Observers are permitted to make presentations to the Board, on the invitation of its Chair. Meetings are, in addition, broadcast live over the Internet without any viewing restrictions and, where EB meetings have coincided with subsidiary body sessions, the EB Chair has briefed observers after the meetings.

Voting

Decisions of the EB are to be taken by consensus where possible but, if all efforts at reaching consensus have been exhausted, decisions may be taken by a three-fourths majority[93] of members present and voting, that is members present at the meeting and casting an affirmative or negative vote.[94] The inter-action of this rule with the quorum requirements described above mean that in practice decisions can only be made by the Board if they have support from the majority of Annex I Parties and also from Parties not included in Annex I. Each member has one vote and alternates may not vote, unless they are acting for the member in his/her absence.[95] Interestingly, the CDM rules of procedure go further in defining an operational meaning of consensus; that is, the Chair ascertains whether consensus has been reached, but is required to declare that there is no consensus 'if there is a stated objection to the proposed decision under consideration'.[96] This definition implies that one member can force the issue to a vote.

The CDM rules of procedure include innovative procedures for electronic remote voting if the Chair judges that a decision cannot wait until the next meeting.[97] A decision proposed by the Chair with an invitation to adopt it by consensus is transmitted to all members (and alternates for information), via the Board's electronic listserv maintained by the secretariat. A quorum of the board must confirm receipt of the message, after which point members have two weeks to formulate comments. Unless any objections are raised, the decision is deemed adopted after that time. However, if an objection is made, the proposed decision is not adopted and is included on the provisional agenda for the next meeting. The electronic voting procedures enable the EB to take decisions on a virtual basis, speeding up its work while keeping the operational costs of the Board low. These requirements also ensure that members do not prejudice the adoption of decisions by the Board by simply failing to turn up.

Panels and liaison with SBSTA

The EB is mandated to establish any 'committees, panels or working groups' that it deems necessary, and may draw on outside expertise, including from the UNFCCC roster of experts, taking into account the need for regional balance among the providers of such expertise.[98] A public call for experts is usually issued via the Internet to obtain a slate of candidates for the panels with 'demonstrated and recognized technical expertise',[99] and members are then appointed by the EB. The Panels are served by a Chair and Vice-Chair (one each from an Annex I and non-Annex I Party) designated from among the EB members. Where the work involved is particularly technical or time-consuming, Panel members are paid fees according to UN procedures.[100] The Board moved quickly to agree General Guidelines for Panels in March 2002.[101] The following panels have since been established:

- CDM Accreditation Panel (CDM-AP), June 2002, which is supported by CDM Assessment Teams (CDM-AT);
- CDM Methodologies for Baselines and Monitoring Plans (Meth Panel), June 2002;
- CDM Small Scale Project Activities (SCC Panel), April 2002.

The work of the EB is closely linked to that of the Subsidiary Body for Scientific and Technological Advice (SBSTA) in several areas including, for example, the SBSTA's work on provisions for LULUCF projects. Interaction between the two bodies was called for in the CDM modalities, and has been ensured by the nomination of certain Board members to follow the work of the SBSTA in these areas, reporting to the EB on developments.[102]

Designated national authorities

The two main functions of a DNA are to:

- provide written approval of the voluntary participation of *each Party* involved in the project; and
- obtain confirmation by the *host Party* DNA that the project activity assists it in achieving sustainable development.[103]

These requirements were drafted to take into account that many Annex I Party countries found the idea of approving individual projects unnecessary and laborious. Developing country Parties on the other hand insisted on formal approval of each project to ensure their officials would be able to better track and control CDM activities to conform to their national sustainable development priorities and strategies. The absence of such a requirement would have meant private and/or public entities in the host country would have determined what was or was not sustainable development for the country as a whole. Letters of approval for each project activity by the host Party must be provided before registration of a CDM project activity by the Board.

Designated operational entities

The CDM project cycle makes DOEs primarily responsible for checking that CDM projects are in conformity with the CDM modalities by specifying that DOEs shall:

- validate proposed CDM project activities and put them forward for registration by the EB; and
- verify and certify reductions in emissions.[104]

Prior to anything being submitted to the EB, therefore, DOEs must be hired on a contractual basis by PPs to undertake one or other of these functions in relation to a particular project (because doing both might create conflicts of interests for a particular project, although in exceptional circumstances and upon request to the EB, a DOE can perform all these functions for a project).[105]

Although the CDM modalities contain provisions allowing the Board to review DEO actions relating to the execution of these functions, the CDM modalities envisage that in the vast majority of cases validation, verification and certification decisions taken by DOEs will be final with the Board only getting involved in examining 'problem' CDM projects (see Box I.1 below on stakeholder involvement in CDM projects which explains the information and reviews relevant for such CDM projects).

Because the quality, consistency and transparency of the work done by DOEs is critical to the CDM, the modalities set out certain standards that must be met by any organization that wants to become a DOE.[106] These standards are referred to as *accreditation standards* and are contained in Appendix A of the CDM modalities. Additional procedures specifying how applicant entities (AEs) should go about applying for DOE status have been agreed by the Board.[107] Designation of an AE can be made by the EB on a provisional basis with COP/MOP reserving the right to confirm this status. Only once the COP/MOP has done so is the entity formally called a DOE.

To maintain standards, the CDM modalities also specify procedures to suspend or withdraw the designation of a DOE if a review pursuant to paragraph 20 of the CDM modalities (which takes place once every three years or is triggered if a spot-check reveals relevant information) finds that the DOE no longer meets the accreditation standards. In such a case, the EB can make a status decision with immediate effect about the status of the DOE which remains in effect until a final decision is made by the COP/MOP. Projects that have been validated, verified or certified by a DOE that has subsequently been suspended are not affected unless 'significant deficiencies' in the DOE's work relating to these were found. In such cases, the EB can decide to appoint a different DOE to review those projects. If this review finds that excess CERs have already been issued, the original DOE shall acquire and transfer the equivalent amount of CERs to the CDM Registry within 30 days of the end of the review. These requirements, in particular making DOEs liable for restoring CERs resulting from their 'bad work', are intended to keep DOEs on their toes.

Additional responsibilities and functions for DOEs are spelt out in the CDM modalities. These specify that a DOE shall:

- be responsible for ensuring that the DOE complies with applicable host-country laws when carrying out its functions;

- demonstrate that it has no real or potential conflict of interest in the CDM project;
- maintain a publicly available list of all CDM projects in which it is involved;
- submit annual activity reports to the EB; and
- make non-confidential information from CDM projects publicly available as required by the Board.

To assess whether applicant entities (AEs) meet the accreditation standards to be recognized as DOEs, the Board has set up a special panel, CDM-AP, which will go through each application, including conducting on-site visits by CDM Assessment Teams, to see if the requisite criteria are met. The Board has confirmed that once the CDM-AT has been undertaken for an AE, the AE can propose a new methodology for baselines which is the first step in getting CDM projects ready for validation (see below).

Box I.1 Information and review provisions and stakeholders and observers in CDM project cycle

Proposing projects will require resources and time, and, while the CDM is in its start-up phase, a certain amount of dedication from the pioneers. Negotiators of CDM modalities had to balance the need for a project cycle that generated legal certainty and reduced transaction costs with the need to ensure that decisions taken by the Board and the DOEs are environmentally credible, of a consistently high quality and are made in a transparent and legitimate manner. The CDM modalities balance these needs in a number of ways. First, the modalities provide for *representation* of UNFCCC accredited observers, stakeholders (discussed under cross-cutting issues) and Parties not on the Board at meetings of the EB (discussed above), and, of course, subject to UNFCCC accreditation procedures, at sessions of the COP.

Second, the CDM modalities mandate that certain kinds of *information* must be made publicly available to allow third parties to subject aspects of the EB to greater scrutiny than might be provided if such information was provided only to fellow time-pressed government delegates who may lack expertise or first-hand knowledge about particular problems. Prior to the project cycle, information provided by an applicant entity to gain DEO status must include publicly available information about its internal procedures for carrying out DEO functions, including procedures for allocation of responsibility and for handling complaints.[108] These Accreditation Procedures adopted by the Board make clear how others actors can help the Board determine whether an AE is qualified to be designated as a DEO. As part of the validation process, the PDD must show that comments by 'local stakeholders' have been invited and were duly taken into account in designing the project;[109] the PDD must contain information provided by stakeholders as well as how it was taken into account.[110]

The PDD and other information provided by a PP related to validation of a particular project must also be made publicly available by a DOE except for those sections which are covered by legitimate confidentialityconcerns (these cannot include matters relating to additionality and environmental impacts).[111] The DOE must allow 30 days to receive comments from Parties, stakeholders and UNFCCC accredited observers, make such comments publicly available and report how it has taken them into account if it decides to go ahead with requesting registration of the project. After a project has been registered another significant provision of information requirement is at the point of verification and certification when the DOE must make the monitoring report and the verification report, which together provide the basis for calculating the amount of CERs that the DEO will request the EB issue, publicly available, as well as the final certification report itself.[112]

The third element is the provision of two sets of *formal reviews* by the EB. NGOs and stakeholders do not have direct standing to trigger these two reviews but can, of course, provide information to and lobby Parties involved in the project and Board Members. The first review relates to validation and registration of CDM projects and is set out in paragraph 41 of the CDM modalities. This provides that registration of a CDM project validated by a DEO becomes final after eight weeks if no request for review has been made by a Party involved in the project or at least three members of the EB. The scope of the review must be related to the validation requirements and the Board must finalize its decision about the review no later than the second meeting following the request for review. The Board has recently adopted procedures which spell out how the review pursuant to paragraph 41 may be undertaken and these are being applied provisionally by the Board pending their adoption by the COP (COP/MOP).[113] The second review relates to issuance of CERs which will be considered final 15 days after the DOE has submitted a certification report unless a review has been triggered by a Party involved in the project or at least three Board members in accordance with paragraph 65 of the CDM modalities. The scope of this review shall be limited to issues of fraud, malfeasance or incompetence of the DEO. The request for a review must be considered by the EB at its next meeting to determine if it has merits. If so, the review must be completed within 30 days following the decision to perform the review. If the EB decides that excess CERs have been issued, it shall request the CDM Registry to transfer an equivalent sum of Kyoto units to its cancellation account and these may not be further transferred or used for compliance purposes.

Validation and registration requirements

This section describes project level requirements that must be met if a project is to be validated by a DOE and then registered by the Board. With the exception of small-scale projects and projects relating to afforestation and reforestation (discussed below), project-level validation and registration requirements are the same for *all* types of CDM projects relating to emission sources.

Although some Parties wanted to exclude certain project activities, such as nuclear projects, based on their views about their inherently unsustainable development credentials, these views did not prevail and, in fact, no project types are excluded from the CDM as such. In the case of nuclear projects the Preamble to Decision 17/CP.7 recognizes that 'Annex I Parties are to refrain from *using* CERs generated from nuclear facilities to meet their commitments under Article 3.1.'[114] Parties agreed, however, that it is the host Party's prerogative to confirm whether a CDM project assists sustainable development. Annex I Parties also have choice over whether to authorize particular legal entities to engage in nuclear CDM projects or to accept certain kinds of projects or CERs.[115]

Standard CDM projects

It is important to remember that a Party and its legal entities can only participate in the CDM if the *government-level participation requirements*, set out above, are met. In the case of the CDM, it is the responsibility of the DOE to check that PPs meet the participation requirements relating to the designation of national authorities and that all PPs meet the 'Party to the Protocol' requirements. The latter requirement has been clarified by the Executive Board and to avoid duplication is discussed under participation in the section on cross-cutting issues.

In addition to these participation requirements, the DOE has responsibility to review the PDD and any supporting information to confirm that a project meets the following requirements:

• Comments by local stakeholders have been invited, a summary of the comments received has been provided and a report to the DOE on how due account was taken of any comments has been received from the PP (see Box I.1).
• PPs have submitted to the DOE documentation on the analysis of the environmental impacts of the project activity, including transboundary impacts, and, if those impacts are considered significant by the PP or the host Party, have undertaken an environmental impact assessment in accordance with procedures required by the host Party.
• The project activity is expected to result in a reduction of anthropogenic emissions of GHGs from sources 'that are additional to any that would occur in the absence of the proposed activity' subject to paragraphs 43–52 of the CDM Modalities (see Box I.2).
• If the baseline and monitoring methodologies used for the project have previously been approved by the Board, they have been made publicly available along with any guidance provided by the Board. If new baseline and monitoring methodologies are proposed, the DOE has submitted these prior to the request for validation for review by the Board and these have

been approved by the Board without subsequent COP/MOP revision (see Box I.2).

- Provisions for monitoring, verification and reporting are in accordance with Decision 17/CP.7, its annex and relevant COP/MOP decisions.
- The project activity conforms to all other requirements for CDM project activities contained in Decision 17/CP.7, its annex and relevant COP/MOP decisions.[116]

Box I.2 CDM ADDITIONALITY, BASELINES AND CREDITING[117]

Additionality, baselines and choice of methodologies are highly technical issues but issues upon which the environmental integrity of the CDM depends. Article 12.5 provides that CERs should be generated from a CDM project if the emissions reductions can be certified on the basis of:

- real, measurable and long-term benefits related to the mitigation of climate change; and
- reductions are additional to any that would occur in the absence of the certified project activity.

Various 'additionality tests' were proposed during the Marrakesh negotiations to prevent CERs going to projects that would happen even without the CDM. Paragraph 43 provides that a project is considered to be additional if anthropogenic emissions by sources are reduced below those that would have occurred in the absence of the registered CDM project activity. Proposals that additionality requirements should include a financial test to ensure that funding for CDM projects is additional to ODA, including Global Environment Facility (GEC) contributions, proved controversial although a reference with this intent is set out in the Preamble to Decision 17/CP.7.[118]

Paragraph 44 provides guidance on the construction of baselines (which refer to estimates of what future emissions would be within the project boundary *without* the CDM project intervention) as these are crucial to deciding how many CERs will be generated by a project. What happens to emissions outside the project boundary is crucial for environmental integrity as CERs can be generated but with increased emissions outside the project boundary – an issue known as 'leakage'. The CDM modalities provide that baselines shall be established in a 'bottom-up' manner by PPs working through DOEs to propose new methodologies for the Board's approval using the 'pre-validation' procedure. Baselines are meant to be transparent, conservative and importantly *project-specific*. They must take national circumstance into account, may include scenarios where projected emissions rise above current levels and must be defined so that CERs cannot be earned for decreases attributable to *force majeure*.

Project-specific baselines increase transaction costs for PPs. Multi-project baselines or where several projects could use the same baseline would reduce these. The CDM modalities try to create some degree of standardization of approach to baseline setting by requiring PPs to select from three approaches set out in paragraph 48, and also by defining how leakage issues should be addressed by PPs. The issue of the crediting period is linked to baselines because the length of time credits can be claimed can 'freeze' the degree of technological and other developments that can be taken into account in constructing a 'reasonable' future. The CDM modalities state that the crediting period must be either a maximum of ten years with no renewal option or a maximum of seven years but which may be renewed twice provided that for each renewable, the DEO informs the EB that the original baseline is still valid.

More detailed provisions on baseline methodologies proved difficult at Marrakesh due to lack of time, conceptual differences and lack of technical expertise. Additional elements were developed in the form of Appendix C to the CDM Modalities but there was insufficient understanding of their full implications. Thus these elements, entitled 'terms of reference for establishing guidelines on baseline and monitoring methodologies' and also included in the CDM Modalities, are not actually mentioned therein. Appendix C is referenced, however, in the 'prompt start' part of Decision 17/CP.7 concerning the COP request to the EB to prepare recommendations on any 'other relevant matter', including but not limited to Appendix C. In response, the EB has undertaken an extensive body of work on baseline and monitoring methodologies.

It has provided guidance and clarification on this issue in four separate guidance/clarification notes.[119] At its eleventh meeting, the Board elaborated procedures for submission and consideration of proposed new baseline methodologies.[120] To date, some 17 proposals for new baseline and monitoring methodologies have been submitted to the Board by PPs, two of which have been approved, nine not approved and the remaining five are to be reconsidered after suggested amendments are taken into account.[121]

These substantive requirements for the DOE are accompanied by additional procedural rules the DOE itself must comply with if it is to transmit a validation report to the Executive Board that is itself in conformity with Decision 17/CP.7. These procedural requirements state that the DOE shall:

- prior to the submission of the validation report, obtain written approval of the voluntary participation of all Parties involved in the project and in addition confirmation from the host Party that the project achieves sustainable development;
- make the PDD publicly available while safeguarding information deemed confidential;
- receive, within 30 days, comments on the validation requirement from Parties, stakeholders and UNFCCC accredited NGOs and make these publicly available;

- after the deadline for receipt of comments, make a determination as to whether, on the basis on the information provided and taking into account comments received, the DOE will validate the project;
- inform PPs of the DOE's decision and date, including, if the project is rejected for validation, the reasons for its rejection; and
- where it decides to validate a project, to make the validation report, comprising the PDD, approval of host Party and comments received, publicly available upon transmissions to the Board.[122]

Any project that has been rejected for validation may be reconsidered provided, after appropriate revisions, it goes through the validation and registration steps, including the public comment requirements.

In sum, to be validated and registered as a standard CDM project, the Parties involved in the project have to meet their participation requirements, the PP and the project itself must conform with the substantive requirements spelled out in the CDM modalities, including, in this early phase, prior acceptance of baseline and monitoring methodologies, and finally the DOE itself must adhere to certain procedural requirements relating to its preparation of the validation report before this can be transmitted to the EB to trigger a valid request for registration.

Small-scale projects

Because Parties realized that the complexity and transaction costs of the 'standard' CDM project cycle might deter small projects and skew the CDM in favour of large-scale projects, COP-7 agreed that as part of the CDM 'prompt start' simplified modalities and procedures should be adopted for small-scale CDM project activities (SCC projects). Based on the Board's work, COP-8 adopted such modalities and procedures as explained in Box I.3.[123]

Afforestation/reforestation projects[125]

The inclusion of sinks in the CDM was controversial at Marrakesh for many political reasons as discussed in section I.1 above.[126] There were also complex technical issues such as measurement uncertainties and how the temporary nature of sinks projects could be reconciled with the creation of CERs that would have long-term value. The Bonn Agreements brokered a compromise that only afforestation and reforestation projects shall be included in the CDM for the first commitment period with a decision on activities to be included in the second commitment period to be negotiated in the context of these negotiations. CDM sinks activities shall be subject to the forest principles set out in Decision 11/CP.7.[127] For the first commitment period, additions of CERs from sinks projects by an Annex I Party are capped at 1 per cent of its base year emissions (times 5). The implications of this cap for each Annex I Party are set out in the Technical Appendix to this book.

BOX I.3 SMALL-SCALE MODALITIES AND PROCEDURES FOR
SMALL-SCALE **CDM** PROJECTS

Decision 17/CP.7 states that the following three small-scale project activities
can take advantage of simplified modalities and procedures:

- renewable energy project activities with a maximum output capacity
  equivalent to up to 15 megawatts (or appropriate equivalent);
- energy efficiency improvement project activities which reduce energy
  consumption, on the supply and/or demand side, by up to the equivalent of
  15 gigawatt/hours per day;
- other project activities that both reduce anthropogenic emissions by sources
  and directly emit less than 15 kilotonnes of carbon dioxide equivalent annually.

COP-8 adopted small-scale simplified modalities and procedures (SCC M&P)
contained in Annex II of Decision 21/CP.8. The EB completed the SCC M&P by
agreeing the three appendices envisaged in Annex II of Decision 21/CP.8 as
follows: simplified PDD (Appendix A), indicative methodologies (Appendix B) and
provisions for avoiding debundling (Appendix C). The completion by the Board of
this work in early January 2003 gives the green light for submission of
applications for potential SCC projects. The Board further stressed that PPs may
propose new SCC project activity categories and amendments/improvements to
the SCC M&P which shall be reviewed at least once a year, and that the CDM
Meth Panel will continue to work on Appendix B in consultation with experts.[124]

Because Article 12 of the Protocol and the CDM modalities themselves refer only to emissions sources and not to sinks, it was not clear how inclusion of afforestation/reforestation would fit into the CDM project cycle, particularly bearing in mind that inclusion of sinks brings in issues that are either unique or have different policy implications. These include issues regarding non-permanence, additionality, leakage, uncertainties and socio-economic and environmental impacts, including impacts on biodiversity and natural ecosystems. Because supporters of sinks inclusion tried at Marrakesh to resist negotiations of an entirely different and new project cycle for sinks, Decision 17/CP.7 specified that inclusion of CDM project activities covering afforestation and reforestation 'shall be in the form of an annex on modalities and procedures … reflecting, *mutatis mutandis*, the annex to the present decision.'

Modalities for afforestation and reforestation projects under the CDM were adopted by COP-9 in December 2003 and are set out in Decision 19/CP.9. The rules for afforestation and reforestation projects under the CDM are identical to those for emission reduction projects with a few exceptions. A CDM sinks project must be implemented on land that was not forested on 1 January 1990. A project may choose a single crediting period of 30 years or a renewable crediting period of 20 years with up to two renewals for a total of

60 years. The project proponents must consider the socio-economic and environmental impacts of the proposed project in accordance with the procedures required by the host Party. The deference to host Party acceptance was criticized by environmental NGOs who argued that this would benefit genetically modified large-scale plantations of non-native monocultures and that local communities and stakeholders will only have a minimal say in national CDM approval processes (Greenpeace, n.d.).

Projects may specify which of the carbon pools – above-ground biomass, below-ground biomass, litter, dead wood and soil organic carbon – are to be included in the project. Project participants may choose not to account for one or more carbon pools if they can provide transparent and verifiable information that the exclusion will not increase the quantity of reductions claimed. Thus if a pool is a sink, it need not be measured, but if a pool is a net source it must be measured since failure to do so would overstate the reductions achieved by the project.[128] Greenhouse gas emissions from activities on the land prior to afforestation or reforestation are not included in the baseline.[129] But emissions associated with the project must be deducted from the increase in carbon stocks.

Since the carbon stored by the trees and soil can be released again by disease, fire, harvesting or other events, special provisions are included to address the non-permanence of these projects. The project proponents must choose one of the two options to address non-permanence. Both options verify and certify the net increase in the carbon stocks due to the project since its inception at regular intervals.

Under one option temporary CERs (tCERs) equal to the certified net increase in the carbon stocks *since the inception of the project* are issued after each verification. The tCERs expire at the end of the subsequent commitment period, so they can only be used for compliance with the commitments for the period in which they are issued. If the trees are still there when the project is next verified, new tCERs will be issued to replace the ones that have expired. No new tCERs can be issued after the end of the project's crediting period.

Under the other option long-term CERs (lCERs) equal to the certified net increase in the carbon stocks *since the previous verification of the project* are issued after each verification. If there has been an increase in the carbon stocks during the period, additional lCERs are issued. If there has been a partial or complete release of carbon since the previous verification, an appropriate share of the outstanding lCERs must be replaced by AAUs, CERs, ERUs, RMUs or lCERs from the same project. If a verification report is not received when required, the lCERs for that project must be replaced. The lCERs for a project expire and must be replaced at the end of the project's crediting period.

The rules define small-scale afforestation and reforestation project activities under the CDM as those that are expected to result in net removals by

sinks of less than 8 kilotonnes of CO_2 per year and are developed or implemented by low-income communities and individuals as determined by the host
Party. If a project is designated a small-scale afforestation or reforestation project activity and then achieves net removals greater than 8 kilotonnes of CO_2
per year, the excess removals will not be eligible for the issuance of tCERs or
lCERs. Simplified rules for small-scale afforestation and reforestation projects
are to be developed for adoption at COP-10 in December 2004.

Monitoring and verification and certification requirements

Monitoring refers to the identification, collection and archiving of information
necessary to design and implement a monitoring plan as required by CDM
modalities.[130] A monitoring plan for a proposed projected activity must be
based on a previously approved monitoring methodology or, if it is new, one
that has been submitted to the Board using the 'pre-validation' procedure set
out in paragraphs 37 and 38 of the CDM modalities (described above in relation to methodologies for new baselines). The proposed plan must satisfy the
DOE it is appropriate and reflects good monitoring practice.

If approved it becomes known as *the registered monitoring plan*. Any revisions to this shall be justified and submitted for validation to the DOE.
Implementation of the registered monitoring plan by the PP is a condition for
the verification, certification and issuance of CERs. The PP shall provide a
monitoring report to the DOE contracted to perform verification in accordance with the registered plan which will normally set out the frequency,
content and timing of such information.

Verification and certification are defined in the section on the CDM project
cycle above.[131] A DOE is responsible for performing verification at regular intervals on the basis of the *monitoring report* which, subject to confidentiality
provisions, it shall make publicly available. A DEO may conduct on-site visits,
review performance records, talk with PPs and stakeholders, examine measurement equipment and use alternative sources of data with a view to establishing
that monitoring methodologies have been correctly applied. It may make recommendations for future improvements. On the basis of the foregoing it shall
calculate the reductions in emissions that would not have occurred in the
absence of the CDM project activity and inform the PP of any concerns it may
have regarding the conformity of the actual project with the PDD, giving the PP
an opportunity to address and correct these concerns.

The DOE's *verification report* shall be provided to the PP, the Parties
involved and the Board and made publicly available. This report will give rise
to a *certification report* by the DEO certifying how many CERs the project has
achieved during a specific time period which shall be transmitted to PPs and
the EB and made publicly available.

Issuance of CERs

The DOE's certification forms the basis of a request by the DOE to the EB for the issuance of CERs.[132] Issuance refers to the instruction by the EB to the CDM Registry Administrator to issue (i.e. create) a specified number of CERs into the pending account of the EB.

Issuance becomes final 15 days after the date of receipt of the request for issuance, unless a Party or at least three members of the EB request a review of the proposed issuance of the project. The scope and procedures for this review are explained in Box I.1 above.

Upon being instructed by the EB to issue CERs, the responsibility to forward CERs to the registry accounts of the PPs rests with the CDM Registry which must also forward, at the same time, the relevant number of CERs to cover the share of the proceeds for administrative expenses (not yet determined) and adaptation (set at 2 per cent) to the appropriate accounts. Thus the EB has no role in the final allocation of CERs to PPs as this is a purely technical act undertaken by the CDM Registry Administrator. A particular merit of these provisions is that the CERs to cover adaptation (and at some future stage for administrative expenses) are collected automatically at the point of issuance leaving no possibility of their non-payment by PPs.

The Executive Board has responsibility for the development, establishment and functioning of the CDM Registry, as defined in Appendix D of the CDM modalities.[133] This Registry creates accounts for non-Annex I Parties and thus allows the accurate tracking of issuance, holding, transfer and acquisition of CERs by non-Annex I Parties. The requirement for the EB to develop and maintain the CDM Registry not only makes possible various approaches to CDM (unilateral, bilateral and multilateral, explained above) but also ensures that non-Annex I Parties are not under an obligation to develop their own national registries. The structure and functioning of the CDM Registry is similar in substance to the national registries to be established by Annex I Parties, including public disclosure requirements.

Where a DOE's accreditation status has been withdrawn or suspended any ERUs, CERs, AAUs and RMUs equal to the excess CERs issued, as determined by the EB, shall be placed in a cancellation account in the CDM Registry and cannot be further transferred or used for the purpose of compliance with Article 3.1 commitments.[134]

Given the work going on in SBSTA to develop Annex I Parties' national registries, the EB has decided it will not establish an interim CDM Registry in 2003.[135] It has agreed instead to issue a public call to Parties and organizations for inputs relating to the development of the CDM Registry and to request the Secretariat to begin development of the CDM Registry with a timeline for continuing work.

Funding issues

ODA/public funds

The issue of whether donors can generate CERs on the back of public funds has given rise to friction between donors and developing countries throughout the negotiations on AIJ and then on the CDM. Developing countries have argued that ODA and public funds should not be used to fund CDM projects because doing so means resources earmarked for the sustainable development of developing countries are used instead to assist Annex II Parties (donors) meet their own climate mitigation commitments, and this has the potential to distort the funding of developing countries' sustainable development as prioritized by them.

This view is widely accepted. Accordingly, Decision 17/CP.7 states that 'public funding for CDM projects from Parties in Annex I is not to result in the diversion of official development assistance and is to be separate from and not counted towards the financial obligations of Parties included in Annex I.'[136] This reference, included in the preamble to Decision 17/CP.7 to reflect its weakened legal status, is not included as a test of additionality to be examined by DOEs. Assessment of whether Annex I Parties are complying with it is a matter for the COP/MOP to review, following its mandate to 'assist in arranging funding of CDM project activities, as necessary.'[137] It should be noted that tracking and separating out funding for sustainable development generally from 'new and additional' resources provided to the Convention's financial mechanism has historically proved problematic for reasons explained elsewhere. Thus broader financial mechanism discussions, in particular whether public expenditures on CDM activities will be defined in non-climate regime institutions such as the OECD Development Assistance Committee (DAC), as climate-related ODA, will have a bearing on monitoring adherence to the 'no diversion' stipulation set out in Decision 17/CP.7.

Assistance in funding and geographic imbalance

Private investment gravitates primarily toward a handful of the larger developing countries. Concerns that the bulk of developing countries would not benefit from the CDM led to inclusion of Article 12.6 in the Protocol. This states that the 'CDM shall assist in arranging funding of certified project activities, as necessary.' The COP/MOP has an explicit function to review the regional and sub-regional distribution of CDM projects with a view to identifying systematic or systemic barriers to their equitable distribution and take appropriate decisions, based, *inter alia*, on a report by the EB.[138]

Concern that, due to lack of capacity and technical expertise, DOEs might only be based in developed countries, which might increase costs and reduce choice for developing country PPs, led to the inclusion of a review function for the COP/MOP of this issue. In this case the EB does have an explicit mandate to report to the COP/MOP on the geographic and regional distribution of DOEs and has already endorsed corrective actions to make the financial payment of fees by developing country based DOEs less onerous (see below). Discussions are also under way regarding a 'phased' approach to accreditation that would allow more developing country DOEs to come forward as applicant entities.[139] Additionally, COP-9 has requested Parties and other organizations to assist with capacity building to encourage a greater spread of DOEs from developing countries.[140]

Share of proceeds

Given the private sector nature of the CDM, Article 12.8 includes a unique international 'levy' on CDM activities by mandating that a 'share of the proceeds' of certified project activities is to be used for two purposes: to cover the CDM's administrative costs and to fund the adaptation needs of developing country Parties vulnerable to the adverse effects of climate change. The wording of Article 12.8 creates no priority between these two uses. The Marrakesh Accords reached a political agreement that the share of the proceeds for adaptation 'shall be two per cent of the CERs issued for a CDM project activity.'[141] CERs will be collected by the EB at the point of issuance of CERs and, in due course, made available to the Adaptation Fund established under the Protocol.[142] It further agreed that CDM projects hosted by LDCs shall be exempt from the share of the proceeds related to adaptation.

So far as administrative expenses are concerned, Parties agreed that the share of the proceeds to cover these shall be determined by the COP upon recommendation of the EB.[143] The EB has in turn stated that it will not consider making a recommendation until 2004 when there is more information about CER prices.[144] Additionally, the Board has pointed out that requiring PPs to pay CDM administrative costs in the early stages of the prompt start phase in which costs are front-loaded due to the need for intensive development of additional rules and procedures could penalize, rather than support, CDM pioneers.

Pending a recommendation by the EB and a decision by the COP, Parties have been invited to provide funding for the EB by contributing to the Trust Fund for Supplementary Activities recognizing that some basic funding for Secretariat support has come from the core budget. COP-7 called for contributions 'in the order of $6.8 million' to support the prompt start of the CDM,[145] on the understanding that, if requested, these would later be reimbursed. By COP-8, however, only a fraction of this total had been received.[146] This led to the preparation of a revised budget for 2002–2003 amounting to US$4.32 million. Actual contributions to date amount to US$1.74 million.[147]

To lessen resource constraints, organizations applying for accreditation as operational entities must also pay a registration fee of US$15,000 with the option for applicants from developing countries to pay this in two instalments.[148] This has raised a total of US$240,000 from the 16 applicant entities to date.[149] In addition, the Board has recently agreed to a system of raising a registration fee as a down payment until a share of the proceeds may be determined.[150] The fee will vary depending on the size of the project, ranging from a minimum of US$5000 to a maximum of US$30,000, with these figures to be reviewed and revised in light of experience. The second report of the EB to COP-9 addresses the financial and budgetary issues in more detail. The costs of operating the CDM prior to the entry into force of the Protocol was part of the contentious negotiations on the budget that took place at COP-9 due to the US position to refuse to contribute towards Protocol-related activities. US intransigence on this matter has prompted COP-9 to invite Parties to urgently make contributions for funding the administrative expenses of the CDM.

I.7 Joint Implementation (Article 6)

Overview

Article 6 of the Protocol allows Parties in Annex I to use 'emission reduction units' (ERUs) resulting from GHG abatement or sequestration projects in any other Annex I Party for the purposes of meeting their Article 3.1 commitments.[151] Joint implementation among Parties with binding quantitative commitments under the Protocol has been succinctly described 'as a specific form of emissions trading related to individual projects rather than a trading of assigned amounts from any source.'[152]

Delegates often find the concept of JI confusing because the term 'JI' straddles elements of emissions trading with project-based forms of trading. Accordingly, the 'hybrid' nature of JI, its history and its institutional implications are explained in more detail in Box I.4.

The Bonn Agreement provided limited guidance on how to institutionalize JI. It clarified that Annex I Parties' eligibility to participate in all mechanisms was dependent on meeting reporting and review requirements. It also recommended that the COP/MOP establish 'a supervisory committee to supervise, *inter alia*, the verification of ERUs generated by Article 6 project activities but without specifying its functions and composition.[153] Negotiations on Decision 16/CP.7 interpreted the Bonn guidance so as to create two 'tracks' for JI. Track 1 and Track 2 are available when a host Annex I Party is in conformity with its reporting and review requirements and allow this Party to validate JI projects and issue and transfer ERUs without additional external scrutiny.[154]

Where the host Annex I Party fails to meet its reporting and review eligibility requirements, it can only participate in JI projects under Track 2. This requires international oversight of JI activities provided by the Article 6

Supervisory Committee (A6SC) whose functions, powers and rules of procedure are very similar to those of the EB. The project cycle under Track 2 is also
very similar to the CDM project cycle incorporating reliance on 'independent
entities' (IEs) rather than on DNAs. IEs are accredited third-party organizations
that perform essentially the same functions as DNAs but in the context of a
streamlined project cycle that merges the distinct steps of the CDM project
cycle. Issues relating, *inter alia*, to participation, eligibility, fungibility, equity
and supplementarity are addressed in section I.3 on cross-cutting issues.

BOX I.4 JOINT IMPLEMENTATION: HISTORY AND CHARACTERISTICS

From Kyoto right through to Marrakesh, negotiators found it hard to address the
particular challenge JI created: how to create institutional arrangements that
would address simultaneously the environmental policy concerns arising from a
cap-and-trade trading scheme with concerns arising from the operation of a
baseline-and-credit one. Where the host Party has a binding target it does not
have long-term incentives to give away ERUs. Although Article 6 provides that
JI projects must generate ERUs additional to any that would otherwise occur,
the zero-sum nature of Annex I Article 3 commitments means additionality is
not critical for the environmental integrity of JI as a mechanism where all Parties
are meeting their reporting and review commitments. By contrast, where
reporting and review commitments are not being met, a host Party can easily
underestimate its emissions and thus oversell its assigned amount through JI.
In these circumstances, its participation in JI requires some form of external
scrutiny if environmental integrity is to be secured.

Article 6 of the Protocol did not help negotiators sort out responses to
these   distinct policy issues because its provisions included additionality and
reporting and review commitments as conditions for JI participation. Finally,
sound environmental reasons, cross-cutting concerns relating to
supplementarity, fungibility, equity and the application of the 'adaptation levy' to
JI and ET led many developing countries to insist on 'institutional parallelism'
for Article 6 – meaning that JI should replicate as closely as possible the rules,
institutional structures and procedures for the CDM.

Substantive progress was only possible at Marrakesh after modalities for
the CDM and ET were finalized for three reasons. First without knowing the
nature of its parentage, the CDM and ET, the 'hybrid' mechanism could not
emerge. Second, mechanism negotiations, with complex linkages to
simultaneous negotiations on compliance and Articles 5, 7 and 8, left delegates
with little time to focus on JI which was accorded the lowest priority out of the
three mechanisms by JUSCANNZ and developing countries. Finally, because
EITs, the principal beneficiaries of JI, do not form a cohesive political bloc in
negotiations and have capacity constraints, a coherent vision of JI modalities
emerged relatively late on in the negotiations.

JI institutions

COP/COP/MOP

The role of the COP/MOP is defined in the annex to Decision 16/CP.7 in one short sentence stating that it 'shall provide guidance regarding the implementation of Article 6 and exercise authority over the Article 6 supervisory committee.'[155] The lack of specificity results from Annex I Parties' overall preference for bilateral approaches to project-based mechanisms. A number of specific tasks and functions for the COP/MOP are set out and/or implied from the section on the A6SC. Additionally the wide scope of this short reference combined with the general functions and power of the COP/MOP under Article 13 mean it could undertake a very wide range of functions on any Protocol matter, including JI.

JI projects starting as of the year 2000 may be eligible as Article 6 projects but these cannot result in ERUs until 2008 as there is no assigned amount before then. Because of this, no prompt start or interim role has been envisaged for the COP prior to the entry into force of the Protocol or of the A6SC.

Article 6 Supervisory Committee

The A6SC was not envisaged in the Kyoto Protocol but was established through the Marrakesh Accords when it became clear some form of independent body was needed to supervise Track 2 JI (see Box I.4).[156] The A6SC will be established at the first session of the COP/MOP. The Marrakesh Accords include provisions governing the institutional and procedural aspects of its operation but a fuller set of procedural rules for the A6SC functioning may be devised for consideration by the COP/MOP, as has happened with the EB/CDM.

When established, the A6SC will function under the authority and guidance of the COP/MOP, to which it will report annually.[157] The Marrakesh Accords include provisions governing the institutional and procedural aspects of its operation but a fuller set of procedural rules for the A6SC functioning may be devised by the A6SC itself for consideration by the COP/MOP.

The main function of the A6SC is to supervise the work of IEs which will be responsible for verification of emission reduction units generated through JI projects hosted by countries that are not fully in compliance with their reporting and review commitments under Track 2.[158] Thus, like the Executive Board, the A6SC's functions include responsibility for the following:

- accreditation of IEs in accordance with standards and procedures contained in Appendix A of the Article 6 Guidelines;
- review of standards and procedures for the accreditation of IEs, giving due consideration to the relevant work of the EB and making recommendations to the COP/MOP on revisions to these standards and procedures;

- review and revision of reporting guidelines and criteria for baselines and monitoring in Appendix B of the Article 6 Guidelines for consideration by the COP/MOP;
- elaboration of a PDD for Article 6 projects for consideration by the COP/MOP, taking into consideration Appendix B of the CDM Modalities and giving consideration to relevant work of the EB as appropriate;
- review of the validation determination made by an IE under paragraph 35 of the Article 6 Guidelines (JI validation review) and review of the verification/ certification determination made by an IE under paragraph 39 of the Guidelines (JI verification/certification review); and
- elaboration of additional rules of procedures for consideration by the COP/MOP.

Given that JI projects take place exclusively among Annex I Parties, the supervisory committee's composition is deliberately skewed to provide for greater representation of Annex I Parties, with EITs granted strong representation. The A6SC is composed of ten members, including three members from the Annex I Party EITs, three from Annex I Parties that are not EITs,[159] three from non-Annex I Parties and one member from the small island developing states.[160] As with the EB, however, each member is accompanied by an alternate from the same region, and both members and alternates are nominated by their constituencies and formally elected by the COP/MOP.[161] As with other elected posts, Parties will be required to give active consideration to the nomination of women for the A6SC to improve the gender balance of FCCC and KP elected institutions.[162] The cost of participation of developing country members and those of other Parties eligible under UNFCCC practices are to be covered by the budget of the A6SC.[163]

The A6SC's procedures parallel those of the EB also in many other respects, including provisions for election of a Chairperson and Vice-Chairperson and the rotation between them, the terms of office for its members, the need for them to serve in their personal capacity, the financial interest and confidentiality provisions to which they are subject and the role of alternates. Again, similar to the EB's procedures, the A6SC's members must have 'recognized competence relating to climate change issues and in relevant technical and policy fields'.

The A6SC is to meet 'at least two times a year, whenever possible in conjunction with the ... subsidiary bodies'. As with the EB, two-thirds of members must be present to constitute a quorum, including a majority of Annex I and non-Annex I Parties, and meetings are 'open to attendance, as observers, by all Parties and UNFCCC accredited observers and stakeholders', unless the A6SC decides otherwise. Decisions are to be taken by consensus unless all such efforts have been exhausted, in which case they may, as a last resort, be taken by a three-fourths majority vote. The full text of all decisions is made publicly available in all six UN languages. As noted above, the A6SC is expected to develop its additional rules of procedure, which are likely to draw heavily on those agreed for the EB.

JI lacks a 'share of the proceeds' provision to fund administrative costs. The issue of who should pay the administrative costs arising from the procedures contained in the Article 6 Guidelines proved too complex to negotiate in detail in the last days of Marrakesh. Accordingly, the Article 6 Guidelines reflect all that could be agreed in terms of principle which is that such costs 'shall be borne by both the Parties included in Annex I and the project participants according to specifications set out in a decision by the COP/MOP at its first session.'[164]

Independent entities

IE functions are not spelled out in a specific section as for the CDM but have to be gleaned instead from the Article 6 Guidelines, from Section E concerning the verification procedure set out in paragraphs 30–45 and from Appendix A which sets out standards and procedures for the accreditation of IEs.

Granting accreditation status to a potential IE rests with the A6SC, in contrast to the CDM (where the COP/MOP designates on the basis of EB recommendations and provisional designation). The standards and procedures set out in Appendix A are almost identical to those used for the CDM, in part, because the actual tasks the IEs will undertake under Track 2 JI are functionally equivalent to those under the CDM.

The provisions for the suspension and withdrawal of accreditation status of an IE who does not meet the accreditation standards set out in paragraphs 42 and 43 are also very similar to the CDM. They give any IE the opportunity for a hearing, provide for the immediate effect of the A6SC's decision and state that the suspension or withdrawal of an IE will not affect verified projects unless 'significant deficiencies' are found, in which case the IE concerned shall acquire an 'equivalent amount of AAUs and ERUs and place them in the holding account of the Party hosting the project within 30 days.'

Participation/eligibility

Participation and eligibility requirements are discussed in detail in section I.3 on cross-cutting issues. For ease of reference, the differences between the eligibility criteria for Track 1 and Track 2 are summarized in Table I.3.

Article 6.4 provides that if a question of implementation of the provisions of Article 6 is identified by the expert review process (Article 8), transfers and acquisition of ERUs may continue to be made after the question has been identified but these units may not be used by a Party to meets its commitments 'until any issue of compliance is resolved'. The Article 6 Guidelines adopted state at paragraph 25 that 'the provisions in Article 6.4 of the Protocol shall pertain, *inter alia*, to the requirements of paragraph 21 above', i.e. the paragraph defining the eligibility conditions for JI. Inclusion of this provision is intended to limit the kinds of questions that can give rise to the restriction of the use of ERUs to questions related to the specified eligibility requirements.

Table I.3 *Difference between Track 1 and Track 2*

	JI Track 1	JI Track 2
Participation requirement to be met by the host country	1. Party to the Kyoto Protocol 2. Has submitted a report for determining their initial assigned amounts (AAUs) 3. Has a national system of evaluation of greenhouse gas emissions from sources and storage using sinks 4. Has a computerised national registry compliant with the international requirements 5. Annually submits a current inventory protocol fully compliant with the Kyoto Protocol requirements	1. Party to the Kyoto Protocol 2. Has submitted a report for determining their initial assigned amounts (AAUs) 3. Has a computerised national registry compliant with the international requirements
Verification	Host country performs the verification of greenhouse gas emissions	An independent entity performs the verification
Transfer of ERUs	Host country transfers the agreed amount of ERUs	Host country can transfer ERUs only after verification by an independent entity

Source: Wollansky and Freidrich (2003).

Paragraph 41 of the Article 6 Guidelines states that 'any provisions relating to the commitment period, reserve or other limitation to transfer under Article 17 shall not apply to transfers by a Party of ERUs issued into its national registry that were verified in accordance with the verification procedure under the Article 6 supervisory committee.'

Track 1 procedure

Although paragraph 23 of the Article 6 Guidelines does not use the term 'Track 1' its contents create a streamlined procedure that has become popularly known by this name. The procedure stipulates that where the *host* Party is considered to meet the eligibility requirements in paragraph 21 of the Article 6 Guidelines it may verify reductions in anthropogenic emissions and removals by sinks from an Article 6 project 'as being additional to any that would otherwise occur, in accordance with Article 6.1(b)'. Upon such verification, the host Party may issue the appropriate quantity of ERUs in accordance with Decision 19/CP.7. This is done by converting AAU into ERUs and transferring them through the system of national registries.

Track 1 does not entail international scrutiny in respect of the JI project by the A6SC and IEs and appears to give discretion to the host Party to choose baseline and monitoring methodologies. Because the reference to Article 6.1(b) of the Protocol refers to the additonality test which also applies in the CDM, it may be that choice of methodologies may, in practice, be influenced by international developments as to what is/is not a reasonable methodology.

A Party which can undertake Track 1 JI projects may at any time elect to use the Track 2 procedure instead. This might be done, for example, to provide greater credibility to the project and the Party than might be possible under Track 1.

Track 2 verification procedure

Track 2 applies when not all the eligibility criteria can be met by a host Party. As Table I.3 makes clear, at a minimum, the host Party must have established its assigned amount and have in place a national registry which meets the requirements under Articles 5 and 7. Such a host Party can then transfer ERUs, provided that an IE validates the project and it is subsequently verified by an IE according to the procedures set out below.

The validation part of the procedure is based on the validation/registration requirements of the CDM. For Article 6 projects, PPs must prepare a PDD (to be tailor-made by the A6SC on the basis of Appendix A) containing information to allow the IE to assess the following:

- approval by the Parties involved;
- whether the project would result in the reduction of anthropogenic emissions by sources or enhancement by sinks additional to any that would otherwise occur;
- has an appropriate baseline and monitoring plan in accordance with Appendix B; and
- has submitted documentation on the analysis of the environmental impacts of the project activity, including transboundary impacts in accordance with host Party procedures, and if the environmental impacts of the project are considered by the PPs or the host Party to be significant, the PPs have undertaken an environmental impact assessment in accordance with host Party procedures.

The PPD and supporting documentation submitted to the IE by PPs must be made publicly available for 30 days to enable the IE to received public comments. The IE then makes its determination as to whether to validate a project and in so doing makes its decision, an explanation of its reasons and how the comments it received were taken into account publicly available. This determination becomes final unless a review is requested by a Party involved in the project or three members of A6SC request it in accordance with paragraph 35 of the Article 6 Guidelines.[165] The A6SC has a maximum of six months from the date of the request to conclude the reviews and its decision is final.

Once the project has commenced, PPs shall submit to 'an' accredited IE a report in accordance with the monitoring plan. The reference to 'an' (rather than to 'the' IE) would appear to suggest that the IE undertaking verification must be different to 'the' IE that determined the project's validation, as would normally be the case under the CDM.[166]

Upon receipt of the report (which is functionally comparable to the monitoring report in the CDM project cycle), the IE shall make a determination of the reductions/enhancements reported by PPs and make its determination publicly available through the Secretariat, together with an explanation of its reasons. This determination shall be deemed final 15 days after the date of it being made public unless a Party involved or three members of the A6SC request a review in accordance with the procedure specified in paragraph 39 of the Article 6 Guidelines.[167] When the final decision is made, the host Party is entitled to issue and transfer the ERUs but only if it is in compliance with its minimum JI eligibility criteria set out above.

LULUCF projects

Article 6 covers projects relating to sink activities which are covered by the Kyoto Protocol as only specific categories may count towards an Annex I Party's Article 3.1 commitments. Article 6 projects aimed at enhancing anthropogenic removals by sinks shall conform to definitions, accounting rules, modalities and guidelines under Articles 3.3 and 3.4.[168] Decision 19/CP.7 on accounting modalities explains how an Annex I Party may issue RMUs in relation to its (domestic) sinks which are being counted towards compliance under Articles 3.3 and 3.4.

Small-scale and nuclear projects

The JI Guidelines allow all types of projects to be JI projects. There are no specific provisions to encourage small-scale projects under JI as is currently the case for the CDM. The Article 6 Guidelines stipulate that Annex I Parties are to refrain from using ERUs from nuclear facilities to meet their commitments under Article 3, as is the case for CERs under the CDM.

I.8 Compliance procedures and mechanisms under the Protocol

Overview

The inclusion of binding targets for Annex I Parties in the Protocol focused attention on how compliance with these can be assured. Because the Kyoto mechanisms provide economic incentives for Parties to engage in trading which would not otherwise exist, the inclusion of the mechanisms creates an additional compliance problem: the possibility that a Party could find itself in a situation of non-   compliance caused by its own excessive transfers. This might be caused by lack of administrative capacity to track mechanism transmissions or failure to forecast emission trends or through cases of overselling by a rogue Party. All of these scenarios would undermine the environmental integrity of the Protocol and also damage confidence in the carbon markets as legal certainty about who could or could not undertake valid transactions might dampen trade and erode the value of Kyoto units. This section provides an overview of the provisions adopted pursuant to the Kyoto Protocol that facilitate compliance by Parties with their international commitments and, where necessary, to correct cases of non-compliance, focusing on mechanism-related issues.[169]

Article 18 provides that the COP/MOP-1 shall 'approve appropriate and effective procedures and mechanisms to determine and to address cases of non-compliance with the provisions of this Protocol, including through the development of an indicative list of consequences, taking into account the cause, type, degree and frequency of non-compliance. Any procedures and mechanisms under this Article entailing binding consequences shall be adopted by means of an amendment to this Protocol.' Shortly after the adoption of the Protocol, and acting on the leadership of the US, Parties began to realize that binding targets would require robust systems to ensure compliance. COP-4, held in Argentina in 1998, agreed to establish a joint working group (JWG) to prepare procedures and mechanisms relating to compliance under the Protocol.[170] These negotiations were highly technical in nature but as the final days of negotiations at COP-6 Part II in Bonn and at COP-7 in Marrakesh demonstrated, high political stakes were involved such that on both occasions compliance issues came to be deal breakers. After intense negotiations, COP-7 agreed the adoption of procedures and mechanisms relating to compliance under the Kyoto Protocol[171] which many observers regard as the most advanced compliance system in international environmental law.[172]

The Kyoto Compliance Procedures aim to prevent non-compliance through the development of an early warning system that can lead to the deployment of facilitative approaches but also to procedures that deal with cases of non-compliance through a quasi-judicial process that is focused on correcting cases of non-compliance and restoring the environment. To ensure that there was as little delay as possible to transactions taking place in the Kyoto carbon markets, fast-track procedures were included to expedite consideration of mechanism eligibility issues, and to reinstate eligibility to use the mechanisms for Parties that had been suspended from using them. This section provides a brief explanation of the key features of the Kyoto Compliance Procedures.

The Compliance Committee

The Compliance Committee will be established pursuant to the procedures and mechanisms relating to compliance under the Kyoto Protocol adopted under the Marrakesh Accords.[173] Establishment of the Compliance Committee will take place at COP/MOP-1. The Committee is a standing body which is likely to meet at least twice each year, probably together with the subsidiary bodies. The Committee itself is composed of four bodies: Plenary, Bureau, Facilitative Branch and Enforcement Branch. The Plenary's functions include reporting to the COP/MOP on the Committee's activities, applying the general policy guidance received from the COP/MOP, submitting administrative and budgetary matters to the COP/MOP and developing any further rules of procedure that may be needed, including rules on confidentiality, conflict of interest, submission of information by IGOs and NGOs and translation. The Plenary of the Compliance Committee is made up of 20 members, ten from each Branch (accompanied by their alternates), with the Chair of each Branch serving as Co-Chairs of the Plenary. The Bureau comprises four persons: the Chair and Vice-Chair of each of the two Branches. The tasks of the Bureau are likely to be akin to those of the COP and subsidiary body bureaux in terms of being restricted to organizational and procedural matters, the most important being deciding to which of the two Branches to allocate questions of implementation upon their receipt by the Committee.

The members of the Compliance Committee are to be formally elected by the COP/MOP. As with the other two Kyoto Protocol bodies, each member can be accompanied by an alternate, elected from the same group. To ensure the independence and quasi-judicial character of the Committee, all members are required to serve in their 'individual capacities' and have 'recognized competence relating to climate change and in relevant fields such as the scientific, technical, socio-economic or legal fields'. The two Branches are each made up of ten members, using the same membership formula as the Executive Board of the CDM-EB (that is, one member from each of the regional groups plus the small island developing states, plus two each from Annex I and non-Annex I Parties). This means a majority of non-Annex I Parties, which was a cause of deep concern for Annex I Parties, given that the work of the Enforcement Branch of the Compliance Committee will be focused on their commitments and not those of non-Annex I Parties.

The basic procedures of the Compliance Committee are set out in the Kyoto compliance procedures and include rules concerning a quorum, the adoption of decisions and the frequency of meetings. A quorum of three-fourths is required, specifically for the 'adoption of decisions'; the rules are silent on whether such a quorum also applies to the convening of meetings. Decisions can be taken by majority voting if necessary. The voting rule for the Compliance Committee incorporates the concept of a double majority. If all efforts at reaching consensus have been exhausted, decisions may be adopted by a three-fourths majority of members present and voting, but decisions of the Enforcement Branch require, in addition, a majority of Annex I and non-Annex I Parties. The safeguard of double majority voting for the Enforcement Branch was included to allay concerns of some Annex I Parties that they might be subject to unfair or politically motivated decisions by that branch given that its membership is based on equitable geographic representation which provides a simple majority for developing countries (Wang and Wiser, 2002, p190).

The Facilitative Branch

The Facilitative Branch is responsible for providing advice and facilitation to Parties in implementing the Protocol and acts as an early warning system for potential non-compliance. The mandate of the Facilitative Branch basically covers everything that is not expressly assigned to the Enforcement Branch. As part of its 'early warning function' the Facilitative Branch will be responsible for providing advice and facilitation for compliance with:

- commitments under Article 3, paragraph 1, of the Protocol, prior to the beginning of the relevant commitment period and during that commitment period;
- commitments under Article 5, paragraphs 1 and 2, of the Protocol, prior to the beginning of the first commitment period; and
- commitments under Article 7, paragraphs 1 and 4, of the Protocol prior to the beginning of the first commitment period.

The Facilitative Branch can apply one or more of the following consequences:

- provision of advice and facilitation of assistance to individual Parties regarding the implementation of the Protocol;
- facilitation of financial and technical assistance to any Party concerned, including technology transfer and capacity building from sources other than those established under the Convention and the Protocol for the developing countries – this is important for EITs as they are not entitled to funding under the Convention's financial mechanism;

- facilitation of financial and technical assistance, including technology transfer and capacity building, taking into account Article 4, paragraphs 3, 4 and 5, of the Convention; and
- formulation of recommendations to the Party concerned, taking into account Article 4, paragraph 7, of the Convention.

In the early years of the regime it is very likely that much of the Facilitative Branch's work will focus on ensuring that Annex I Parties that are EITs are able to meet their reporting and review requirements. This may mean the Facilitative Branch having to consider how to arrange 'facilitation' of financial resources 'outside' the financial mechanism of the climate regime.

The Enforcement Branch

The Enforcement Branch is responsible for determining whether Annex I Parties are in compliance with:

- their quantified emission limitation or reduction commitment under Article 3;
- the methodological and reporting requirements under Articles 5.1, 7.1 and 7.4; and
- the eligibility requirements for the flexible mechanisms under Articles 6, 12 and 17.

It is also responsible for authorizing the application of adjustments to Annex I inventories in the event of a dispute between a Party and the ERT. The compilation and accounting databases for accounting of assigned amounts, which is to be maintained by the secretariat according to Decision 19/CP.7, will take into account any corrections to be made as a resolution of questions of implementation raised by ERTs once these have been determined by the Enforcement Branch.

The degree of discretion given to the Enforcement Branch to impose consequences has been limited to ensure legal certainty and decrease the chance of political interference. Thus the consequences to be 'automatically' applied by the Enforcement Branch are defined tightly to suit the type of commitment that has not been fulfilled. Thus failure to meet Article 5.1, 5.2 and 7.1 requirements will lead to a declaration of non-compliance and a requirement to submit an action plan indicating how and by when a party intends to remedy the failing.

Failure to meet mechanism eligibility requirements means a general suspension from mechanism eligibility but the precise limitation on what kinds of transactions are not allowed depends on the eligibility requirement for particular mechanisms (discussed below). Non-compliance with Article 3.1 will require the following: deduction of 1.3 tonnes for every tonne over-emitted from a Party's assigned amount for the next commitment period, plus a detailed compliance plan indicating how the Party will meet its new target and inability to transfer under Article 17 emissions trading. The application of all such consequences must be aimed at the restoration of compliance to ensure environmental integrity and to provide an incentive to comply.

Mechanism-relevant procedures

The most significant trigger source for the Enforcement Branch is likely to be reports containing questions of implementation raised by ERTs. The ERTs have the right to send these to the Compliance Committee without political intervention from a Party. The procedures provide that the Committee shall receive through the secretariat questions of implementation indicated in the reports of ERTs under Article 8 of the Protocol, together with any written comments by the Party which is subject to the report. Additionally, questions of implementation can also be submitted by any Party with respect to itself or any Party with respect to another Party, supported by corroborating information. The Committee shall also receive all other final ERT reports.

The general procedures of the Compliance Committee are designed to secure due process, transparency and legal certainty, including by defining, *inter alia*, upon what basis a Branch may make a determination, allow a Party to be represented, give it opportunities to comment, allow competent NGOs/IGOs with relevant information to make factual and technical information available, and make information available to it publicly available, bearing in mind any confidentiality rules. Because of the gravity of the commitments it is dealing with, and the consequential need for efficiency and timeliness, time-bound general procedures have been developed for the Enforcement Branch.

A determination of non-eligibility with participation/eligibility conditions (explained in section I.4 above) by the Enforcement Branch leads automatically to the suspension of eligibility to use the flexibility mechanisms by the Party concerned.[174] Such a determination cannot be appealed to the COP/MOP in any circumstance.

The consequence of ineligibility is that the ineligible Annex I Party (and any legal entities it had authorized to participate in the mechanisms under its own authority) cannot undertake transactions dealing with Kyoto units. The range of transactions that an ineligible Party is barred from undertaking is mechanism-specific because Decision 24/CP.7 states that the suspension 'is to be in accordance with the relevant provisions' under Articles 6, 12 and 17. Thus an Annex I Party that does not meet the eligibility criteria for JI is still able to issue and transfer ERUs using the Track 2 procedure provided it meets some of the eligibility criteria set forth in Decision 16/CP.7. For IET, a suspension of eligibility appears to mean that no transactions relating to any Kyoto units can be undertaken until eligibility is reinstated.[175]

Because the effects of being barred from the use of the mechanisms could be very significant for Parties that place high reliance on the use of the mechanisms to achieve their Kyoto commitments, special expedited procedures were included to speed up the overall assessment of eligibility by the Enforcement Branch to enable it to make positive determinations of mechanism eligibility as speedily as possible.

Section X of the Annex to Decision 24/CP.7 itself defines expedited procedures for the consideration of mechanism eligibility by the Enforcement Branch. It should be noted that the initial eligibility requirements for all three mechanisms (rather than the reinstatement requirements) state that an Annex I Party will be deemed to be eligible once 16 months have elapsed from the date on which it submitted the information needed to establish its assigned amount and demonstrated its capacity to account for its emissions and assigned amount under Article 7.4 unless the Enforcement Branch finds that it does not meet eligibility conditions, or alternatively there is earlier confirmation by the Enforcement Branch that it does not meet these conditions or that it is not proceeding with any question of implementation with respect to that Party, and has informed the Secretariat accordingly.[176] The rules further provide that an Annex I Party will continue to meet the eligibility criteria specified for each of the mechanisms 'unless and until the Enforcement Branch ... decides that the Party does not meet one or more of the eligibility requirements, has suspended the Party's eligibility, and has transmitted this information to the Secretariat.'[177] The 16-month period was chosen to take into account the timeframe for the completion of the annual review of inventories and supplementary information due under Kyoto – a process which is supposed to take approximately 12–13 months, if everything proceeds according to schedule. The availability of adequate funding and resources for the ERTs to complete their work to schedule is thus of critical importance.

Agreement that eligibility is to be presumed unless rebutted by an actual decision of the Enforcement Branch was a contentious issue in the mechanism negotiations. The approach that the 'green light is on unless switched red' by the Enforcement Branch was included because some JUSCANNZ countries were concerned that bureaucratic delays in the Article 8 review processes might unduly limit their (and hence their legal entities') participation in the mechanisms. Other Parties, such as the EU and AOSIS, were concerned that presumed eligibility potentially could allow participation in the mechanisms to proceed without any of the environmental integrity eligibility criteria being met, even when fundamental 'questions of implementation' have been raised by ERTs. The integrity of Enforcement Branch procedures and the ability of its member to act as independent experts will thus be critical to the environmental integrity of the Protocol. Additionally, depending on resources and workload, the tight timetables for assessment of mechanism eligibility may set difficult challenges for the Enforcement Branch, although it is likely in practice that many cases of potential non-eligibility will already have been referred to the Facilitative Branch as part of its early warning functions.

Since the adoption of the Marrakesh Accords, the expedited procedures for assessing mechanism eligibility have been supplemented by additional procedures to deal with cases of suspension. These procedures were elaborated because some JUSCANNZ countries, notably Japan, considered the existing procedures might be unnecessarily time-consuming for Parties that wanted to reinstate their eligibility. The additional expedited review procedures are found in the guidelines on the annual review of inventories and other information requirements due under Article 8 of the Protocol which were adopted by COP-8 to deal specifically with cases concerning reinstatement of mechanism eligibility.[178] The expedited reinstatement procedures can be triggered at any time by the concerned Party and will take no longer than 21 weeks. They allow the suspended Party to reapply directly to the Committee or to an ERT. The COP-8 guidelines provide that such a review should be expedited only by restricting it to the issue that caused the suspension and not by adopting a less rigorous approach.

Notes

1. The CDM is in a different legal position: it is already in legal existence because it is being implemented under the authority of the FCCC and is further discussed below.
2. JUSCANNZ is a loose negotiating coalition comprising Japan, the US, Canada, Australia, Norway and New Zealand.
3. See Lefevere, Part II, and Anderson and Bradley, Part III, Chapter 4, in this volume for an extended discussion of these concerns.
4. See Lefevere Part II, on the evolution of market-based approaches in EU environmental policy.
5. See Wilder et al, Part III, Chapter 5, in this volume on the development of the Chinese trading scheme addressing SO_2 which demonstrates that while developing countries are learning they are also dealing with environmental problems more developed countries have long legislated for. The capacity-related needs of developing countries are also important and discussed by Michaelowa, Part III, Chapter 8.
6. FCCC/AGBM/1997/MISC.1. See also Oberthur and Ott (1999, p188).
7. See Anderson and Bradley, Part III, Chapter 4, in this volume on EITs and their approach to the mechanisms.
8. Ibid.
9. See Haites in the Conclusion to this book who discusses the role of the mechanisms in the context of future commitments by developed and developing countries.

10. By early 1999, proposals for the UK and Danish trading schemes were advanced and the EU Commission had decided to investigate the policy framework for the design of a GHG trading scheme for the EU, work on which was coordinated by the author as part of a multi-partner consortium headed by FIELD. For details, see Lefevere, Part II, on the evolution of the EU ETS. The final report and background papers are available from FIELD at www.field.org.uk.

11. See Blanchard, Part III, Chapter 2, for economic analysis of who trades what with whom in different kinds of carbon markets.

12. For details, see Anderson and Bradley, Part III, Chapter 4, and Haites in the Conclusion to this volume.

13. See Michaelowa, Part III, Chapter 7, on the issues raised by CDM baselines and additionality.

14. See, for example, the report for the Pentagon on the national security implications of climate change on the US, summarized in *Fortune* magazine in February 2004.

15. *Faking It*, Climate Action Network Report on the US climate policy under the Bush Administration, December 2003, available from Climate Action Network US (CAN-US), Washington, DC.

16. See Lefevere, Part II.

17. See Michaelowa, Part III, Chapter 7, for a discussion of these.

18. Decision 10/CP.3.

19. Decision 20/CP.8.

20. UNFCCC website: www.unfccc.org.

21. *JI Quarterly*, available http://www.northsea.nl/jiq/.

22. Decision 17/CP.7, paragraph 13.

23. Decision 18/CP.9, paragraph 1.

24. Decision 16/CP.7, Draft Decision -/CMP.1 (Article 6), paragraph 5.

25. Decision 16/CP.7, Draft Decision -/CMP.1 (Article 6), paragraph 8. Decision 17/CP.7, Draft decision -/CMP.1 (Article 12) paragraph 4 and Decision 18/CP.7, paragraph 2.

26. Decision 15/CP.7, preamble, paragraph 5 and Draft Decision -/CMP.1 (mechanisms), Principles, nature and scope of the mechanisms pursuant to Articles 6, 12 and 17 of the Kyoto Protocol, preamble paragraph 5.

27. Decision 15/CP.7, preamble, paragraph 6 and Draft Decision -/CMP.1 (mechanisms), preamble paragraph 6. See also Decision 5/CP.6, Section VI.I, paragraph 4.

28. Decision 22/CP.7, Draft Decision -/CMP.1, Guidelines for the preparation of information under Article 7 of the Protocol, paragraph 4.

29. Decision 15/CP.7, paragraph 4 and Decision 24/CP.7, annex, paragraph 5(b).

30. Decision 11/CP.7, annex, paragraph 14.

31. Decision 15/CP.7, paragraph 6.

32. Decision 19/CP.7, annex, section F.

33. Decision 19/CP.7, Draft Decision -/CMP.1 (modalities for accounting of assigned amounts), paragraph 2 and Annex, paragraph 9.

34. Decision 16/CP.7, Annex, paragraph 1(e) and Decision 17/CP.7, Annex, paragraph 1(e).

35. See, for example, CDM Watch, which is monitoring all CDM projects to see if these conform with the Marrakesh Accords and is also monitoring broader trends useful for policy-makers such as what types of CDM projects are being favoured as well their geographic location and whether relevant affected communities are being consulted (available at http://www.cdmwatch.org/).

36. Decision 19/CP.7, annex, paragraph 44 and Decision 24/CP.8.

37. Decision 24/CP.8. See also FCCC/TP/2002/3, Registries under the Kyoto Protocol and FCCC/TP/2002/2, Treatment of Confidential Information by International Treaty Bodies and Organizations and Decision 21/CP.9 concerning issues relating to the implementation of Article 8 of the Kyoto Protocol.
38. Decision 16/CP.7, Annex, paragraph 29, Decision 17/CP.7, Annex, paragraph 33 and Decision 18/CP.7, Annex, paragraph 5.
39. For emissions trading, a Party must maintain an up-to-date list of such entities and make it available to the Secretariat.
40. The policy rationale for linking such schemes and concrete proposals for how these linkages might be implemented are discussed by Haites in the Conclusion to this volume.
41. Decision 17/CP.7, paragraph 3.
42. Decision 17/CP.7, Annex, paragraph 30.
43. Decision 16/CP.7, Annex, paragraph 20. Decision 17/CP.7, Annex, paragraph 29.
44. CDM Monitor, Point Carbon, 12 February 2004. For a more detailed discussion of some DNAs see Wilder in this volume.
45. Decision 20/CP.7, Guidelines for national systems under Article 5.1 of the Kyoto Protocol and draft decision -/CMP.1 of the same title (KP national system guidelines).
46. For a discussion of the national systems and their place in the Protocol's system of reporting and review see Chapter 11 in Yamin and Depledge (2004).
47. Decision 24/CP.8, Technical standards for data exchange between registry systems under the Kyoto Protocol and FCCC/SBSTA/2003/L.20, SBSTA conclusions on issues relating to national registries under Article 74 of the Protocol.
48. Decision 19/CP.7, Modalities for the accounting of assigned amount and draft decision -/CMP.1 of the same title.
49. See Chapter 11 Yamin and Depledge (2004) for a detailed explanation of when inventories under the Protocol are due and how these will be reviewed.
50. Decision 22/CP.7, Draft Decision -/CMP.1 (Article 7), paragraph 3(a)–(f).
51. The reference to assistance from the Facilitative Branch is part of its 'early warning' and proactive assistance functions designed to ensure the FB helps Parties fix problems to stop non-compliance issues from arising. See section I.8.
52. Decision 20/CP.9 Technical guidance on methodologies for adjustments under Article 5.2 KP.
53. Decision 21/CP.7, paragraphs 2–4.
54. See Anderson and Bradley, Part III, Chapter 4, who discuss the problems EITs may face in meeting the mechanisms eligibility conditions.
55. Decision 19/CP.7, Annex, paragraph 26.
56. These provisions were included at the insistence of OPEC countries. The use of the mechanisms as a whole will reduce these potential impacts as they will lower Annex I Parties compliance costs.
57. Decision 15/CP.7, paragraph 8 and Draft Decision -/CMP.1 (mechanisms), paragraph 7.
58. Chapter 5 explains the difference between the term 'Annex I Parties' and Annex B Parties.
59. Decision 18/CP.7, Annex, paragraph 2.
60. See section I.1 above and also the contributions by Lefevere, Part II, and Anderson and Bradley, Part III, Chapter 4.
61. Decision 5/CP.6.
62. Decision 18/CP.7, Annex, paragraph 8.

63. Decision 19/CP.7, Annex, section II.D.
64. Chapter 11, Yamin and Depledge (2004).
65. On CERs and ERUs generated by nuclear facilities see validation and registration section below. See also discussions of the Linking Directive Proposal to the EU ETS in Part II which will exclude certain kinds of projects such as CDM sinks and nuclear credits.
66. On linking national and international trading schemes, see Haites in this volume.
67. Article 12 and Decision 17/CP.7 setting out the modalities and procedures of the CDM ('CDM Modalities and Procedures') refer to CDM 'project activities' rather than to 'projects'. The former term is conceptually broader as 'activities' could cover policies and measures unrelated to physical projects. For convenience this part uses the two terms interchangeably as, presently, current proposals for the CDM seem only to involve physical projects.
68. See section I.4 on participation above.
69. Decision 17/CP.7, Annex, paragraphs 64–66, and Appendix D.
70. Decision 21/CP.8, Annex II, Simplified modalities and procedures for small-scale CDM project activities.
71. CDM-PDD, Version 01, in effect 29 August 2002, available at: http://cdm. unfccc.int/Reference/Documents/cdmpdd/English/cdmpdd.doc.
72. Decision 17/CP.7, Annex, paragraph 35.
73. Decision 17/CP.7, Annex, paragraphs 35–52.
74. For an extended discussion of baselines see Michaelowa, Part III, Chapter 7.
75. Decision 17/CP.7, Annex, paragraphs 53–60.
76. Decision 17/CP.7, Annex, paragraphs 61–63.
77. Decision 17/CP.7, Annex, paragraph 64–66.
78. Decision 17/CP.7, Annex, paragraphs 2–4.
79. Decision 21/CP.8, paragraph 1.
80. Decision 21/CP.8, Annex I, Rules of Procedure of the EB, hereafter 'CDM Rules'.
81. Decision 17/CP.7, Annex, paragraph 6.
82. Decision 36/CP.7.
83. To enhance continuity, Decision 18/CP.9 adopted by COP-9 amended rules 4 and 12 of the CDM Rules to allow the Chair and Vice-Chair being in office between the election of new members and alternates and the first meeting of the EB in a calendar year.
84. Service as an alternate does not count, however, towards the term of office of a member. An alternate may, therefore, serve two consecutive terms and then be elected as a full member for a further two consecutive terms, and vice versa. This allows alternates to 'train up' as well as providing an important measure of continuity, although it does mean Parties not represented on the Board may have to wait a longer time to get a chance to be elected to the Board.
85. CDM Rules, Rule 10.2 for the text of the oath of service.
86. This provision aims to ensure the effective functioning and full representation of the Board and to ensure that Board members know what is expected to them and are fully committed to their role from the start. It reflects problems encountered in other intergovernmental arenas (e.g. the IPCC), where Bureau/committee members have routinely failed to attend meetings.
87. Decision 17/CP.7, paragraph 8(c).
88. CDM modalities and procedures, paragraph 14, and CDM rule 28.

89. CDM modalities and procedures, paragraph 17, CDM rule 31.
90. FCCC/CP/2003/2 and Add.1, Annual Report of the EB to COP (2001–2002) ('First CDM Report'). FCCC/CP/2003/2, Annual Report of the EB to COP (2002–2003).
91. Decision 18/CP.9 and Decision 19/CP.9.
92. CDM Rule 2.14 and Rules 26 and 27. The draft rules of procedure originally proposed by the EB to COP-8 were amended by the COP to add this point. This is, of course, of particular concern to the US.
93. CDM modalities and procedures, paragraph 15, and CDM Rule 29, paragraphs 1.
94. CDM Rule 29, paragraph 3.
95. CDM Rule 29, paragraph 4.
96. CDM Rule 29, paragraph 2.
97. CDM Rule 30.
98. CDM Modalities and Procedures, paragraph 18, CDM Rule 32.
99. CDM Rule 32, paragraph 2.
100. See First CDM Report, paragraphs 29(c), 23(a) and 21(b).
101. See First CDM Report, section III.D.
102. Decision 17/CP.7, paragraph 6(e). See First CDM Report, section III.D and Second CDM Report, section III.F.
103. Decision 17/CP, 7, Annex, paragraph 40(a).
104. Decision 17/CP.7, Annex, paragraphs 27.
105. Decision 17/CP.7, Annex, paragraph 27(e).
106. Decision 17/CP.7, Annex, paragraphs 20–25.
107. Report of the Seventh Meeting of the EB, Annex 2, Procedures for accrediting operational entities by the EB, Version 03, January 2003, as clarified by the EB at its eleventh meeting which issued additional clarification regarding cost implications of changes to an application made by an applicant entity, set out in Annex 1, of the Report of the Eleventh Meeting of the EB/CMD, and clarification agreed by the Board regarding witnessing opportunities, set out in Annex II of the same Report.
108. Decision 17/CP.7, Annex, Appendix A.
109. The PP has to observe all relevant domestic laws and other international laws that may generate such a requirement. Paragraph 27(c) states it is the responsibility of the DOE to check a CMD project meets these legal requirements.
110. Decision 17/CP.7, Annex, Appendix B.
111. Decision 17/CP.7, Annex, paragraph 6, paragraph 27(h), and paragraph 40.
112. Decision 17/CP.7, Annex, paragraphs 62–63.
113. Decision 18/CP.9 contains a draft recommendation to COP/MOP setting out procedures for the reviews under paragraph 41. See also Second Annual Report of the EB, FCCC/CP/2003/2, Annex.
114. Similar language is found in JI modalities and procedures.
115. See Part II for a discussion of the Linking Directive proposal which will restrict ERUs and CERs from certain kinds of projects from entering the EU ETS.
116. CDM Modalities and Procedures, paragraph 37.
117. Baseline and additionality CDM issues are discussed by Michaelowa, Part III, Chapter 7.
118. See below on the funding of CDM.
119. Eleventh Meeting of the EB, Annex I, Further clarifications on methodological issues, Ninth Meeting of the EB, Annex 3, Further clarifications on methodological issues, Eighth Meeting of the EB, Annex I, Clarifications on issues relating to

baseline and monitoring methodologies and Fifth Meeting of the EB, Annex 3, Guidance by EB to the Panel on guidelines for methodologies for baseline and monitoring plans.

120. Eleventh Report of the EB, Annex Version 04.
121. Second Annual Report of the EB, FCCC/CP/2003/2, paragraphs 31–39.
122. Decision 17/CP.7, Annex, paragraph 40.
123. Decision 21/CP.8, Annex II, Simplified modalities and procedures for small-scale CDM project activities (hereinafter 'SCC M&P').
124. Second Annual Report of the EB, FCCC/CP/2003/2, paragraphs 27–30).
125. The author is grateful to Erik Haites for clarifying the rules on CDM sinks.
126. For a summary of these see Greenpeace (n.d.).
127. See below.
128. So project proponents may choose to exclude a pool that is a small sink that is costly to measure.
129. This yields a conservative estimate of the reductions achieved. It also avoids the possibility of earning credits for reducing the emissions associated with displacing activities on the land prior to it being planted with trees. For example, if the land was used for cattle grazing prior to being planted, credits cannot be earned for reducing those emissions.
130. Decision 17/CP.7, Annex, paragraphs 53–60.
131. Decision 17/CP.7, Annex, paragraphs 61–63.
132. Decision 17/CP.7, Annex, paragraph 64–66.
133. Decision 17.CP.7, Annex, paragraph 5(l) and 66.
134. Decision 17/CP.7, Annex, Appendix D, paragraph 8.
135. Second Annual Report of the EB, FCCC/CP/203/2, paragraph 46.
136. Decision 5/CP.1 on AIJ contains similar wording but the issue was nuanced because no credits could be gained in the pilot phase.
137. Decision 17/CP/7, Annex, paragraph 3.
138. Decisions 17/CP.7, Annex, paragraph 4. See also Second Annual Report of the EB, FCCC/CP/2003/2.
139. Second Annual Report of the EB, FCCC/CP/2003/2.
140. Decision 18/CP.9, paragraph 1.
141. Decision 15/CP.7, paragraph 15.
142. For details of the Adaptation Fund and other funding issues under the climate regime see Yamin and Depledge (2004).
143. Decision 15/CP.7, paragraph 15.
144. Second Annual Report of the EB.
145. Decision 38/CP.7, paragraph 14.
146. First CDM report, paragraph 34.
147. Second CDM Report, paragraph 74.
148. First CDM report, paragraphs 19(c) and 35.
149. Second Annual Report of the EB, FCCC/CP/2003/2, paragraph 69.
150. Second EB Report, FCCC/CP/2003/2, paragraph 69.
151. For convenience, the terms 'JI' and 'Article 6 projects' are used interchangeably in the main body of the text because the mechanism defined by Article 6 continues to be popularly known as 'joint implementation'.
152. European Commission, 1998.
153. Decision 5/CP.6, section 2.
154. The reporting and review requirements for Annex I Parties mean that expert review teams will already have provided one layer of external scrutiny (see Yamin and Depledge, 2004, Chapter 11).

155. The reference to 'Article 6 Guidelines' refers here to the Annex to Draft decision -/CMP.1 (Article 6) Guidelines for the Implementation of Article 6 of the Kyoto Protocol which is appended to Decision 16/CP.7.
156. Decision 16/CP.7, Article 6 Guidelines. Unless otherwise specified, references to rules of the supervisory committee are taken from the Annex to draft decision -/CMP.1, which sets out the guidelines (section C).
157. Article 6 Guidelines, Annex, paragraphs 2 and 3(a).
158. Article 6 Guidelines, Annex, paragraph 3.
159. These include not only the Annex II Parties, but also Annex I Parties that are neither EIT nor included in Annex II (namely, Liechtenstein, Monaco and Turkey).
160. Article 6 Guidelines, Annex, paragraph 4.
161. Article 6 Guidelines, Annex, paragraph 5.
162. Decision 36/CP.7.
163. Ibid., paragraph 10(a).
164. Draft Decision -/CMP.1 (Article 6), paragraph 7.
165. The comparable CDM procedure under paragraph 40 of the CDM modalities (see Box I.3).
166. CDM Modalities, paragraph 27(e).
167. This corresponds to the review at post-certification/pre-issuance stage of the CDM under paragraph 64 of the CDM Modalities.
168. Decisions 16/CP.7, Draft Decision, paragraph 4.
169. For a more detailed account of the Kyoto compliance procedures see Chapter 12 in Yamin and Depledge (2004).
170. Decision 8/CP.4.
171. Decision 24/CP.7, Procedures and mechanisms relating to compliance under the Kyoto Protocol containing an Annex of the same title (Kyoto Compliance Procedures). The word mechanisms in the title of Decision 24/CP.7 does not refer to the Kyoto flexibility mechanisms but refers to institutions referred to in Decision 24/CP.7.
172. On the non-compliance with MEAs and the Kyoto Protocol, see Werksman (2004), Wang and Wiser (2002) and Lefeber (2001).
173. Decision 24/CP.7, Procedures and mechanisms relating to compliance under the Kyoto Protocol. Unless otherwise stated, all references relate to the Annex to this decision, which sets out the procedures and mechanisms.
174. Decision 24/CP.7, Annex, Section XV, paragraph 4.
175. Decision 18/CP.7, Annex, paragraph 2.
176. Decision 16/CP.7, Annex, paragraph 22(a), Decision 17/CP.7, Annex, paragraph 32(a) and Decision 18/CP.7, Annex, paragraph 3(a).
177. Decision 16/CP.7, Annex, paragraph 22(b), Decision 17/CP.7, Annex, paragraph 32(b) and Decision 18/CP.7, Annex, paragraph 3(b).
178. Decision 22/CP.8, Additional sections to be incorporated in the guidelines for the preparation of the information required under Article 7 and in the guidelines for the review of information under Article 8 of the Kyoto Protocol.

References

Bodansky, D. (1993) 'The United Nations Framework Convention on Climate Change: a commentary', *Yale Journal of International Law*, vol 18, no 2, pp451–558.

Depledge, J. (2000) *Tracing the Origins of the Kyoto Protocol: An article by article textual history*, a technical paper for the UNFCCC Secretariat, FCCC/TP/2002/2, 25 November.

Greenpeace (n.d.) *Sinks in the CDM: After the climate, biodiversity goes down the drain, an analysis of the CDM sinks agreements at COP-9*, by Malte Meinhausen and Bill Hare. Available from Greenpeace International.

Grubb, M. and Yamin, F. (2001) 'Climate collapse at The Hague: what happened, why, and where do we go from here?', *International Affairs*, vol 77, no 2, pp261–76.

Grubb, M., Vrolijk, C. and Brack, D. (1999) *The Kyoto Protocol: A Guide and Assessment*, London, Earthscan/RIIA.

Lefeber, R. (2001) 'From the Hague to Bonn to Marrakesh and beyond: a negotiating history of the compliance regime under the Kyoto Protocol', *Hague Yearbook of International Law*, 17 25.

Oberthur, S. and Ott, H. (1999) *The Kyoto Protocol: International Climate Policy for the 21st Century*, Berlin, Springer Verlag.

Wang, X. and Wiser, G. (2002) 'The implementation and compliance regimes under the Climate Change Convention and its Kyoto Protocol', *Review of European Community and International Environmental Law*, 11.

Werksman, J. (1998) 'Compliance and the Kyoto Protocol: building a backbone into a "flexible" regime', *Yearbook of International Environmental Law*, vol 9.

Werksman, J. (1999a) *WTO Issues Raised by the Design of an EC Emissions Trading System*, paper prepared for the European Commission, available from FIELD.

Werksman, J. (1999b) 'Greenhouse gas emissions trading and the WTO', *Review of European Community and International Environmental Law*, vol 8, no 3, pp251–64.

Wilkins, H. (2002) 'What's new in the CDM?' *Review of European Community and International Environmental Law*, vol 11: 2.

Wollansky, G. and Freidrich, A. (2003) *A guide to carrying out Joint Implementation and Clean Development Mechanism Projects*, Version 1.1, produced for and published by Austrian Federal Ministry of Agriculture, Forestry, Environment and Water Management, Vienna.

Yamin, F. (1993) 'The use of joint implementation to increase compliance with the Climate Change Convention: international legal and institutional questions', *Review of European Community and International Environmental Law*, vol 2: 4.

Yamin, F. (1999) 'Equity, entitlements and property rights under the Kyoto Protocol: the shape of things to come', *Review of European Community and International Environmental Law*, vol 8: 3.

Yamin, F. and Depledge, J. (2004) *The International Climate Change Regime: A Guide to Rules, Institutions and Procedures*, Cambridge, Cambridge University Press.

Yamin, F., Burniaux, J.-M. and Nentjes, A. (2001) 'Kyoto mechanisms: key issues for policy-makers', *International Environmental Agreements: Politics, Law and Economics*, 1, pp187–218.

Part II:

The EU Greenhouse Gas Emission Allowance Trading Scheme

Jürgen Lefevere

II.1 Introduction

On 4 March 2002, the 15 Members of the European Union decided to ratify the Kyoto Protocol before 1 June 2002, in line with their previous commitment to allow its entry into force at the World Summit on Sustainable Development in Johannesburg in August/September 2002. Accordingly, the EU adopted its ratification decision on 31 May 2002 and the European Community and its Member States jointly submitted the ratification instrument to the UN.[1]

The Kyoto Protocol will require the European Union to reduce its aggregate emissions of a 'basket' of six greenhouse gases[2] (GHGs) by 8 per cent over the period 2008–2012 compared to its 1990 emissions. Although in the year 2000 community-wide greenhouse gas emissions stabilized in relation to 1990 emissions, recent inventories have shown a rise in emissions to 2.1 per cent above the Kyoto target by the end of 2001 (see Figure II.1 below). Member States' projections furthermore suggest that existing policies and measures, both at national and EU levels, will not be sufficient to continue the EU-wide reductions of total EU greenhouse gas emissions.[3] Instead, progress made so far will be outweighed by further increases unless further measures are taken. The 'business-as-usual' scenario, with existing measures, suggests that in 2010 EC-wide emissions will have decreased by only 0.5 per cent, which leaves a significant gap of 7.5 per cent to be achieved through new measures.[4]

In 1998 the need to reinvigorate the debate on the development and adoption of effective policies and measures to reduce the EU's GHG emissions led
the Commission to focus on the introduction of an innovative instrument to
tackle the EU's greenhouse gas emissions: emissions trading. This discussion
was in particular inspired by the inclusion of emissions trading as one of the
flexible mechanisms under the Kyoto Protocol. As explained in Part I, Article
17 of the Kyoto Protocol lays the foundation for international emissions trading (IET) among Parties, who can, if they so wish, establish domestic trading
schemes and/or authorize legal entities to participate in IET under Article 17
itself. The development of an EU-wide emissions trading scheme can be seen
as a domestic (regional) trading scheme which links in with IET under the
Protocol and with the two Kyoto project-based mechanisms. The EU ETS will
be operational before the start of Kyoto trading and could function independently of the Kyoto Protocol's entry into force.

Since the start of the discussions on EU-wide emissions trading, the instrument has rapidly gained support across a broad range of stakeholders within
the European Union, representing a variety of interests. Together with the
Commission's eagerness to establish a trading regime that could serve as the
'flagship' of the Community's strategy to implement the Kyoto Protocol, this
momentum led to the adoption in October 2001 of the Proposal for a
Directive establishing a scheme for greenhouse gas emission allowance trading
within the Community.[5] After an unusually short decision-making process, the
final text of the Directive establishing a scheme for greenhouse gas emission
allowance trading within the Community (henceforth the 'ET Directive') was
adopted by the European Parliament and the Council in July 2003.[6] Shortly
after agreement was reached on the Directive by the European Parliament and
Council in July 2003, the Commission proposed the first amendment to the
ET Directive.[7] This Proposal for a Directive of the European Parliament and
of the Council amending the Directive establishing a scheme for greenhouse
gas emission allowance trading within the Community, in respect of the Kyoto
Protocol's project mechanisms (henceforth 'the Linking Directive'), expand the
EU ETS by allowing the use of credits from the Kyoto Protocol's project-based
mechanisms, JI and CDM, for compliance with the targets set under the ET
Directive. Both the ET Directive and Linking Directive are set out in
Appendices 2 and 3 of this book.

This section discusses the EU ETS and describes the background and negotiations of the ET Directive. It starts by giving an overview of the EU's
greenhouse gas emission reduction objectives as set out in the EU burden-sharing agreement. The development of the ET Directive took place against the
background of existing EU policy instruments concerning climate change and
other environmental policy problems, and because the ET Directive has had to
'fit' into the existing framework, this section describes this background. It also

addresses certain ethical issues raised by the ET Directive in the EU context. The bulk of Part II then continues with an explanation of the evolution of the EU ETS setting out some of its key design issues and ends with a discussion of the key components of the Commission's Linking Directive.

II.2 The EU burden-sharing agreement

The EU's burden-sharing agreement is the backbone of the EU's implementation of the Kyoto Protocol and the EU climate change target for the years 2008 to 2012. The burden-sharing agreement divides up the EU's overall reduction target of 8 per cent of 1990 emissions between the years 2008 and 2012 by setting individual emission targets, in percentages of 1990 emissions, for each of the 15 Member States that jointly ratified the Kyoto Protocol in May 2002. The burden-sharing agreement was made legally binding under Community law through its inclusion in Annex II to the Council Decision concerning the approval, on behalf of the European Community, of the Kyoto Protocol to the United Nations Framework Convention on Climate Change and the joint fulfilment of commitments thereunder, adopted on 25 April 2002.[8]

The origins of the idea of burden-sharing can be traced back to the elaboration of the EU negotiating position in preparation for the third Conference of the Parties (COP-3) in December 1997 in Kyoto, Japan, at which the Kyoto Protocol was adopted.[9] The EU Bubble was found necessary to allow the Community to adopt a strong common negotiating position for a challenging target under the Kyoto Protocol. Identical targets for each Member State were not seen as feasible, in view of the widely different energy-generation infrastructure, economic development and energy consumption patterns in each Member State. Coming up with a differentiated set of targets at the international level would have significantly complicated and even jeopardized the success of the international negotiations. Rather than negotiate an individual target for each Member State under the Kyoto Protocol, the EU sought to negotiate a target for the EU as a whole, and subsequently redistribute this target internally among the EU Member States through a burden-sharing agreement.

By December 1995, the Environment Council had already decided that:

The equitable sharing of the objective within the Community should be discussed and agreed in parallel; in this respect, account should be taken of cost-effectiveness and the aspects included in the Berlin Mandate, such as differences in starting-points, approaches, economic structures and resources, capital, the need to maintain strong and sustainable economic growth, available technologies and other features specific to each case with each Member State having, moreover, to contribute substantially to the fulfilment of the obligations in the protocol.[10]

This decision followed calls from the EU's less developed Member States, Greece, Ireland, Portugal and Spain (the so-called 'cohesion countries', thus named because of the support they receive for development from the EU's cohesion fund), who were concerned that the EC's climate policy could negatively affect their development. While other Member States insisted on improving energy efficiency in the cohesion countries, more advanced Member States acknowledged that these countries had fewer resources and a legitimate need for economic development. The difficulty was, however, to translate this agreement into concrete differentiated targets.

Although the EU's ability to agree upon internally differentiated targets was initially viewed with scepticism, the Environment Council, under Dutch leadership, managed to reach agreement on an internal EU burden-sharing agreement at its meeting on 2 March 1997.[11] The environment ministers agreed to propose a 15 per cent cut in emissions of a basket of three greenhouse gases (carbon dioxide (CO_2), methane (CH_4) and nitrous oxide (N_2O)) by 2010, as the EC negotiation position in the talks under the UNFCCC.

The EU burden-sharing agreement was the result of a series of discussions among Member States, concluded under the Dutch presidency. The basis for the agreement was a 30-page report on the so-called 'Triptych Approach' which was presented for the first time at an informal workshop of the EU's Ad Hoc Group on Climate on 16 and 17 January 1997 in Zeist, the Netherlands.[12] The Triptych Approach, developed by experts from the University of Utrecht in the Netherlands, separated the national economy into three sectors: the domestic sector, the energy-intensive, export-oriented sector, and the electricity generation sector. CO_2 emissions per capita from the domestic sector did not vary significantly across Member States, although the cohesion countries in general emitted less. In the case of the heavy industry and in particular the power generation sectors, however, considerable differences existed across the EC. Sweden and France for example depended greatly on carbon-free nuclear and hydro-power, whereas Danish and German electricity production was heavily coal-based. By using this sectoral approach, the Triptych Approach clearly demonstrated and justified the need for distinguishing between various countries when setting targets. The Triptych Approach analysed emissions from the various sectors in each Member State, taking into account economic growth, population changes and climate-adjusted energy use (heating and cooling). On the basis of this analysis it set out four variants for Member State targets, depending on reductions to be achieved in the domestic sector and energy efficiency improvements in the energy-intensive sector.

The Triptych variants played a central role in the agreement on the burden-sharing in the March 1997 Environment Council by providing a technical justification for differentiating targets between Member States and demarcating the boundaries for the ensuing political discussions. The Triptych Approach also formed the basis for the Dutch proposal for a burden-sharing agreement. This proposal, setting the stage for the formal negotiations leading to the agreement in March 1997, was put forward in a letter from the Dutch minister for the environment at the end of January 1997.[13] While the ensuing negotiations had a more political character, the burden-sharing agreement that was reached in March 1997 is still firmly rooted in the Triptych Approach.

The EU's initial burden-sharing agreement set specific emission targets for the year 2010, as a percentage of 1990 emissions, for each of the 15 Member States (see Table II.1).[14] The success of this victory was somewhat overshadowed by the fact that the total emissions on the basis of the agreed burden-sharing amounted to only two-thirds of the 15 per cent reduction agreed for the Community as a whole. Also, the agreement was conditional on what countries would agree upon at the meeting in Kyoto in December 1997, including whether the EU's negotiating partners recognized the 'joint fulfilment' proposed by its burden-sharing agreement. The conclusion of the EU burden-sharing agreement in anticipation of the negotiations at COP-3 in Kyoto, however, significantly strengthened the EU's negotiating position and challenged other developed countries to come up with proposals for targets.

Table II.1 *Member State targets under the initial EU burden-sharing agreement (before the finalisation of the Kyoto negotiations)*

Austria	−25%
Belgium	−10%
Denmark	−25%
Finland	0%
France	0%
Germany	−25%
Greece	+30%
Ireland	+15%
Italy	−7%
Luxembourg	−30%
The Netherlands	−10%
Portugal	+40%
Spain	+17%
Sweden	+5%
United Kingdom	−10%

During the negotiations in Kyoto, the EC succeeded in getting its proposal for 'joint fulfilment' into Article 4 of the Protocol, to allow a group of developed countries to agree on a common reduction target and subsequently redistribute this target among the different countries (see Part I). The burden-sharing agreement must be notified to the secretariat on the date of ratification and remains in place for the duration of the first Kyoto Protocol commitment period. When the group of countries as a whole fails to meet its common target, each of its members will be held to their individual target under the burden-sharing agreement. Article 4 does not allow alterations in the burden-sharing agreement due to changes in the EU's membership. This means that the accession of 10 new EU Member States on 1 May 2004 leaves the EU burden-sharing agreement unaffected.

The outcome of the Kyoto negotiations was, however, different from the EU's assumptions for the agreement on the first burden-sharing agreement. The scope of gases was expanded from the three gases originally proposed by the Community to six gases. The reduction target that was finally agreed upon was only 8 per cent over the period 2008–2012 rather than 15 per cent by 2010. Therefore, to allow the Community to use its burden-sharing agreement, the initial agreement of 23 March 1997 had to be adapted to the results of the Kyoto negotiations. Following difficult political negotiations, Member States agreed at the Environment Council meeting on 16 and 17 June 1998 to divide the 8 per cent emission reduction for the European Community as a whole over the Member States as set out in Table II.2.

Table II.2 *Member State targets under the final EU burden-sharing agreement*

Austria	−13%
Belgium	−7.5%
Denmark	−21%
Finland	0%
France	0%
Germany	−21%
Greece	+25%
Ireland	+13%
Italy	−6.5%
Luxembourg	−28%
The Netherlands	−6%
Portugal	+27%
Spain	+15%
Sweden	+4%
United Kingdom	−12.5%

As already mentioned above, the burden-sharing agreement was made legally binding under Community law through its inclusion in Annex II to the Council's Decision of 25 April 2002. Article 2 of that Decision states that 'the European Community and its Member States shall take the necessary measures to comply with the emission levels set out in Annex II, as determined in accordance with Article 3 of this Decision'. The Council Decision is a self-standing and legally binding Decision, which applies irrespective of the entry into force of the Kyoto Protocol. While the calculation of the precise target is based upon rules developed pursuant to the Kyoto Protocol, the obligation for Member States to meet these targets under the burden-sharing agreement is not made conditional upon the entry into force of the Kyoto Protocol. The burden-sharing agreement therefore binds the Member States to their targets irrespective of the timing of the entry into force of the Kyoto Protocol. As will be discussed below, each Member State's targets under the burden-sharing agreement plays an important role in the discussion on the allocation of emission rights in the form of EU 'allowances' under the ET Directive.

II.3 EU environmental policy: from command and control towards market-based mechanisms

EU environmental legislation is traditionally based on so-called 'command and control' legislation. Typical command and control legislation functions through a permitting regime, under which a regulated activity is prohibited unless the operator of the activity has a permit. The permit determines the conditions under which the activity is allowed to take place and includes, in particular, limit values for emissions into various aspects of the environment, as well as monitoring and reporting provisions. A modern example of a command and control instrument is the Community's Directive concerning integrated pollution prevention and control (henceforth 'the IPPC Directive'), described in more detail in Box II.1.[15]

BOX II.1 COMMAND AND CONTROL LEGISLATION: THE 1996 INTEGRATED POLLUTION PREVENTION AND CONTROL DIRECTIVE

The IPPC Directive seeks to regulate emissions in the air, to water and land by requiring an integrated permit for the categories of industrial activities listed in its Annex I. Annex I lists a large number of industrial installations, including, *inter alia*, energy industries such as combustion installations with a rated thermal input exceeding 50 MW, mineral oils and gas refineries, coke ovens and coal gasification and liquefaction plants.

Each IPPC permit must contain emission limit values (ELVs) for pollutants that are emitted from the particular installation. ELVs are to be included in particular for the substances included in Annex II of the Directive, which do not explicitly list any of the greenhouse gases regulated under the KP, but also for other substances which are emitted in 'significant quantities' (Article 9(3) IPPC Directive). The ELVs may also be supplemented or replaced by equivalent parameters or technical measures.

IPPC permits are to be issued by Member State authorities. These authorities also decide on the ELVs to be included in the permits, except when these ELVs

---

have been harmonized at the Community level.[16] The Directive requires the ELVs and other parameters in the permit to be based on 'best available techniques' (BAT). The interpretation of what constitutes BAT for a particular installation is in principle left to Member State authorities, but these need to take into account a number of considerations listed in Annex IV of the IPPC Directive. Among these considerations are 'the consumption and nature of raw materials (including water) used in the process *and their energy efficiency*', 'the need to prevent or reduce to a minimum the overall impact of the emissions on the environment and the risks to it', 'comparable processes, facilities or methods of operation which have been tried with success on an industrial scale' as well as 'technological advances and changes in scientific knowledge and understanding'. Member State authorities must impose ELVs stricter than BAT if this is needed to comply with environmental quality standards.

Member State authorities must include ELVs on greenhouse gases in the IPPC permit when these are emitted 'in significant quantities'. When determining the BAT as the basis for the ELVs to be included in the IPPC permit, Member State authorities must also take into account the energy efficiency of the specific technology. Depending on the installation, IPPC permits can thus contain ELVs for greenhouse gases as well as energy efficiency measures.

Since the beginning of the 1990s, there has, however, been a growing interest within the EU to start using more flexible, market-based mechanisms. A first indication of the EU's interest in the use of market-based mechanisms can be found in its 5th environmental action programme.[17] In this programme, under the heading 'Broadening the Range of Instruments', the EU proposed to use a broader mix of instruments, which would include 'Market-based instruments, designed to sensitize both producers and consumers towards responsible use of natural resources, avoidance of pollution and waste by internalising of external environmental costs (through the application of economic and fiscal incentives and disincentives, civil liability, etc.) and geared towards "getting the prices right" so that environmentally-friendly goods and services are not at a market disadvantage vis-à-vis polluting or wasteful competitors'.

Although emission allowance trading was not specifically mentioned in the 5th environmental action programme, it is often seen as one of the prime examples of a 'market-based mechanism'. The use of market-based mechanisms has for some time now been promoted by academics and critics of the current regulatory system as an alternative to the 'outdated' 'command and control' type legislation like the IPPC Directive. Proponents of market-based mechanisms argue that the current approach, which brings emissions under government control through permitting and stringent monitoring requirements, is too rigid, fragmented, costly and bureaucratic, is not transparent and fails to stimulate innovation.[18] Market-based mechanisms, they argue, would achieve the same level of pollution reduction as traditional instruments, but do so at lower costs, reduce government intervention, provide incentives for technology development and in some cases even generate additional revenue for the government. (Johnson, 2001) This potential for cost-saving plays an important role in the growing interest in the instrument of emissions trading (see Box II.2).

Box II.2: Emissions trading: the ultimate market-based instrument?

There are various approaches to emissions trading, but all are based on the same concept: a target, which can be either fixed or performance-related, is given to each source. In most trading regimes these targets are set by the regulator, not the market. If a source does better than its target it can trade its overachievement, usually in the form of emission rights or 'allowances' with other sources. If it does worse, then it has to buy from other sources on the market. The source will base its decisions on whether to buy or to sell allowances on the market price of the allowances and its marginal costs of abatement. If the market price is higher than the marginal costs to reduce emissions at the source, the source will choose to reduce its emissions further and sell the allowances that are freed up by doing so. If the market price of allowances is lower than the marginal costs to reduce emissions at the source, then the source will choose not to reduce its emissions, but maintain its emissions or even buy allowances on the market to increase its emissions.

A well-functioning trading regime will level the marginal reduction costs across all sectors of industry, by allowing sources with high marginal reduction costs to invest in reductions in sources with lower marginal reduction costs through buying allowances freed up by these sources. By allowing sources to optimally use all cheap abatement options, emissions trading can significantly lower compliance costs and ease the achievement of targets. The potential benefit depends upondifferences in the marginal cost of reducing emissions among participating sources, due to the ability to use different control options, remaining life of the facility, or other reasons. To realise the potential savings, the trading programme must include enough buyers and sellers to create a competitive market.

A well-designed emission allowance trading programme shifts the location and the timing of the emission reductions, but, provided there is effective compliance and enforcement, ensures that the target is achieved. The programme design must ensure that such shifts do not create environmental problems, such as local pollution 'hot spots'.

Discussions on the introduction of market-based mechanisms in the EU have in large part been fuelled by the experiences in the United States.[19] In Europe, both Member States and the EU have, however, until recently, been slower in introducing these mechanisms. Attempts to do so in the last few years have largely focused on an increasing use of environmental taxation as an instrument to reduce fossil fuel and energy consumption and reduce greenhouse gas emissions.[20] Proposals to introduce environmental taxation at the Community level have, however, for a long time been thwarted by the EC Treaty's requirement that these measures must be adopted on the basis of unanimity among the Member States.[21] This does not mean that no regulatory innovation has taken place. More popular 'second-generation' regulatory instruments have been the 'environmental', 'voluntary' or, better, 'negotiated agreements', which are increasingly used at both national and EU levels (see also Box II.3).[22] Negotiated agreements, however, cannot be classified as 'market-based instruments', but are rather attempts to build more flexibility into the current regulatory framework, in particular through more direct involvement of the regulated sector in the design of the instrument and its application and enforcement through contract law rather than through legislation. They could thus be labelled as 'negotiated command and control instruments' (Stewart, 2001, p60).

Box II.3   Negotiated agreements in the EU

In 1998 the European Commission and the European Automobile Manufacturers Association (ACEA) reached agreement on the reduction of CO_2 emissions from passenger cars, known as the 'ACEA Agreement'. Under this agreement car manufacturers have committed themselves to substantially reduce CO_2 emissions from new passenger cars through technological innovation. The agreement requires car manufacturers to reduce CO_2 emissions from new cars by 25 per cent by 2008. Cars produced in 2008 should emit an average of 140 g/km CO_2, equivalent to a fuel consumption of 5.71 l/100 km. In addition, ACEA has agreed to introduce some vehicles that emit 120 g/km by 2000. The agreement is currently being reviewed with the aim to propose further reductions in order to reach the target of 120 g/km by 2012. The Commission endorsed the agreement in a Communication to the Council and the European Parliament.[23] Following the agreement with ACEA, similar agreements have been concluded with the Japanese (JAMA) and Korean (KAMA) car manufacturers.[24]

The voluntary agreements were closed with the understanding that they would provide a 'complete and sufficient substitute' for other Community regulatory measures for the car manufacturers aimed at limiting fuel consumption or reducing CO_2 by cars. Should the car manufacturers fail to meet the targets set out in the agreement, the European Commission has undertaken to propose legislation to achieve the objective of 120 g/km by 2012.

Negotiated agreements are also used in the Member States. On 6 July 1999, the Dutch government concluded the Energy Efficiency Benchmarking Covenant with industry. In it, the energy-intensive industry pledged to be among the world leaders in terms of energy efficiency for processing installations by no later than 2012. In exchange for this undertaking, the government has agreed not to impose any extra specific national measures governing energy conservation or CO_2 reduction on the participating companies.[25]

Recent experience with the implementation of market-based mechanisms in both the US and Europe[26] has, however, shown that simply replacing existing 'old' regulatory instruments with new market-based instruments is often not an option, for a number of reasons. The investment in designing and implementing the 'old' instruments is often considerable; these regimes have been in place in most Member States since the early 1970s. These regimes have been considerably improved over the years, in particular through the introduction of the concept of integrated permitting in a number of Member States from the mid-1980s,[27] and the Community-wide consolidation of this concept in the IPPC Directive has made these regimes more sensitive to cross-sector pollution and introduced attention to a source's impact on the environment as a whole. The combination of permitting with general binding rules, allowed on the basis of Article 9(8) of the IPPC Directive, and the increased use of negotiated agreements[28] have introduced more flexibility and reduced the costs of achieving the targets. In many cases both the regulated sectors and the regulator are of the opinion that the system works and therefore do not see the benefit of a major overhaul of the rules.

It is thus not surprising that recently proposed and introduced market-based instruments can mostly be found in 'new' areas in which behaviour was previously either unregulated or in areas where the current set of regulatory tools has been found insufficient to reach the goals set by the regulator. Climate change is one of those 'new' areas. It is therefore in this area that, within the EU, the instrument of emissions trading has also found its broadest application. Before the adoption of the ET Directive emissions trading had already been implemented in Denmark and the United Kingdom, and trading regimes were in various stages of development in other Member States.[29]

But perhaps the most important argument that shows that simply replacing existing 'old' regulatory instruments with new market-based instruments is often not an option follows from recent experience with the design and implementation of emissions trading regimes. It is increasingly clear that market-based mechanisms are no substitute for legal controls on conduct, backed up by effective government enforcement and sanctions. Instead, these mechanisms are designed to complement, rather than substitute for, command and control measures. Most market-based mechanisms indeed rely for their success upon an underlying programme of government regulatory control (Johnson, 2001, p422; Schwarze and Zapfel, 2000, p293). Emissions allowance trading is a prime example of this, as will become clear from the rest of this chapter. To function well, emissions allowance trading is in practice often built on top of existing permitting regimes. Emissions trading may thus even be described as a 'command and control *plus*' instrument, with often even stronger government oversight, in particular in relation to the monitoring and reporting of emissions and high non-compliance sanctions. The true value of this market-based instrument therefore lies not in replacing the existing command and control regimes, but in building flexibility, cost-effectiveness and incentives for technology development into those regimes.

II.4 The concept of emissions trading

This section gives an overview of the various approaches to the concept of trading that are relevant in the area of climate change. A clear view of these approaches is essential to understand the background to, and implications of, the approach chosen by the ET Directive. Although the issues involved in the design of a trading regime are numerous, this section focuses on the three most important: the choice between an absolute or a relative trading regime; the issue of allocation; and the question of coverage.

Absolute v. relative regimes

As briefly described in Part I, there are two main approaches to emissions allowance trading: baseline-and-credit, and cap-and-trade regimes. In the discussions on emissions trading in the EU, the cap-and-trade approach came to be referred to as the 'absolute target approach', whereas the baseline-and-credit approach came to be referred to as the 'relative target approach'.

Cap-and-trade regimes set a total cap, an absolute quantity of emissions measured over a specific period of time, on all emissions from the sources covered by the regime. This total is subsequently allocated free or by auction in the form of a right to emit a specific quantity, usually in the form of 'allowances', to the various sources under the regime. After the allocation sources can choose to reduce their emissions and sell their allowances, maintain their emissions, or increase their emissions and buy allowances. Choices to buy or sell are made on the basis of the market price of the allowances and the marginal costs of emission reductions at the source. At the end of the trading or compliance period, sources have to match up their actual emissions with sufficient allowances. Sources that are unable to do so must buy allowances, whereas sources that hold an excess of allowances can sell these. Article 17 of the Kyoto Protocol establishes a trading regime with absolute targets. It lays down the amount of emissions during the first commitment period for each of the developed countries and allows them to trade surplus amounts. The ET Directive is also an absolute regime – it requires Member States to set absolute targets for the trading sectors and subsequently allocate the allowances to sources.

Unlike absolute regimes, relative trading regimes do not set a fixed absolute cap on the emissions from the sectors covered by the regime, although the regulator usually has an absolute target in mind when setting the relative target. The relative target is usually set through defining a baseline, which is expressed in the emissions efficiency in relation to the activity of the source, measured in weight per unit of input, output or activity. The same baseline can often be set across a wide range of similar installations. Installations that can reduce their emissions more cheaply than the market price of allowances, will reduce their emissions and obtain allowances which they can sell. Installations for which the reduction of emissions is more expensive than the market price of allowances will maintain or increase their emissions and buy additional allowances on the market. A key difference with an absolute trading regime is that allowances are not allocated up front but when a source demonstrates that it performs better than its baseline. It is possible to use relative targets for the free-of-charge allocation under a regime with absolute targets. The relative targets are used to calculate the share of the total quantity of allowances allocated to each participant under the cap.

Absolute trading regimes are attractive to both policy-makers and environmental groups since they give certainty on the environmental outcome of the trading system. The regime's cap determines the total amount of emissions from the sectors covered by the regime, while the market determines where the necessary reductions in emissions will take place. The fact that the Kyoto Protocol itself sets an absolute cap and the international emissions trading under the Protocol is based on absolute units are reasons for policy-makers to choose an absolute system in the design of their domestic or regional trading regime.

Relative regimes are usually more attractive for regulated sources, as the absolute effects of most abatement techniques are usually dependent on the activity of the installation. Any reductions in emissions caused by the installation of new technology can often easily be outdone by an increase in activity. Industry usually argues that production growth itself should not be punished, but that the efficiency of production should be increased. Relative targets do exactly this. An additional reason for industry to support trading with relative targets is that a number of Member States, including Germany, the Netherlands and the United Kingdom, have already negotiated agreements on CO_2 emissions with industry sectors, most of which contain relative targets (Barth and Dette, 2001). Both industry and regulator are reluctant to reopen these agreements and replace them with a trading regime with absolute targets. The downside of relative targets, however, is that the environmental outcome cannot be predicted with full certainty. This risk, usually determined by economic circumstances, is borne by the regulator.

Trading regimes with absolute and relative targets are not necessarily incompatible, and relative trading regimes could be used to attain an absolute target such as the one set under the Kyoto Protocol. Predictions can be made of the emissions in absolute numbers from a relative trading regime, the accuracy of which is dependent on the predictability of the behaviour of the sources. It is also possible to periodically adjust relative targets to increase the chance that an absolute objective is met. The domestic nitrous oxides (NO_x) trading regime that is currently being developed in the Netherlands, for instance, will set a relative target in amounts of NO_x emissions per unit of energy consumed. If it becomes clear that through the application of this relative target the absolute target that the Dutch government has set itself for NO_x emissions in 2010 will not be met, the relative target can be adjusted downwards in 2006 by a maximum of 20 per cent.[30]

Relative and absolute regimes can also be linked, although these links are in practice often complex and difficult to operate. At the end of the compliance period of a relative regime, when the volumes of emissions or the activity rate of the sources is known, the results of the relative regime can be translated into absolute numbers and a link can be created with an absolute trading regime. The greenhouse gas allowance trading regime in the United Kingdom, for instance, allows trading with sectors that have closed negotiated agreements with the government, many of which are based on relative targets.[31] The negotiated agreement sector is linked with the absolute trading sector through a 'gateway'. Sources with absolute targets can sell their allowances to sources with relative targets, but only when there has been a net flow into the relative sector will any relative sector participant be permitted to transfer allowances out to the absolute sector. The gateway thus allows the relative sector to use allowances generated by the absolute sector to facilitate its compliance, but it cannot increase the total amount of allowances available to the absolute sector.

Allocation of allowances

Through allocation, the total amount of emissions allowed under a trading regime is distributed in the form of emission 'rights', 'allowances' or 'permits'. In most trading regimes these are allocated free or auctioned to the sources covered by the regime. These emission rights could, for instance, also be allocated free to a state's citizens or other legal entities. The method of allocation is not important for the operation of the emissions trading regime itself. Allocation is often the most politically controversial part of a trading regime as allocation determines who gets the economic value of the allowance. It is because of this that discussions on allocation often tend to take up the largest part of the negotiating time needed to establish a trading regime, and can often be the reason for significant delays in the entry into force or even failure of the adoption of a trading regime.[32]

The allocation issue is usually a key argument for regulators to opt for a baseline and credit regime. In such a regime the allocation is not done through transferring allowances to sources up front but through defining the baseline.

Using a baseline-and-credit approach usually eliminates auctioned allowances as an option, as well as free allocations to entities that are not sources of the regulated emissions and an absolute cap on total emissions. Since the baseline for a source determines the free allocation it receives, industry in general also prefers a baseline-and-credit approach. The baseline can often be determined for a broad range of sources within a regime and is independent of a source's activity, which makes these baselines significantly easier to negotiate. This in turn is likely to increase the political acceptability of a regime and to significantly reduce the time needed for its adoption. Following the same reasoning used by supporters of a baseline-and-credit regime to argue that it is not incompatible with an absolute target, the setting of baselines in combination with predictions of the sector's activity can, however, also be used for allocation under a regime with absolute targets.

Two main allocation methods are usually distinguished: free allocation and auctioning. A free allocation may be based on historic emissions (or other historic variables) – grandfathering – or on a baseline related to current activity as in a baseline-and-credit system. Auctioning usually stands for the auctioning of the allowances to the sources covered under the regime, where the amount of allowances allocated to a source depends on the price it is willing to pay for those allowances.

Grandfathering is usually industry's preferred allocation method. Under grandfathering sources are unlikely to incur much extra cost from the allocation but will hold a valuable asset which they can use or sell on. Grandfathering is, however, also the most complicated allocation method. Many different interests will need to be accommodated, including the need not to disadvantage sources that significantly reduced their emissions in the past ('early movers') or benefit those that have not done so. Because of the value of the allowances, each of the sources will seek to maximize the amount of allowances allocated to it through the allocation process, regardless of whether it actually expects to need these allowances to cover its emissions. If very different types of sources are covered by the regime, the selection of the baseline or benchmark used to help set more objective grandfathering criteria may be even more complicated. Grandfathering may also have important implications under the Community's state aid rules: valuable assets are given free of charge to industry. These state aid aspects of allocation will be further discussed below.

Auctioning is often favoured by economists as the most efficient and easiest allocation method, but in practice it is often not applied due to opposition by industry. This opposition can be explained by the fact that use of a resource which was previously free of charge must now be paid for through an auction, often at prices unknown at the time of the design of the regime. During the stakeholder discussions on the ET proposal, a number of industry representatives even stated that auctioning, by requiring industry to pay for allowances, would be taking away industry's 'right to pollute' and therefore require financial compensation to industry. This argument is unfounded. Industry, certainly under Community legislation, does not have a 'right to pollute', just as countries under the Kyoto Protocol do not have a right to pollute.[33] Although in many cases these emissions have not been regulated under either Community or national law, this cannot mean that industry's emissions cannot be further reduced without compensation.[34] To compensate for industry's objection to the large cost burden that auctioning may impose, it has been suggested that the revenue from the auctioning process might be redistributed to sources after the auction.[35] This would, however, bring a difficult distribution debate into the allocation discussion, which would take away one of the key advantages of the use of auctioning in the first place.

In practice, combinations of these allocation methods are sometimes used – the government may decide to grandfather part of the allowances and hold regular auctions to sell off the remaining part.[36] A variation on the auctioning approach was used by the UK government in the establishment of the targets and coverage of its trading regime. The UK government made available £215 million in incentive money to the sources willing to participate in the trading regime. This incentive money was paid to sources in relation to their committed reductions per tonne of CO_2 equivalent. The allocation took place through a 'descending clock auction'. In this auction the government announced an opening price per tonne, to which firms could offer commitment to reduce a specific amount of tonnes by 2006. As long as the total tonnage offered times the price was larger than the amount of the incentive money, the government revised its price downwards, with bidding rounds continuing until the clearing price was reached. The firms that remained in the auction are those included in the trading regime.[37]

Coverage

The question of coverage relates to the gases and sources that are covered by the emission limitations of the trading regime. Although the Kyoto Protocol includes a range of six greenhouse gases from a wide range of sources, as listed in Annex A to the Protocol, a domestic trading scheme could cover fewer (or more) gases and sources.

The choice of the coverage of gases by a trading regime usually depends on the coverage of the sources and the measurability of the emissions of gases by those sources, which is determined by the diffuse nature of the source of the emissions and the uncertainty related to the estimation or measurability of the quantities of those emissions. Some greenhouse gases are only emitted in small quantities by the sources covered by a regime – inclusion of these gases would therefore not make much sense. Some greenhouse gas emissions from specific sources also have a considerable degree of uncertainty in relation to the measurement or estimation of their emissions. While CO_2 emissions from energy can be measured with uncertainty of around 10 per cent, the uncertainty in measuring CH_4 emissions from the combustion of biomass has been estimated at around 100 per cent and for N_2O emissions from agricultural soils even around 200 per cent (Denne, 1999). Bringing emissions that have a high degree of measurement uncertainty attached to them within a trading system could complicate the trading system and undermine its practical feasibility and political acceptability. It is for those reasons that the ET Directive, discussed below, initially limits the coverage of gases to CO_2 only. The Danish trading regime is similarly limited to CO_2 only.[38] The UK trading regime, on the other hand, leaves the choice to sources to either include only CO_2 emissions, or to include their emissions from all six greenhouse gases. The inclusion of all six gases is, however, subject to the source being able to demonstrate that it can actually monitor the emissions in a sufficiently accurate manner.[39]

Discussions on the coverage of sources are usually characterized by defining a trading regime as either upstream or downstream (Hargrave, 1999). This characterization can be unclear as different authors use different interpretations of these terms. The two terms can, however, be used to indicate the extremes of a spectrum of choices related to coverage, rather than single options, and they do provide a useful aid to illustrate the various options. A typical upstream regime for fossil-fuel CO_2 emissions would require fuel producers or importers to hold allowances equal to the carbon content of the fossil fuel sold or produced. Rather than limiting the amount of emissions, the regime would limit the carbon content of fossil fuels used in a country. The advantage of this regime is that virtually all fossil-fuel combustion CO_2 emissions would be covered, no matter how diffuse the actual sources of emissions actually are. The regime would be administratively easy to manage, in view of the limited amount of producers or importers of fossil fuel. Although the allowance requirement would not be linked with the actual CO_2 emissions, those emissions can be easily determined on the basis of the carbon content of each fuel and the allowance holders would translate their requirement to purchase permits for each unit of fossil fuel sold into a price signal towards the final consumers. A disadvantage is that the permit requirement may be too distanced from the actual emissions, and the price signal passed through by the producers or importers may be seen as a fuel tax. In some countries the number of fossil-fuel producers and importers may be too small to create a competitive market. The upstream design also presumes an auction of allowances, since the participants have virtually no options for reducing the carbon content of the fuels, so the result is higher prices for fossil-fuel products. A free allocation would provide the fossil-fuel producers and importers with a windfall profit. An auction captures this windfall and allows the government to use the revenue to mitigate the impact of the higher prices.

A purely downstream regime includes the actual sources of the emissions. A downstream regime with full coverage would not only include household emissions from natural gas boilers and emissions from cars, but also methane emissions caused by the digestive processes of ruminant animals. It is obvious that a purely downstream system with extensive coverage would be administratively too difficult to manage, because of the enormous amount of sources and the often tiny amount of emissions from each source. It is for this reason that most of the trading regimes that are currently being designed are actually hybrids of these two extreme approaches. These regimes cover only larger industrial sources, and mostly only their direct emissions, although the peculiarities of the UK trading regime include indirect emissions caused by electricity consumption as well.[40]

The upstream–downstream choice as a point of regulation exists for many but not all sources of GHG emissions. It exists for CO_2 emissions from fossil-fuel combustion. In principle it exists for manufactured gases such as HFCs, but in practice the number of downstream sources is too large, so the only practical option is an upstream design. Process emissions such as CO_2 emissions from cement, N_2O emissions from adipic acid and PFCs from aluminium production can only be regulated at the point of production.

II.5   The ethical dimension of emissions trading

Although the ET Directive received a cautious welcome from European environmental NGOs, the concept of emissions trading is by no means widely accepted.[41] Former French Minister for the Environment Voynet is even rumoured to have called the concept of emissions trading 'diabolique'.[42] This section explores some of the moral and ethical arguments for and against emissions trading.[43]

Arguments in favour of emissions trading

Emissions trading, more than other instruments, turns pollution reduction from a 'burden' into a 'business'. This can boost the development and distribution of new technologies generating flows of capital and the transfer of technology into regions with cheaper, generally older technologies, particularly in less developed economies, promoting not only emissions reductions but also positive feedbacks across the whole economy.

Emissions trading regimes with sufficiently strict caps and a good compliance regime are also likely to gain support from environmental NGOs because of the certainty of environmental outcomes. Emissions trading can thus provide an important benefit over environmental taxation, which does not usually provide this certainty. While environmental taxation allows the polluters to pay the tax and continue to pollute, the finite availability of emission allowances under a cap-and-trade regime, combined with a strong enforcement regime, ensures that it will be unattractive for sources to pollute more than the total cap.

Arguments against emissions trading

Arguments against emissions trading have a moral and ethical basis. Emissions trading 'turns pollution into a commodity to be bought and sold' and by doing so it 'removes the moral stigma that is properly associated with it', which 'makes pollution just another cost of doing business, like wages, benefits, and rent'.[44] As long as the permit price is paid, the behaviour is allowed. The argument that emissions trading is morally wrong can be illustrated by explaining the moral difference between a fee and a fine (Sagoff, 1999). Payment of a fee exempts the behaviour for which the fee was paid from being classified as morally wrong. Payment of a fine on the other hand confirms the morally wrong character of the behaviour by confirming it as a violation of a society's rules. Emissions trading allows the polluter to pay a fee, rather than a fine, exempting him from the guilt associated with the behaviour.

Whether emissions trading is perceived as 'morally wrong' thus hinges on whether all emissions of greenhouse gases should be stigmatized as wrong behaviour, or whether society should accept some level of pollution (Scott, 1998). The Convention and the Protocol are silent on this issue. But the Decision on the principles, nature and scope of the Kyoto Protocol mechanisms in the Marrakech Accords contains an interesting phrase, inserted at the insistence of G-77 and China, which states that the UNFCCC's supreme body, the Conference of the Parties, recognizes that 'the Kyoto Protocol has not created or bestowed any right, title or entitlement to emissions of any kind' on developed countries.[45] Both the Convention and the Protocol also recognize the responsibility of emitters of greenhouse gases for the damage caused by their behaviour, by requiring them to 'assist the developing country Parties that are particularly vulnerable to the adverse effects of climate change in meeting costs of adaptation to those adverse effects'.[46] It can thus be argued that the climate regime does not exempt developed countries from being morally wrong – it merely recognizes the need to reduce those emissions and lays down a first step for doing so.

Another argument against emissions trading, linked to the previous argument, relates to possible equity concerns. When applied on a global scale, emissions trading could allow the rich to buy their way out of their obligations thus sanctioning their wasteful lifestyles. Emissions trading could also consolidate the economic power of the rich, by allowing them to buy emission rights from the poor. A counter-argument is that the emission rights from the poor are not simply taken away, but are bought at a price, which can be used to 'buy' more pressing developmental benefits.[47] The other part of the argument, the need for the rich to change their lifestyles and reduce their emissions at home, is precisely the reason why the European Union and others insisted on the elaboration of the supplementarity principle during the negotiations on the Marrakech Accords, as discussed in Part I.

An argument related to the need to require 'domestic action' is that emissions trading, rather than promoting technology development, may actually provide a disincentive to technological innovation. Polluters with relatively advanced technology will no longer have an incentive to explore further reduction opportunities 'at home'. Instead they will invest 'abroad', using existing technology to clean up polluting sources, and by doing so merely 'pick the low-hanging fruits'. The validity of this argument is linked to the incentives or requirements given to polluters to reduce their emissions 'at home', as well as to the stringency of the 'cap'. If the cap is sufficiently stringent, there may not be sufficient reduction options available to buy, which will increase the price to a level where it becomes attractive to reduce emissions domestically. Without a sufficiently strict cap, a trading regime may lead to adverse environmental effects, such as the creation of 'hot air', i.e. allowances vastly in excess of anticipated needs.[48]

A final argument against emissions trading relates to the issue of 'hot spots'. If, as a result of trading, one source significantly increases its emissions, this may have severe local impacts and lead to the creation of pollution 'hot spots'. Research in the United States has shown that these pollution 'hot spots' tend to be located in low-income or minority areas, which in turn raises serious environmental justice concerns (Stewart, 2001, p101). The concentration of emissions in hot spots can be addressed in the design of a trading regime, through limitations on the use of allowances by sources located in specific areas or through local environmental quality standards.[49] More importantly, the 'hot spots' argument does not apply to the emission of substances which do not have a local effect, such as most greenhouse gas emissions, and in particular CO_2. To avoid the indirect creation of hot spots, it is important that other emissions that do have a local impact are covered by sufficiently strict emission requirements.

On the whole, it thus seems that strong arguments can be made in favour of emissions trading as an instrument to combat greenhouse gas emissions. Care, however, must be taken with the design of emissions trading regimes to ensure that they actually deliver a positive environmental result and minimize negative equity and environmental justice implications.

II.6 The development of emission allowance trading in the EU

This section will give a brief background to the development of the EU ETS as well as some of the instruments that preceded or influenced the Directive.[50] It starts with a discussion of the origins of the idea of an EU trading scheme in the Commission, followed by a discussion of the Green Paper. It then gives an overview of the discussions on the Green Paper in Working Group 1 of the European Climate Change Programme, followed by a description of the process leading to the adoption of the proposal and the adoption of the ET Directive itself. An overview of key dates in the development of the Directive is given in Table II.3.

Table II.3 *Overview of key dates and documents relating to the development of the ET Directive*

- *8 March 2000*: Commission Green Paper on Greenhouse Gas Emissions Trading within the European Union, COM(2000)87.

- *23 October 2001*: Commission proposal for a Directive establishing a scheme for greenhouse gas emission allowance trading within the Community and amending Council Directive 96/61/EC, COM(2001)581.

- *10 October 2002*: First reading opinion European Parliament.

- *18 March 2003*: Council Common Position (agreed on 9 December).

- *2 July 2003*: Adoption of second reading agreement by the European Parliament.

- *22 July 2003*: Adoption of second reading agreement by the Council.

- *13 October 2003*: Directive 2003/87/EC of 13 October 2003 establishing a scheme for greenhouse gas emission allowance trading within the Community and amending Council Directive 96/61/EC, [2003] OJ L275/32.

- *23 July 2003*: Commission Proposal for a Directive of the European Parliament and of the Council amending the Directive establishing a scheme for greenhouse gas emission allowance trading within the Community, in respect of the Kyoto Protocol's project mechanisms, COM(2003)403.

- *27 October 2004*: Directive 2004/101/EC of the European Parliament and of the Council amending Directive 2003/87/EC establishing a scheme for greenhouse gas emission allowance trading within the Community in respect of the Kyoto Protocol's project mechanisms.

From conception to Green Paper

The specific interest in emission allowance trading developed quickly after the adoption of the Kyoto Protocol in December 1997. A first reference to an 'EC-wide approach to emissions trading' can be found in the Commission's Communication on the EU's post-Kyoto strategy from 3 June 1998.[51] In this Communication the Commission asked the Council to:

- endorse the introduction of the flexible mechanisms in a step-by-step and coordinated way within the Community; and
- endorse the objective of the gradual inclusion of private entities over time, and that, as national use of the flexible mechanisms will have to respect the Community law, it would be desirable to have a Community framework to safeguard the internal market.

In the same Communication, the Commission suggested that the 'Community could set up its own internal trading regime by 2005 as an expression of its determination to promote the achievement of targets in a cost-effective way'. In its Communication 'Preparing for the Implementation of the Kyoto Protocol' of 19 May 1999, the Commission expanded this idea and suggested that the Commission should adopt a Green Paper on EU greenhouse gas emissions trading and organize a wide consultation on the basis of this issue in 2000.[52] In preparation for this Green Paper, a study was commissioned by a group of consultants, led by the London-based Foundation for International Environmental Law and Development (FIELD) in autumn 1998. The results of this study (FIELD, 2000) provided the foundation for the Commission's Green Paper on greenhouse gas emissions trading within the European Union, which was released in little more than three months after the finalization of the study. On 8 March 2000 the Commission published the Green Paper on Greenhouse Gas Emissions Trading within the European Union (the Green Paper),[53] together with a Communication on EU Policies and Measures to Reduce Greenhouse Gas Emissions: Towards a European Climate Change Programme (ECCP).[54]

Within less than two years the Commission moved from a first interest in the instrument to the adoption of a Green Paper on the issue. The quick development of the Commission's interest in trading, and the very short time-span within which the Green Paper was produced, are remarkable. It is interesting to note that shortly after the adoption of the Kyoto Protocol most of the staff of the Commission's Climate Change Unit, including the head of unit, changed. Key members of the new staff had previously worked in the economic instruments unit and had participated in the preparation and negotiation of the Commission's CO_2 taxation proposals. This may explain their eagerness to embrace this new market-based mechanism and make it a key feature of the Community strategy to implement the Kyoto Protocol.

The aim of the Green Paper was to start a discussion within the European Union on the merits and design of a greenhouse gas emissions trading system, and the relation between such a system and existing and future policies and measures. The ECCP started a stakeholder dialogue to prepare further common and coordinated policies and measures. It created a number of Working Groups, each of these providing a platform for stakeholder discussions on a specific issue. The ECCP and the Green Paper were mutually supportive: Working Group 1 of the ECCP addressed the Flexible Mechanisms under the Kyoto Protocol, focusing on developing the Community's GHG emissions trading regime. The ECCP set out the larger policy context and created the platform for discussion, while the Green Paper focused on the further development of one specific instrument. Both processes are discussed in more detail in the following paragraphs.

The Green Paper on emissions trading

The Green Paper made a strong case for the introduction of GHG emissions trading in the European Union. It pointed out the economic benefits and advantages for the internal market of applying emissions trading at the EU level rather than only at the national level. It explored design options for a regime with a degree of harmonization at the EU level able to reap these benefits, but with sufficient freedom for the Member States to adapt the specifics of the regime to their national circumstances and preferences.

The Green Paper advocated a step-by-step approach. Starting in January 2005, a Community regime could initially confine itself to large fixed-point sources of carbon dioxide. The Green Paper set out a list of possible industry sectors to include in the emissions trading regime, which covered in total 45.1 per cent of all Community CO_2 emissions. Although it pointed out the advantages of having an identical coverage of sources across the Community, it recognized the need for flexibility. It proposed a number of alternative approaches for doing so, including variations where Member States choose to either opt in or opt out of certain sectors from an EU list of participating sectors and installations.

The Green Paper also addressed the central issue of the allocation of allowances between sectors and for individual sources. It set out the various options available, including auctioning and different approaches to free allocation, such as grandfathering, and discussed the case of new market entrants. It also addressed the question of whether the allocation process and choice of allocation methods should be left to the Member States or whether some degree of harmonization would be appropriate.

Both the EU and its Member States already have a body of environmental legislation in place, including measures addressing climate change. The Green

Paper discussed the relation between a possible EU trading system and existing and planned policies and measures. It specifically addressed the trading scheme's relation with the IPPC Directive,[55] negotiated agreements, energy taxation and points at potential links with the Monitoring Mechanism.[56]

Robust compliance and enforcement are the basis for any GHG emissions trading regime. Recognizing this, the Green Paper described compliance and enforcement at both the Member State level (vis-à-vis companies) and the EU level (vis-à-vis the Member States). It asked whether elements of national compliance and enforcement regimes need to be harmonized at the Community level to avoid gaming by participants.

The Green Paper laid the foundations for the development of an EU-wide GHG emissions trading regime. Rather than proposing a specific regime, the Green Paper's aim clearly was to stimulate a focused debate. It set out the available design options and some of their implications. To stimulate discussion within the EU, it contained a list of specific questions addressed to stakeholders and interested Parties, and invited interested parties to give their reactions and opinions to the Green Paper by 15 September 2000.

The Green Paper on Emissions Trading was discussed at the Environment Council Meeting on 22 June 2000.[57] In its conclusions, the Environment Council welcomed the Green Paper and encouraged the Commission to take the process forward, recognizing that emissions trading could play an important role in the Community's strategy to reduce its GHG emissions. The European Parliament discussed the Green Paper on 26 October 2000. In its resolution in reaction to the Green Paper, Parliament supported the introduction of an EU-wide trading regime by 2005. Parliament also urged that the trading regime be based on quantified greenhouse gas abatement targets, set in advance, by country and by sector.[58] The consultation process launched by the Green Paper triggered an unexpectedly large number of reactions. In total 90 reactions were submitted by a wide range of interested parties, including Member States, industry organizations and NGOs, virtually all of whom supported the development of an EU-wide trading regime.[59]

European Climate Change Programme, Working Group 1

In parallel to the discussion of the Green Paper in the Council and the European Parliament, the Commission organized the stakeholder meetings under the European Climate Change Programme, Working Group I on Flexible Mechanisms (WG I). Ten meetings in total were held between July 2000 and May 2001. These meetings served not only as a platform for different stakeholders to exchange their views on the development of the trading regime, but also as a capacity-building exercise, as the exchange of opinions between the various participants helped clarify different concepts and approaches and generated new ideas.

The final report of WG I was presented at the ECCP conference held in July 2001. Even though the stakeholder group participants came from very different backgrounds and represented a broad range of interests, the degree of consensus on a number of key design issues was encouraging. The group was unanimous on the need for emissions trading to be introduced as soon as practicable. Differences of opinion did, however, remain on key design issues, such as the role of relative targets in the regime and the degree of harmonization of emissions trading at the EU level.[60]

The adoption of the proposal

At the end of January 2001, the Commission started with the drafting of the ET proposal, again helped by a small group of consultants led by FIELD (FIELD, 2001). The US rejection of the Kyoto Protocol in March 2001 caused the Commission to significantly advance its agenda for bringing out a proposal. Environment Commissioner Wallström decided that there was a need for the Community to give a clear signal to the outside world that it was still taking the Kyoto Protocol seriously and was preparing for its implementation. The Commissioner instructed the Climate Change Unit to prepare a draft proposal to be adopted by the Commission before the start of the international Climate Change meeting in Bonn on 16 July 2001. A draft proposal was sent to other Commission services for informal consultation in mid-May 2001, and went into inter-service consultation at the end of May.

The widely-leaked draft of the ET Proposal, only a few weeks before the ECCP conference, caused uproar among a number of industry lobby groups. Industry had not expected the proposal to be presented so soon, and had anticipated another round of stakeholder consultations before a final proposal would be presented.[61] The short deadline for the adoption of the ET proposal caused a panicked reaction from a small number of key industry representatives, who heavily pressured the Commission to delay the adoption of the proposal. Faced with the heavy industry lobbying and growing opposition within the Commission, Wallström decided on 28 June 2001, while senior international negotiators where discussing the progress of the negotiations on the implementation of the Kyoto Protocol at a high-level meeting in The Hague, to postpone the proposal.

The success of the international climate change negotiations in Bonn in July 2001, which led to the adoption of the Bonn Agreements,[62] reinvigorated the Commission's determination to push for the adoption of the ET proposal before the end of 2001. After an extra round of consultations on 4 September 2001 (with industry and NGOs) and on 10 September 2001 (with Member States, EEA and accession countries), the Commission published on 23 October its proposal for a Directive establishing a scheme for greenhouse gas emission allowance trading within the Community.[63] The ET Proposal was accompanied by a proposal for a Council Decision to ratify the Kyoto Protocol[64] and a Communication on the implementation of the first phase of the ECCP.[65] The adoption was timed to occur just before the start of the final act in the international negotiations on the completion of the package of measures necessary to implement the Kyoto Protocol in Marrakesh on 29 October 2001.

The adoption of the Directive

The Directive was negotiated and adopted by the Council of the European Union and the European Parliament under the so-called 'co-decision procedure' (see Box II.4). Within little over a year after the proposal's adoption by the Commission, the Council agreed its Common Position on 9 December 2002, which was adopted on 18 March. Only six months after the Council's agreement, and twenty months after the original Commission proposal, the final text of the Directive was agreed in second reading between the European Parliament and the Council on 23 June 2003. The compromise was approved by the European Parliament on 2 July 2003 and the Council on 22 July 2003, and formally entered into force on the date of its publication, 25 October 2003.

BOX II.4   THE EU'S CO-DECISION PROCEDURE[66]

The EU's co-decision procedure, set out in Article 251 of the European Community Treaty, provides for 'qualified majority voting' in the Council, composed of relevant Ministers of the Member States, and allows Member States to be outvoted. Member States are accorded votes based on their size, and proposals are adopted when around two-thirds of the votes of the Member States are in favour. Parliament has the right to propose amendments and can veto the adoption of an entire proposal if it feels that its amendments have not sufficiently been reflected in the final text. The adoption of a proposal usually goes in two so-called 'readings'. Parliament can propose amendments in each of these readings. If Council accepts all Parliament's amendments in the first reading of the text, the proposal can be adopted in first reading. If Council does not accept all Parliament's amendments, then it adopts a 'common position' and the proposal goes into second reading. In second reading, Parliament can propose new amendments to the Council's common position. If no agreement is reached between Parliament and the Council on these amendments, a conciliation committee is convened, which seeks agreement between the two institutions. Without agreement on the final text, Parliament can vote to veto the adoption of a proposal.

The unusually fast adoption of a Directive on such a complex topic can be explained by the high priority given to this proposal by the Commission, combined with the political pressure to adopt a meaningful measure at the Community level to prepare for the implementation of the Kyoto Protocol, and almost unprecedented widespread support for the proposal from key sectors of industry and NGOs. Although the key elements of the original Commission proposal have remained intact, the adoption process did not go without difficulty, and the final text of the Directive shows the scars of the negotiations that took place in Parliament and Council. The key issues in these negotiations included the binding nature of the trading regime, its coverage of sectors and gases, its link with the Kyoto Protocol flexible mechanisms, the allocation of allowances (method and total quantity) and the issue of penalties.

The most intensive discussions on these various issues took place before the adoption of the Council's common position on the proposal on 9 December 2002. Parliament proposed in total 73 amendments,[67] and the discussions in Council had to be postponed due to negotiations within Germany on its position on the Directive. Unlike virtually all other major pieces of environmental legislation that have been recently adopted by the European Community under the co-decision procedure, the ET Directive did not go into conciliation. Uniquely, Parliament was lobbied by a broad range of stakeholders, including industry, environmental NGOs, the European Commission and most Member States, to avoid proposing too many amendments and accept the Common Position as it was adopted by the Council in December 2002. Had the Directive gone into conciliation, its adoption would have been delayed at least until the end of 2003, which would not only have jeopardized the start of the trading regime by 1 January 2005, but could even have undermined the adoption of the Directive altogether. The unique compromise between the Council and Parliament on 23 June 2003 avoided the conciliation procedure and allowed the Directive's timetable to remain unchanged. As part of this compromise, the Council did, however, have to make a number of concessions to the European Parliament relating to the coverage of sectors and gases, the method and amount of allocation and the link with the Kyoto Protocol's project-based mechanisms.

The issues that proved key during the negotiations on the adoption of the Directive are elaborated in the next section, describing the contents of the Directive.

II.7 Core elements of the ET Directive

This section provides an overview and background of the key aspects of the ET Directive as it was finally agreed upon in June 2003. When analysing the ET Directive it should not be overlooked that this is a 'Directive'. The ET

Directive provides the backbone of the European trading regime. For it to become effective it will need to be implemented in the domestic legislation of the Member States, each of which has a certain amount of freedom to choose the methods of implementation, as long as the results prescribed by the Directive are actually achieved. By choosing the Directive as the backbone of the European trading regime there will not actually be one single trading regime but a series of 15 and, from 1 May 2004, 25 connected domestic trading regimes with common elements and elements that are different.

The following part of this section will give an overview of the core elements of the ET Directive, starting with a discussion of the timing and nature of the regime, followed by a discussion of its coverage, the instrument's 'permit' and 'allowance', the allocation of allowances, enforcement and the relation with the IPPC Directive. The role of the project-based mechanisms will be discussed in the next section. Table II.4 below provides an overview of key dates in the implementation of the ET Directive.

Table II.4 *Key dates for the implementation of the ET Directive*

Implementation	
13 October 2003	Adoption of the Directive
25 October 2003	Entry into force of the Directive
31 December 2003	Deadline for implementation in Member States
Reviews	
31 December 2004	Commission may propose to amend Annex I to include further gases and activities
30 June 2006	Commission report on the Directive, accompanied by proposals as appropriate
First trading period (January 2005 – December 2007)	
31 March 2004	Deadline for submission of Member State national allocation plans
30 June 2004	Deadline for Commission rejection of Member State national allocation plans that are submitted on 31 March 2004
1 October 2004	Member State decision on the total quantity of allowances for the first trading period
1 January 2005	Start of the first period of the EU emissions trading scheme (until 31 December 2007)
28 February 2005	Issue of a proportion of the allowances to each operator
28 February 2006	Issue of a proportion of the allowances to each operator
31 March 2006	Operators to submit emissions report for previous year
30 April 2006	Operators to surrender allowances to cover emissions over previous year

28 February 2007	Issue of a proportion of the allowances to each operator
31 March 2007	Operators to submit emissions report for previous year
30 April 2007	Operators to surrender allowances to cover emissions over previous year
31 March 2008	Operators to submit emissions report for previous year
30 April 2008	Operators to surrender allowances to cover emissions over previous year

Second and subsequent trading periods (January 2008 – December 2012 and for five-year periods thereafter)

30 June 2006	Deadline for submission of Member State national allocation plans [*every five years*]
30 September 2006	Deadline for Commission rejection of Member State national allocation plans that are submitted on 30 June 2006 [*every five years*]
31 December 2006	Member State decision on the total quantity of allowances for the second trading period [*every five years*]
1 January 2008	Start of the second period of the EU emissions trading scheme (until 31 December 2012) [*every five years*]
28 February 2008	Issue of a proportion of the allowances to each operator
28 February 2009	Issue of a proportion of the allowances to each operator [*annually*]
31 March 2009	Operators to submit emissions report for previous year [*annually*]
30 April 2009	Operators to surrender allowances to cover emissions over previous year [*annually*)

Compulsory participation from 1 January 2005 onwards

The ET Directive lays down the framework for a Community-wide compulsory greenhouse gas emission allowance trading scheme in all EU Member States from 1 January 2005 onwards. Although the original Commission proposal put forward a compulsory regime that must commence on 1 January 2005, it was not until shortly before the adoption of the Common Position that it became clear that participation in the regime would indeed be compulsory right from the start. The discussions on the binding nature of the regime focused mostly on the period from 1 January 2005 to 31 December 2007. While the Commission and the environmental NGOs[68] favoured a binding regime, a number of industry groups argued that EU emissions allowance trading regime should be voluntary, at least before the first commitment period under Kyoto (2008).[69] Experience with voluntary trading regimes in Canada[70] and the United Kingdom,[71] however, had shown that these regimes would not bring the large-scale reductions needed to meet the Kyoto Protocol targets, as turn causing an oversupply of allowances on the market. The Commission had in addition argued in its original proposal that a pre-2008 trading regime would benefit the Community in allowing it to gain important experience with trading prior to the commencement of IET under the Kyoto Protocol.

The final compromise on the ET Directive keeps the trading regime binding from 1 January 2005 onwards. To allow for this, a number of concessions had to be made. The first concession is laid down in a new Article 27 of the ET Directive. This provision, inserted at the insistence of the United Kingdom, in particular allows Member States to request permission from the Commission to exclude certain installations during the period up to 31 December 2007, subject to a range of conditions. A further concession is laid down in the new Article 29 of the ET Directive. This provision allows Member States to apply to the Commission for permission to issue additional, non-transferable, allowances to certain installations in limited cases of *force majeure*.[72]

But perhaps the most interesting addition can be found in the form of a new Article 28 on 'pooling'. This article is the result of strong lobbying by German industry and ensuing pressure from the German government during the negotiations. German industry has since the beginning of the discussions on trading strongly opposed a mandatory regime.[73] The reason for this is that in Germany agreements were negotiated between industry and government to reduce greenhouse gas emissions from key industry sectors with the shared understanding that no further requirements would be imposed upon the sectors included in those agreements (see Box II.5). Industry feared that mandatory participation in an emissions trading regime could not only lead to a revision in the domestic legal framework and the administrative burdens attached to that, but also to reduction obligations going beyond those already agreed in the climate change agreements. Under pressure from the BDI, the main German industry association, Germany proposed during the negotiations leading up to the Common Position that the possibility be created for sectors of industry, or even all industry within a country, to pool their allowances together and allow for a single trustee to manage the allowances held by the pool. The original idea behind this was that Germany would make participation in this pool mandatory for all sectors covered by the trading regime and the pool would be managed by the German government as the trustee. Under this scenario there would therefore be no emissions trading within Germany, but the German government would hold all allowances and do the buying and selling to other Member States. The original German pooling proposal would have effectively taken German industry out of the internal EU emissions trading market. It would have turned an instrument that was intended to bring the Kyoto Protocol's emissions trading down to the level of individual installations back to state-level trading as introduced under the Protocol, thus significantly undermining the viability and operation of the EU trading scheme. During the discussions on the pool proposal, the proposed text was, however, significantly changed. The compromise text as set out in the ET Directive no longer allows for mandatory pools and it requires that if a trustee fails to comply with the penalties under the trading regime, each operator of an installation remains responsible in respect of emissions from its own installation. The current provisions on pooling have been weakened to the extent that it is now unlikely that German industry will use this possibility in practice.

Box II.5 The German climate change agreements[74]

Voluntary agreements have played an important role in Germany's climate change policy. First agreements between the German government and industry were reached in 1995, and these were renewed in November 2000. Under the current agreement German industry commits itself to reduce emissions of all six greenhouse gases referred to in the Kyoto Protocol by a total of 35 per cent by the year 2012 compared to 1990 levels. In addition to this, German business agrees to make additional efforts to reduce its CO_2 emissions by 28 per cent as compared to 1990 levels. In return for this commitment, the German government has undertaken not to introduce domestic legislation to achieve further greenhouse gas emission targets. It has also undertaken to ensure that ecological tax reform will not cause a competitive disadvantage for German industry. The German government will also 'endeavour' to ensure that German business will not suffer any competitive disadvantages at the international level as a result of the Kyoto obligations and the instruments involved and EU Burden-Sharing.

To incentivize over-achievement and create legal certainty, the ET Directive is designed to continue after 2012. From 2008 onwards the Directive works with five-year periods, for each of which Member States are required to draw up a national allocation plan. The second 'commitment period' under the ET Directive overlaps with the first commitment period of the Kyoto Protocol. After 2012 a new five-year commitment period starts, regardless of the outcome of the forthcoming negotiations on the second commitment period of the Kyoto Protocol.[75] The ET Directive could, however, be amended to be brought in line with any future international agreements if these negotiations result in a commitment period different from the five-year period chosen under the Directive.

Coverage of sectors, gases and Member States

The ET Directive contains in its Annex II a list of the six Kyoto greenhouse gases, but it initially applies only to CO_2 emissions from activities set out in its

Annex I. The sources listed in Annex I are a subset of the installations listed in Annex I of the IPPC Directive, limited to installations that emit large quantities of CO_2. One addition is that combustion installations over 20 MW, rather than the 50 MW threshold set in the Annex of the IPPC Directive, are covered by the ET Directive. The ET Directive is thus expected to cover approximately 46 per cent of the estimated EU CO_2 emissions in 2010. Initial Commission estimates were that 4000 to 5000 installations would be covered.[76] More recent estimates, however, have shown that, in particular in light of the planned accession of ten new Member States in May 2004, up to 15,000 installations may be covered by the regime, of which almost 1500 installations are in the UK and 2600 in Germany alone.[77]

Article 30, paragraph 1 of the ET Directive states that 'the Commission may make a proposal ... by 31 December 2004' to expand the sectors and gases covered by the regime. Paragraph 2 requests the Commission to draw up a report on the application of the Directive, 'accompanied by proposals as appropriate' by 30 June 2006, which is to include the coverage of the regime. In the original Commission proposal, expansion of the list of activities and the gases emitted by those activities was possible only through amendment of the Directive, through the normal co-decision procedure, on the proposal of the Commission (Article 26). The Commission had, in the explanatory memorandum to the original proposal, defended this approach on the basis of the need to reduce monitoring uncertainties for other gases and the need to limit the number of sources that would initially enter the regime, to make the regime more administratively practical given the lack of experience of large-scale trading schemes in the environmental sector in most Member States. This essentially pragmatic approach was also supported by the environmental NGOs.[78] Industry, however, lobbied heavily for an expansion of the coverage of the trading regime to include all six Kyoto gases, arguing that this would provide greater flexibility and prove more cost-effective.[79] The European Parliament also urged the Commission to include other gases at the earliest possible stage, although it recognized that this should not go at the cost of jeopardizing the simplicity of the regime and that expansion to other gases and sources should only be possible if the quality of monitoring and measurement of these emissions could be ensured. Parliament therefore requested that the Commission develop the methodologies for monitoring these emissions to be finalized in time for the Directive's revision in 2004.[80]

The result of the discussions in Parliament and Council is the addition of a number of new provisions. The first relevant provision, already mentioned above, was included more as a result of the discussions on the compulsory nature of the trading regime in the pre-Kyoto period. A new Article 27 was included in the ET Directive, which allows for the temporary exclusion of installations up to 31 December 2007, subject to strict conditions. Member States must apply for such temporary exclusion with the European Commission. In the Common Position a temporary exclusion could be given for 'certain installations and activities', giving the impression that entire categories could be opted out. Due to pressure from the European Parliament during the second reading of the Directive, temporary exclusion in the final text of the ET Directive is limited to specific installations. To apply for a temporary exclusion a Member State has to list and publish details of each installation for which it seeks exclusion. A temporary exclusion is only given if the installations will limit their emissions as much as they would have under the trading regime, and are subject to equivalent monitoring, reporting and verification requirements and equivalent penalties.

The ET Directive, however, also allows for expansion of the coverage of activities and gases beyond those included in Annex I. As a result of the negotiations on coverage, a new Article 24 on 'procedures for unilateral inclusion of additional activities and gases' was included in the final text of the ET Directive. This provision allows Member States, with the approval of the Commission, to expand the coverage of the regime to other activities and gases from 2008 onwards, and to expand coverage to activities below the thresholds in Annex I from 2005 onwards. Expansion of coverage under this provision is approved by the Commission following the 'comitology' procedure (see Box II.6) referred to in the Directive's Article 23(2), 'taking into account all relevant criteria, in particular effects on the internal market, potential distortions of competition, the environmental integrity of the scheme and reliability of the planned monitoring and reporting system'. Article 24(3) requires the Commission, on request by a Member State, to adopt monitoring and reporting guidelines for these activities 'if the monitoring and reporting can be carried out with sufficient accuracy'.

Box II.6 The EU's Comitology procedure[81]

'Comitology' refers to a legislative procedure where the adoption of the legislation has been delegated to the Commission, assisted by a committee composed of Member State representatives. A comitology procedure is normally used for the elaboration of more technical, less politically controversial legislation and allows for a significant speeding up of the decision-making process compared to the adoption of regular legislation.

The ET Directive uses the so-called regulatory procedure. Under this procedure the Commission can only adopt implementing measures if it obtains approval by a qualified majority of the Member States meeting within the committee. In the absence of such support, the proposed measure is referred back to the Council which takes a decision by qualified majority. However, if the Council does not take a decision within three months, the Commission finally adopts the implementing measure provided that the Council does not object by a qualified majority.

The committee used under the ET Directive is the Climate Change Committee (formerly known as the monitoring mechanism committee), set up under the monitoring mechanism Decision.[82]

In addition to these new provisions, preambular paragraph 14 was amended to point out the possibility of extending the coverage of the regime to 'emissions of other greenhouse gases than carbon dioxide, *inter alia*, from aluminium and chemicals activities' in the final compromise between the Council and Parliament. Similarly, Article 30(a) was amended to explicitly mention chemicals, aluminium and transport as sectors to be included in the 2006 review. These amendments were included at the insistence of the European Parliament, which had argued for the inclusion of the aluminium and chemicals sectors in the EU trading regime in both first and second reading.[83] Both sectors were left out in the Commission's original proposal and in the final text of the Directive because of their relatively low contribution to the EU's CO_2 emissions and in particular also because of the high number of chemical installations (estimated in the Commission's original proposal at 34,000 plants).[84] It should, however, be noted that combustion installations over the 20 MW threshold set out in Annex I to the Directive in both the chemicals sector and in the production of aluminium are included in the scheme, thus covering the bulk of CO_2 emissions in these sectors.

The original Commission proposal briefly states that it will apply to the then current 15 EU Member States and to new Member States.[85] None of the accession agreements concluded between the EU and the ten new accession states includes transitional measures for the ET Directive. The ET Directive will therefore by default become applicable to those new Member States at their accession on 1 May 2004 as part of the so-called 'acquis communautaire' encompassing the total body of Community law, including legislation and related instruments such as judgments of the European Court of Justice, that is in place in the Community. Interestingly, a number of the new accession states have negotiated transitional measures for the closely related IPPC Directive.[86] The absence of such extended deadlines for the new accession states is due to the fact that the Directive can be seen as an important incentive to exploit their generous allocations under the Kyoto Protocol and use their tremendous potential for further emission reductions. Their full participation under the ET Directive will, however, require greater attention to capacity building.

At the time of writing it was still unclear whether the Directive would also apply to the European Economic Area (EEA) states (Norway, Iceland and Liechtenstein). A decision on their participation would need to be taken by the three countries jointly under the EEA agreement. It is also unclear whether Switzerland is interested in participating in the framework of its bilateral agreements with the EU. On 1 January 2005 there will therefore be at least 25, and potentially even 29, states participating in the trading regime.

Permits and allowances

The ET Directive distinguishes between a 'greenhouse gas emissions permit' and 'allowances'. The permits are the prerequisite framework for installations to participate in the trading regime, as they set out the conditions for the monitoring and reporting of installation emissions. Permits are, however, not the units that are traded – those are the allowances.

Article 4 of the Directive requires Member States to ensure that from 1 January 2005 no installation undertakes any activity listed in Annex I of the Directive unless it holds a greenhouse gas emissions permit. Article 5 of the Directive sets out the information that needs to be included in the permit application, which includes a description of the installation, the raw and auxiliary materials used that are likely to lead to greenhouse gas emissions, the sources of greenhouse gas emissions and the planned monitoring and reporting measures. Article 6 of the Directive contains the conditions for and contents of the greenhouse gas emissions permit. A permit may only be issued if the operator is capable of monitoring and reporting emissions. The permit must include monitoring requirements, specifying monitoring methodology and frequency, reporting requirements and, importantly, the obligation to surrender allowances equal to the total emissions of the installation in each calendar year. As the procedure for granting the permit is based on that of the IPPC permit, Article 8 of the ET Directive requires the permitting procedure to be coordinated with the granting of the IPPC permit and allows Member States to fully integrate the two permitting processes.[87]

Article 3(a) of the ET Directive defines 'allowance' to mean 'an allowance to emit one tonne of carbon dioxide equivalent during a specified period, which shall be valid only for the purposes of meeting the requirements of this Directive and shall be transferable in accordance with the provisions of this Directive'. Allowances are initially allocated to the operators of installations covered by the Directive, but can be transferred between natural and legal persons within the Community and between persons within the Community and persons in third countries if a bilateral agreement under Article 25 of the ET Directive exists with those countries. According to Article 13 of the ET Directive, allowances are only valid to offset emissions during the period for which they are issued. Four months after the end of the pre-Kyoto period, the allowances issued for this first period are automatically cancelled. Member States *may*, but are not obliged to, issue new allowances for the period starting in 2008 to replace the allowances that were cancelled. The reason for this 'may' is that a large surplus in allowances from the pre-Kyoto period could jeopardize a Member State's compliance with its Kyoto targets if it is used to offset emissions during the first Kyoto Protocol commitment period. Pre-2008 allowances allocated under the ET Directive would not be valid Kyoto

Protocol units, and therefore any emissions offset by pre-2008 EU allowances that take place within the Kyoto Protocol's first commitment period would need to be offset by reductions elsewhere. The 'may' thus gives the flexibility to Member States to disallow the banking of pre-2008 allowances into the first Kyoto commitment period (2008–2012). The situation is different for surplus allowances at the end of 2012. As the Kyoto Protocol also allows for the banking of most of its tradable units, Article 13(3) of the ET Directive requires that Member States *shall* issue allowances to replace those cancelled in May 2012. This means that once Kyoto enters into force the EU trading scheme will fit 'hand in glove' with the international trading regime under Kyoto.

With the introduction of the term 'allowances', the EU has introduced its own tradable unit. An important question is what is the nature of the relationship between the EU allowances and the units created under the Kyoto Protocol. This question is addressed below in the section on the links to the project-based mechanisms.

Allocation of allowances

The allocation of allowances is one of the core parts of the Directive. The ET Directive does not specify the total amount of reductions that each Member State individually or all Member States collectively must achieve. Instead, it requires Member States to decide upon their own targets (the 'cap', an essential part of a 'cap-and-trade regime') for the sectors covered by the ET Directive. The lack of a 'cap' in the Commission proposal was criticized by environmental NGOs, who would have liked more certainty on a robust environmental outcome of the operation of the trading regime.[88] The call from environmental NGOs for either an EU level or Member State specific cap that would set out the exact contribution each Member State would require its covered sectors to make was supported by the European Parliament. In the first reading of the ET proposal, Parliament put forward a number of amendments to place a ceiling on emission allowances to be allocated by each Member State, representing 50 per cent of the emissions forecast annually for each Member State on a linear curve converging with the Kyoto commitments.[89] This was not accepted. In the negotiations on the second reading agreement between the Council and Parliament was reached instead to include two new sentences in paragraph 1 of Annex II of the ET Directive, which point out that 'the total quantity of allowances to be allocated shall not be more than is likely to be needed for the strict application of the criteria of this Annex' and that 'prior to 2008, the quantity shall be consistent with a path towards achieving or over-achieving each Member State's target' under the Kyoto Protocol.

It is difficult to see how the ET Directive could have been adopted within such a short period of time if Member State-specific caps had been included.

At the time of the drafting of the Directive there was very little information available on the number of installations that would be covered by the regime, let alone their emissions and their potential for emission reductions. The composition of the sectors and the distribution in the types of installations furthermore differs too much between the various Member States for the Commission to define EU-wide or Member State-specific caps on an *a priori* basis. An EU-wide debate on caps for specific industry sectors would have been much more complicated than the discussions on the burden-sharing agreement, as it would have exposed the participating governments to much more intensive lobbying from these industry sectors, which could have made the adoption of the Directive altogether impossible. Although certainly not perfect, the approach taken in the text of the ET Directive, set out below, succeeds in setting boundaries for the national target setting process, while avoiding protracted discussions on specific targets at the EU level.

As anticipated by the Commission, the issue of allocation turned out to be one of the most hotly-debated during the discussions on the adoption of the ET Directive.[90] The result of the heated debate is that the ET Directive provides that Member States caps and criteria for allocation must be laid down in an 'Allocation Plan', which is submitted to the commission and other Member States and can be rejected by the Commission. The ET Directive requires Member States to allocate at least 95 per cent of the allowances free of charge for 2005–2007 and at least 90 per cent for the period 2008–2012. It also contains a number of requirements for the Allocation Plan in its Annex II. But before this result was reached, a number of key questions had to be addressed.

A first question relates to the division of competence between the Member States and the Community. To what extent is it desirable or necessary for the Community to prescribe not only allocation methods, but also the exact allocation to individual sources at the Community level, or what control can the Community exercise over this allocation by Member States? To provide an answer to this question it is interesting to take a closer look at which aspects of allocation have a Community dimension under current Community law.

The overall reduction target for the Community and all Member States, covering emissions between 2008 and 2012 of all six greenhouse gases under the Kyoto Protocol, was agreed upon in the Kyoto Protocol, which sets an 8 per cent reduction for the EU as a whole and for each Member State individually. The EU burden-sharing agreement subsequently redistributed this reduction target among the Member States. During the stakeholder discussions in the ECCP Working Group I, German industry stated on several occasions that it did not want Member States or the Commission to directly translate the burden-sharing target into the allocation for the sectors under the ET regime. It clearly feared that Germany's 21 per cent reduction target would be translated into a similar reduction target for particular sectors or each individual source.[91] While this is a fanciful extrapolation, it does reflect the fear within a number of Member States that their targets under the burden-sharing agreement may directly reflect upon the competitiveness of their industry under the ET regime.[92] After all, a number of Member States are allowed large increases in their emissions, whereas others must make significant reductions. As seen above, the burden-sharing targets do not fully reflect a Member State's marginal reduction costs but are to an extent based on a political redistribution of the Community's overall target.

Related to the burden-sharing agreement is the 'Dutch situation'. The Netherlands has indicated that it will meet half of its reduction effort domestically and the other half through the use of the Kyoto Protocol's flexible mechanisms. Since May 2000, the Dutch government has started two major investment programmes, the ERUPT programme and the CERUPT programme.[93] Using these programmes the government, through public tendering procedures, is purchasing CDM and JI credits. The government has set aside a total of US$450 million for buying credits, representing between 20 and 25 million tonnes of CO_2 emissions annually between 2008 and 2012. The question is to what extent this large-scale purchasing of credits will have an impact on the allocation of allowances to installations covered under the ET Directive in the Netherlands and, through this allocation, an impact on EU-wide competition issues. A Member State could choose to buy Kyoto credits using public funds to allocate larger amounts of allowances to its industry sectors, thus in effect exempting these sectors from significantly reducing their emissions under the Community trading regime, or at least significantly reducing the emission reduction burdens these sectors would otherwise have to carry. The next question is, however, how the distribution of these 'extra' credits by a Member State government differs from decisions taken on the stringency of caps for trading and non-trading sectors to comply with a Member State's burden-sharing target. But this does not take away the fear within other Member States that the Dutch government could be indirectly subsidizing its industry participating in the trading regime.

A government's potential to confer a competitive advantage on an industry sector or a specific enterprise raises questions about the relation between the allocation of allowances and the Community's state aid rules.[94] The application of the Community's state aid rules to the various aspects of emissions trading is not entirely straightforward. The issue was not addressed in detail in the Community Guidelines on state aid for environmental protection.[95] Accordingly, requests for more clarity on the relation between Community competition law and emissions trading were made on various occasions throughout the stakeholder discussions in ECCP Working Group 1.[96] The Commission decided, however, that it would be too early to draft guidelines for the application of state aid rules under the ET proposal. Instead, it preferred to gather more experience with the application of the current rules, in particular by examining the kinds of issues that might practically arise as a result of pilot emissions trading schemes launched by the UK and Denmark.

Subsequent Commission Decisions on the state aid implications of the UK and Danish domestic trading regimes and domestic energy taxation in combination with the state aid guidelines have helped provide some guidance on the application of those rules. These decisions have confirmed that the Commission will regard the free allocation of allowances to entities as constituting state aid.[97] The Commission has reached a similar conclusion for the UK's payment of incentive money.[98] The Commission has, however, found the state aid aspects of both the Danish and the UK trading regimes to be permissible state aids under Article 87(3)(c) of the EC Treaty. The main reason for allowing the UK incentive money was that it allows companies to go beyond existing Community standards and provides a net environmental benefit. The free allocation in the UK scheme is matched by further reduction of the source's emissions.[99] The Commission similarly allowed the free allocation under the Danish trading scheme because of the further emission reductions sought by the trading regime. The Commission did, however, require Denmark to change its regime to ensure that new entrants on the market will receive allowances 'based on criteria that are objective and non-discriminatory in relation to those applied to incumbent producers'.[100] More recently, in June 2003, the Commission also decided on the state-aid aspects of the Dutch NO_x trading scheme. In this Decision, the Commission found that systems where a tradable emission allowance is considered an intangible asset, representing a market value which the authorities could also have sold or auctioned therefore leading to foregone revenues, would be qualified as state aid. The Dutch NO_x trading scheme was, however, considered permissible state aid since the participating companies have to reduce their emissions further than EU target levels in return for the potential aid received from their free of charge allocation (Könings, 2003).

Initial internal Commission proposals for the ET Directive said very little about the allocation methodology, as the Commission wanted to await further clarity on the applicability of state aid rules following the screening of the Danish and UK schemes. It became clear during the preparation of the proposal that carefully-balanced guidance on allocation was needed to address a number of the issues above.[101] The general expectation was that the Commission would provide this clarification through its Communication on guidance to assist Member States in the application of the allocation criteria set out in the Directive, discussed below.[102] Annex II to the ET Directive contains a criterion (5) which states that the Member State national allocation plans 'shall not discriminate between companies or sectors in such a way as to unduly favour certain undertakings or activities in accordance with the requirements of the Treaty, in particular Articles 87 and 88 thereof'. The

Commission's guidance on this criterion is, however, extremely short, limiting itself to stating that 'normal state aid rules will apply'.[103] The precise applicability of the Community's state aid rules to the allocation process under the ET Directive therefore still remains unclear.

The final version of the ET Directive contains a number of elements that are directly relevant to the issue of allocation. A first element is the virtually complete harmonization of the method of allocation. The Commission's original proposal stated that for the period between 2005 and 2008 allowances were to be allocated free of charge and required the Commission to specify a harmonized method of allocation for 2008–2012, using the comitology procedure. By doing so, the Commission addressed the fear existing with many industry groups that the ET Directive, or individual Member States in the implementation of the Directive, would opt for auctioning as the allocation method, which is strongly opposed by industry.[104] Free of charge allocation of valuable emission allowances significantly complicates the allocation debate. This is not only because it sets governments the challenging tasks of finding allocation criteria that are acceptable to the sectors involved, but also because of the danger of judicial challenges to the allocation decisions. It can, however, be argued that auctioning also brings a difficult allocation debate as the government needs to decide on the allocation of the revenue from the auction.

Environmental NGOs strongly favoured auctioning as an allocation methodology because: it rewards companies that have taken early action; it follows the 'polluter pays principle' as companies do not receive 'windfall profits' in the form of valuable allowances; and it allows the recycling of the revenues of the auction, for example to the development of renewable energy sources.[105] Parliament equally argued strongly for compulsory auctioning of at least part of the allowances.[106] As a result of Parliament's insistence that at least part of the allowances should be auctioned, the final text of Article 10 of the ET Directive now requires that for the three-year period beginning 1 January 2005 Member States shall allocate at least 95 per cent of the allowances free of charge, and from 2008 onwards at least 90 per cent of the allowances are allocated free of charge. At the insistence of the European Parliament, Article 30(c) was also amended in second reading to include the consideration of auctioning as the method of allocation after 2012 in the Commission's review of the Directive in 2006. While this is not a commitment to auction after 2012, it does signal expectations that auctioning will play a greater role post-2012.

The second element of the guidance on allocation given in the ET Directive is the 'national allocation plan' (NAP). Article 9 of the ET Directive requires each Member State to submit to the commission a National Allocation Plan by 31 March 2004 for the period starting on 1 January 2005 and at least 18 months before the start of each subsequent period. In its plan, the Member

State has to indicate the total quantity of allowances that it intends to allocate for that period and how it proposes to allocate these allowances to individual installations. Member States are required to base their plans on objective and transparent criteria, and the Directive lists a number of those criteria in its Annex III.

The ET Directive also sets out a procedure for the assessment of NAPs. Member States must submit to the commission their NAPs to the Commission and to other Member States. The NAPs are subsequently 'considered' in the Climate Change Committee, set up under the Community's Monitoring Mechanism.[107] The Commission may reject the Member State's NAP within three months after its submission, on the basis of its incompatibility with Annex III of the ET Directive or the Community's prescribed allocation method, set out in Article 10 of the ET Directive. Member States are only allowed to allocate allowances to the installations once proposed amendments to the NAP have been accepted by the Commission.

In February 2003 DG Environment of the European Commission issued a non-paper for discussion in the Monitoring Mechanism Committee, addressing various process-related allocation issues.[108] In this non-paper DG Environment sets out six steps to be taken in order to draft a NAP. The first step is a 'top-down' analysis to define the share of domestic emissions covered by the ET Directive, compared to the share of emissions from other sectors. The second step is a 'bottom-up' exercise to collect data from installations covered under the Directive. The third step involves a consolidation of the top-down and bottom-up information, in which the Member State has to decide how to treat any discrepancies between the two approaches. Where the bottom-up analysis results in a larger share of emissions than the top-down analysis, Member States could adopt further measures in the non-covered sectors or reduce the number of allowances that would have been allocated following a bottom-up approach. The fourth step is the setting of allocations for sectors and installations through sharing out the total number of allowances available under the trading scheme. The fifth step is a decision on the treatment of new entrants, in particular, whether new entrants should buy their allowances on the market, whether regular auctions of allowances should take place or whether new entrants should be given allowances from a new entrants reserve. The sixth and final step is the finalization of the NAP through addressing the remaining elements contained in Annex III of the ET Directive.

Article 9(1) of the ET Directive requires the Commission to develop, by 31 December 2003 at the latest, guidance on the implementation of the allocation criteria listed in Annex III. The Commission issued an informal draft of this guidance in November 2003 and adopted its final guidance on 7 January 2004.[109] The guidance not only provides further assistance to Member States in the interpretation and application of Annex III, but it also gives more clarity on the Commission's approach in testing the application of the Annex III criteria in its assessment of the national allocation plans under Article 9(3) of the Directive. The guidance also contains a common format for Member States to use for the notification of their NAPs to the Commission.

The Commission's allocation guidance qualifies the Annex III criteria on the basis of their mandatory or optional nature, as well as on the level (total, activity/sector, installation) at which the criteria apply. Underlying the Commission's Communication is the 'ex ante principle': the Commission is of the opinion that the Directive requires that, once the allocation has been decided by the Member State for that commitment period, it can no longer be adjusted. The only exception to this is where so-called *'force majeure'* allowances are created following Article 29 of the ET Directive, or where an installation's permit is revoked following its closure.

Central in the Commission's allocation guidance is the application of criterion 1 of Annex III to the ET Directive and, in particular, its requirement that 'the total quantity of allowances to be allocated shall not be more than is likely to be needed for the strict application of the criteria' in Annex III. The Commission considers that in order to satisfy this requirement and fulfil all mandatory criteria and elements of Annex III, a Member State should not allocate more than is needed for or allowed by the most constraining of criteria (1) to (5) in Annex III, and the application of any of the 'voluntary' criteria may not lead to an increase in the total quantity of allowances.

The Commission's allocation guidance also demonstrates the key role that the ET Directive and its NAP will play in the Member States' efforts to achieve their targets under the burden-sharing agreement and the Kyoto Protocol. Preparation of a NAP requires a detailed analysis of a Member State's strategy to comply with these targets, going beyond the actions covered by the ET Directive. In order to implement criterion (1) in Annex III to the ET Directive, Member States will need to determine their 'path' towards achieving their target. While this path does not necessarily have to be a straight line, it must be one that is leading towards or goes beyond the reductions required under the burden-sharing agreement and the Kyoto Protocol. Member States also have to decide how to divide their total emissions budget under this 'path' among all sectors in their economies in order to determine the proportion that is to be allocated to the sectors covered under their trading scheme. This requires a Member State to project emissions in the non-covered sectors, including projections of the effects of any policies to reduce emissions from those sectors. Importantly, it also requires Member States to substantiate any intentions to use the Kyoto mechanisms. This requirement in particular refers to the situation where Member States, such as the Netherlands, indicate that they will use the Kyoto mechanisms to alleviate the burden on their economy, including the sectors covered under the ET Directive. These Member States must not only substantiate the extent to which they wish to use the Kyoto mechanisms, but also provide sufficient evidence of whether this objective can also be realized in practice.

It remains to be seen how the Commission is going to apply its allocation check in practice. The original Commission proposal required that the NAPs set out the total quantity of allowances that a Member State intends to allocate and how it proposes to allocate them. While this language was retained in the ET Directive, a new paragraph 10 was added to Annex III, which requires that the plan 'shall contain a list of installations covered by this Directive with the quantities of allowances intended to be allocated to each'. It is unclear whether the Commission's allocation check will be limited to how Member States propose to allocate, or whether it will apply additional state aid checks on allocations to individual installations. Although the emphasis is likely to be on the first, the Commission is unlikely to exclude the possibility of scrutinizing allocation on an installation-by-installation basis in order to test the compatibility of that particular allocation with the Community's state aid rules.

NAP allocation checks will also raise capacity issues for the Commission. Fifteen Member States will have to submit their allocation plans by 31 March 2004, and another ten Accession States will follow in May 2004. This means that the Commission will have to check at least 25 allocation plans between 1 April and 31 July 2004. It is therefore possible that the Commission's checks may not be as in-depth as they could be, given more time. Comments and information from stakeholders and other Member States, including through the Climate Change Committee's 'consideration' of all national allocation plans, are therefore likely to provide important inputs for the Commission's assessment.

The ET Directive is also likely to raise important questions in relation to the implementation of the ET Directive's allocation provisions at the national level. Most Member States have elaborate administrative appeal procedures in place, which allow entities covered by the trading scheme and other stakeholders to challenge specific types of government decisions. Should Member States also allow appeal against allocation decisions, or is it even possible to exclude such appeals? Appeals may cause delays in the allocation process, which in turn may delay the start of the regime and potentially bring Member States into violation of Directive requirements. Challenges may be prevented through a wide acceptability of the allocation methods, but these may in turn take a long time to negotiate. Constitutional law could prevent Member States from excluding the allocation decision from appeal. To solve this, Member States may choose to institute special fast-track appeal procedures.

The allocation provisions in the ET Directive are an innovative attempt to create an allocation framework which addresses the various issues raised above. Perhaps the biggest advantage of the ET Directive's approach is that it has managed to bring the main components of the allocation decisions down to the Member State level. By doing this it has avoided lengthy subsidiarity and competence discussions at the EU level and has set time limits for the debate within the

Member States, thereby limiting the impact of the allocation discussion on the entry into force of the trading regime. One important disadvantage of this process, however, is that Member States are deciding in their allocation process on both the total cap for emissions from the sectors covered under the trading scheme and on the allocation to individual installations. This has the potential to put pressure on the total cap as a consequence of discussions on individual allocations. Had these two decisions been separated, the setting of the total cap for the trading sector might have been easier. Another potential threat to the allocation process may be the application of the allocation criteria in practice and the Community's scrutiny of this application. The strength of the total cap on emissions from the sources covered under the trading scheme will depend on a combination of the willingness of Member States to use the NAP process to set a credible path towards their target under the burden-sharing agreement, peer pressure from other Member States and industry sectors in those Member States and, importantly, the Commission's ability to actually (threaten to) turn down NAPs that do not set out a credible path towards a Member State's burden-sharing target.

As the ET Directive is the first instrument of its kind, and as a result of the scale of its application, there are still a number of important allocation issues that will need to be resolved and lessons to be learned in the implementation of the ET Directive.

Enforcement

The ET Directive breaks new ground as regards enforcement measures and pushes outwards the limits of what Member States usually find acceptable in an environmental Directive. It can, however, be argued that the enforcement issue is still insufficiently addressed, and that the package of measures currently on the table to address the implementation of the Kyoto Protocol and the EU trading regime leaves a number of important gaps.

Four categories of compliance issues can be identified in the ET Directive. The first three of these relate to the behaviour of the installations and their operators:

1. Compliance with the monitoring, reporting and verification requirements.
2. Compliance with the requirement to surrender allowances for each tonne of emissions controlled under the regime.
3. Compliance with other, more general requirements set out in national legislation implementing the ET Directive, including the duty to undertake transfers in accordance with these provisions and the prohibition from committing fraud.

The fourth relates to the behaviour of the Member State and concerns:

4. The correct implementation of the Directive.

In relation to the first compliance issue, Article 14 of the Directive requires Member States to ensure that emissions from installations covered by the Directive are monitored and reported in accordance with harmonized monitoring and reporting guidelines. The Commission adopted these guidelines in January 2004.[110] The guidelines contain precise technical requirements for the monitoring and reporting of emissions from those installations. Before the report is submitted to the competent authority it must be verified in accordance with the list of criteria set out in Annex V of the Directive. Article 15 of the ET Directive requires Member States to 'ensure that an operator whose report has not been verified as satisfactory in accordance with the criteria set out in Annex V by 31 March each year for emissions during the preceding year cannot make further transfers of allowances until a report from that operator has been verified as satisfactory'. This sanction bans an operator from selling his allowances until the installation's emissions have been verified in accordance with the requirements of the Directive. The provision does not preclude the operator from buying allowances.

It is now generally accepted in EU legislation that Directives can include provisions which require that 'Member States shall determine the sanctions applicable to breaches of the national provisions' adopted pursuant to the Directive. Since the European Court of Justice's judgment on a case involving the enforcement of Community Funds in Greece in 1989,[111] it is also general practice to include language in Community Directives requiring that 'sanctions determined must be effective, proportionate and dissuasive'. More recent Community Directives have also included language requiring Member States to 'take all necessary measures' for the implementation or application of sanctions, and requiring Member States to notify their provisions on sanctions, as well as any amendments to them, by a certain date. Article 16(1) of the ET Directive follows this trend by requiring Member States to lay down rules on penalties applicable to infringements of the national legislation implementing the Directive and ensure that these rules are implemented. It also requires that the penalties must be effective, proportionate and dissuasive and Member States must notify these provisions to the Commission by 31 December 2003. This standard provision serves as a 'catch-all' for different types of non-compliance with the national legislation implementing the ET Directive, as required to address the third compliance issue above.

This standard requirement is, however, not sufficiently strong to address non-compliance with the requirement to surrender allowances for each tonne of emissions controlled under the regime. This is where the ET Directive pushes the boundaries of what Member States have in the past found acceptable in EC environmental legislation. The Directive requires Member States to impose a minimum financial penalty on operators that do not hold sufficient allowances to cover the emissions from installations under their control.

Experience in the US has shown that a high penalty for excess emissions is important to ensure compliance with the trading regime, and through that its success.[112] Although Community environmental law has in the past never prescribed specific enforcement measures, it is important that all Member States apply a minimum level of penalties. Doing otherwise would result in companies shifting their allowances away from the Member State in which the compliance penalty was low and allowing their operators in that Member State to be in non-compliance and pay the low penalty, rather than buying more expensive allowances elsewhere. This could in turn have serious consequences for that Member State's compliance with its Kyoto targets.

The ET Directive's unique penalty regime is set out in Article 16, paragraphs (2), (3) and (4). For the period between 2005 and 2008 it sets a minimum penalty rate of 40 euro for each tonne that an installation is in non-compliance with its obligation to cover its emissions by sufficient allowances. This minimum penalty rate is increased to 100 euro per tonne from 2008 onwards. On top of the requirement to pay the penalty, operators of installations have to compensate for their excess emissions in the following compliance period. The Directive also requires Member States to publish the names of those operators who are in breach of their obligation to surrender allowances.

The final version of Article 16 differs on a number of points from the original Commission proposal, but, in view of the innovative nature of these provisions, these differences are marginal. The original proposal provided for a penalty of 50 euro or twice the average market price during the preceding year, from 2005 to 2007, and 100 euro or twice the average market price of the preceding year, from 2008 onwards. Industry and a number of Member States objected to the link between the penalty and the average market price, which they argued would cause too much uncertainty on the risks of non-compliance.[113] A number of Member States felt that a penalty of 50 euro during the trial period from 2005 to 2007 was too high. During the negotiations on the Common Position the link to the market price was removed and the penalty from 2005 to 2007 was lowered to 40 euro. The Requirement for the operator to surrender sufficient allowances to compensate for the excess emissions, however, remains. The original proposal furthermore contained the Requirement for Member States to publish the names of operators who are in non-compliance with their obligations in general, which was limited to cases of non-compliance with the obligation to surrender sufficient allowances to compensate for excess emissions in the final version of the Directive.

The fourth compliance issue concerns the correct implementation of the Directive by the Member States. For the operation of the EU trading regime it is vital that the Directive is fully implemented and on time. Experience with Member State implementation of Community environmental Directives has shown that the track-record of most Member States continues to be deplorable.[114] The Commission's current enforcement tools, set out in Articles 226 and 228 of the EC Treaty, have proven insufficient and especially too time-consuming to provide for the effective enforcement tool necessary to ensure the correct operation of a Community-wide greenhouse gas emissions allowance trading regime.[115] The final version of the ET Directive leaves the various implementation deadlines that were included in the Commission's proposal untouched. This means that Member States have to implement the ET Directive into domestic legislation by 31 December 2003 at the latest. The usual time given to Member States for the implementation of Community environmental legislation is two years. The time for the implementation of the ET Directive is likely to be little over three months after its formal adoption. In this time Member States have to adopt the domestic legislation needed to make the ET Directive operational – a deadline which is unlikely to be met by most Member States, especially if this legislation has to be approved by national parliaments. This situation is further complicated by the fact that all Accession States will be required to have implemented the Directive on the moment of their accession on 1 May 2004.

One option which the Commission could have chosen would have been to require Member States to impose caps on installations in their territories, but then make the participation of installations in a Member State conditional on the Member State's compliance with a set of eligibility criteria. This approach is similar to the approach chosen in the rules elaborated under the Kyoto Protocol. The eligibility of installations in a Member State to participate in the trading regime would be determined through a periodic assessment of whether each Member State is fulfilling its obligations under the Directive and not undermining the environmental integrity or reducing the efficiency of the Community scheme. The advantage of this approach is that it would have separated the requirement to cap emissions from the operation of the trading regime. This approach would not only maintain the integrity of the regime's objective, the cap, but also give industry a clear incentive to pressure government to ensure that the necessary rules for the operation of the trading regime are in place – because the trading component is after all what makes the emission caps palatable for industry.

Now that the ET Directive is adopted, its implementation is likely to become a major issue. Implementation will pose challenges for the Member States and the Commission and will be decisive for the regime's success.

Relation with the IPPC Directive

Defining the relationship between the IPPC Directive (described in detail in Box II.1 above) and the ET Directive was one of the Commission's key challenges in its development of the ET proposal. An underlying procedural issue which had to be resolved was whether emissions trading was to be introduced as a separate proposal or as an amendment to the IPPC Directive. Although the latter could, for reasons of legislative consistency, have been preferable, this was not achievable in practice. Introducing the trading proposal by amending the IPPC Directive could have undermined the IPPC Directive's own implementation process. Introducing emissions trading into the IPPC Directive would have required a significant amendment, and because a considerable part of the IPPC Directive has not yet entered into force,[116] this might have interrupted the implementation process currently ongoing in most Member States. Instead, the Commission chose to design a parallel instrument which very closely   follows the approach and language of the IPPC Directive.

Because of this strategic choice, the Commission had to deal in its ET proposal with the relationship between the ET proposal and the IPPC Directive. Although the ET proposal addresses this relationship in a number of places, as will be described below, industry and Member States raised questions on the Commission's approach during the negotiations on the Directive. In January 2002, in reaction to these questions, the Commission's Environment Directorate General released a non-paper on the synergies between its ET proposal and the IPPC Directive.[117] The Commission's non-paper clarifies a number of these questions, specifically in relation to the overlaps in the use of terminology, coverage and the permitting procedures, as well as on the relationship between the energy efficiency requirements and the emission limits in the IPPC Directive and in the ET proposal.

The close link between the two instruments is most relevant in relation to three issues: the overlap in coverage of installations, coverage of greenhouse gas emissions (leading to the fear of 'double regulation') and the relationship between the permitting procedures.

The first issue, the overlap in coverage of installations, has already been discussed above. Almost all sectors regulated by the ET Directive are also regulated by the IPPC Directive. The only exception relates to combustion installations, where the ET Directive lowers the threshold for inclusion in the regime from a rated thermal input of 50 MW – the threshold used in Annex I of the IPPC Directive – to 20 MW. The overlap between the sectors covered by both instruments allows the ET Directive to address the largest industrial sources of CO_2 emissions and creates additional flexibility for reducing greenhouse gas emissions which, as the Commission argues, were already covered by the IPPC Directive.

The second overlap between the ET Directive and the IPPC Directive relates to the extent to which greenhouse gas emissions are already regulated under the IPPC Directive. The IPPC Directive addresses greenhouse gas emissions in a number of places. Article 3(d) imposes a general obligation upon Member States to ensure that energy in the installations covered by the Directive is used efficiently, a requirement that indirectly regulates energy-related greenhouse gas emissions, in particular CO_2. Article 6 requires that the application for an IPPC permit include a description of 'the raw and auxiliary materials, other substances and energy used in or generated by the installation' as well as 'the nature and quantities of foreseeable emissions from the installation into each medium'. Article 9(3) requires that the IPPC permit includes 'emission limit values for pollutants, in particular, those listed in Annex III'. The IPPC Directive defines in Article 2(2) 'pollution' as the 'direct or indirect introduction as a result of human activity, of substances, vibrations, heat or noise into the air, water or land which may be harmful to human health or the quality of the environment, result in damage to material property, or impair or interfere with amenities and other legitimate uses of the environment'. In view of the contribution of greenhouse gases to global warming it can be argued that greenhouse gases fall within the definition of pollution in the IPPC Directive. While Annex III of the IPPC Directive does not specifically list any of the greenhouse gases, it does include 'volatile organic compounds', which covers CH_4, 'oxides of nitrogen', which covers N_2O and 'fluoride compounds' which covers HFCs, PFCs and SF_6. Pollutants not included in the indicative list should also be subject to emission standards if they are 'likely to be emitted from the installation concerned in significant quantities'. It can thus be argued that the IPPC Directive requires that significant emissions of greenhouse gases, including CO_2, be covered by an emission limit value or by 'equivalent parameters or technical measures'. Support for this argument can furthermore be found in the Directive's definition of 'best available techniques' (BAT). The IPPC Directive requires that Member States use BAT to determine the emission limit values and the equivalent parameters and technical measures. Annex IV of the IPPC Directive lists among the considerations to be taken into account in determining BAT 'the consumption and nature of raw materials ... used in the process and their energy efficiency'. Energy efficiency criteria are indeed included in a number of the BAT reference documents (BREFs).[118] The final, important, argument that the IPPC Directive covers greenhouse gas emissions is that all six gases are included as 'pollutants' in the Commission's decision on the European Pollutant Emission Register (EPER). This register, elaborated under Article 15(3) of the IPPC Directive, provides public access to information of all major pollutants from sources covered under the IPPC Directive.[119] In its non-paper the Commission confirms this interpretation of the IPPC Directive.[120]

The Commission could have opted to allow emissions trading to take place within the limits of the IPPC Directive, i.e. installations would have had to fulfil their IPPC requirements but would be allowed to trade any reductions over and above the minimum emission limit values set in their IPPC permits. This is in fact the approach that is chosen explicitly by the Dutch NO_x trading regime (FIELD and IEEP, 2002) and implicitly by the UK greenhouse gas trading regime. One of the reasons that the Commission approved the UK's incentive payments to the sectors participating in the regime was that it requires those sources to achieve a level of protection higher than the Community standards, including those based on the IPPC Directive.[121]

It should, however, be noted that Member States have an important element of discretion in applying the IPPC Directive, and in particular its 'best available techniques' criterion. This discretion could lead to the imposition of different emission limit values for similar types of installations in different Member States. This would in particular be the case if some Member State authorities were to make full use of the possibilities offered by the IPPC Directive to apply strict emission limit values for greenhouse gas emissions, while in other Member States these emission limit values are deliberately loosened to increase the margin for the trading scheme. Different application of the discretion given to national authorities in the IPPC Directive could therefore have disadvantaged certain sectors or installations in different Member States.[122] Perhaps more importantly, trading within the IPPC framework could have significantly reduced the scope of the trading regime, which could have defied the objective of introducing the instrument in the first place.[123]

The ET Directive therefore excludes emission limit values for greenhouse gas emissions covered under its Annex I from the scope of the IPPC permit. Article 26 of the ET Directive amended Article 9(3) of the IPPC Directive by adding a number of subparagraphs that guarantee that only the emission limit values for direct emission of greenhouse gases are removed from the scope of the IPPC Directive, unless local environmental quality standards require that a minimum standard is set. By doing so it only intervenes to the minimum extent possible in the scope of the IPPC permit, thus leaving its integrated approach as much as possible intact. It also guarantees that emissions from other substances that are not covered under the ET Directive remain covered by the IPPC regime.

In its original proposal, the Commission had opted to explicitly retain the requirement in the IPPC Directive to set targets related to energy efficiency for installations covered under the ET proposal. This requirement, included in Article 2(2) of the ET proposal, set compliance with energy efficiency requirements as a clear baseline below which trading cannot take place. In its non-paper the Commission explained that it did not expect this to be problematic, and that both the UK and Danish trading regimes apply the same minimum baseline.[124] In addition it can be argued that BREF documents that do contain energy efficiency standards mostly do not set those standards far beyond generally available technology levels. Combined with the significant margin of interpretation afforded to national authorities in the application of BAT, this requirement does indeed seem to set a sensible minimum level, which would prevent emissions trading that would allow installations to operate in a very energy-inefficient manner, while not unreasonably restricting the possibilities for trading. Opposition by the Member States to maintaining the energy efficiency requirement as a baseline for trading led to the requirement being dropped in the final version of the ET Directive. Article 2(2) of the ET Directive now only reads 'this Directive shall apply without prejudice to any requirements pursuant' to the IPPC Directive. The amendment to the IPPC Directive now explicitly allows Member States not to impose requirements relating to energy efficiency.

The third area, the relation between the two permitting procedures, is explicitly addressed in Article 8 of the ET Directive. This Article states that:

> *Member States shall take the necessary measures to ensure that, where installations carry out activities that are included in Annex I to Directive 96/61/EC, the conditions of, and procedure for, the issue of a greenhouse gas emissions permit are coordinated with those for the permit provided for in that Directive. The requirements of Articles 5, 6 and 7 of this Directive may be integrated into the procedures provided for in Directive 96/61/EC.*

In its non-paper, the Commission stressed that this allows Member States to combine the permitting procedures for the ET permit with the IPPC permit. Although the Commission expects that Member States will make use of this, it stresses that this is not obligatory.[125] In view of the similarity and complementarity of the two permitting procedures, it is even likely that Member States will implement the ET Directive in the same legal framework they use to implement the IPPC Directive. By doing so, the two procedures could be fully integrated and any disadvantages created by the Commission's opting for a separate proposal rather than the amendment of the IPPC Directive would be fully removed. It is even imaginable that Member States may choose to merge the ET permit and the IPPC permit into a single permit. This is not precluded by the ET Directive and would indeed further promote integration between the two regimes.

Other issues

The ET Directive furthermore lays down the basic requirements for the monitoring and verification of greenhouse gas emissions by the installations covered under its Annex I. It also sets out the public participation requirements and basic requirements for a Member State's registry to track the trade in allowances, as well as various reporting requirements for Member States and the Commission. The proposal allows the linking of the EU trading regime with regimes in non-Member States on the basis of bilateral agreements between the Community and these other States. It furthermore delegates a wide range of tasks to a Committee under a comitology procedure. These tasks include the elaboration of more detailed monitoring and reporting guidelines, a Regulation on the standardization of the national registries, criteria for mutual recognition of allowances from non-Member State regimes, the consideration of national allocation plans, the revision of the allocation criteria in Annex III and establishing a harmonized allocation method to be used from 1 January 2008 onwards.

II.8 The ET Directive and the Kyoto project-based mechanisms

The Commission has been criticized for not including an explicit link to the Kyoto project-based mechanisms, JI and CDM, in its original ET proposal. The reason for this exclusion was that when the Commission issued its original proposal, the negotiations on the Marrakesh Accords, which spell out the rules for the functioning of the Kyoto flexible mechanisms, had not been finalized. These rules were only adopted in Marrakesh in November 2002, a month after the Commission issued its proposal.[126] The Community used the ET proposal, together with the proposal for a ratification Decision and the ECCP communication,[127] as an important political signal before the commencement of the negotiations in Marrakesh to show the world that the European Union was serious about its intentions to ratify and implement the Kyoto Protocol. Waiting to issue the proposal for the ET Directive until after the conclusion of the Marrakesh Accords would have significantly weakened this important signal.

The more important substantive reason for not including a direct link in the ET Directive was that the ET Directive is in the first instance a domestic implementation measure. It was felt that creating a direct link in the original proposal could complicate and delay the adoption of the ET Directive because of the differing views between stakeholders on the desirability of linkage, but also because of the uncertainty on the future of the Kyoto Protocol. Environmental NGOs, from the start of the emissions trading debate, strongly objected to any links with the Kyoto Protocol's flexible mechanisms.[128] Industry, on the other hand, has traditionally been a strong proponent of a direct link with the Protocol's mechanisms as linking provides access to cheaper sources of offsets than may be available within the EU.[129] The European Parliament has throughout the negotiations maintained that any link with the Kyoto mechanisms would only be acceptable after 2008 and as long as there is no link with credits from projects covering carbon sinks or nuclear power. Although the Commission approved of a link with the Kyoto flexible mechanisms in principle, such a link was to be 'subject to the satisfactory resolution of outstanding issues regarding their environmental integrity'. The Commission therefore preferred to discuss the link with the Kyoto mechanisms in the context of the elaboration and adoption of a separate legislative instrument.[130] This proposal (the Linking Directive) was published by the Commission on 23 July 2003.[131] It is not a stand-alone directive because its main effect is to amend and supplement the ET Directive.

This section will give a background to the Linking Directive. It will first discuss the linking provisions in the ET Directive and then provide a more in-depth analysis of its contents.

Linking provisions in the ET Directive

Although the original Commission proposal did not include linkages with the Kyoto flexible mechanisms, it did refer to linking the EU scheme to these mechanisms in a number of places. The first reference was included in the review clause in the proposal's Article 26 (Article 30 in final text), which included 'the use of credits from project mechanisms' as one of the issues to be considered in the review.[132] The second reference was included in the proposal's Article 24 (Article 25 in the final text) which allowed the Community to conclude agreements with third countries to link with their trading regimes.

With both industry and a number of Member States strongly advocating a direct link with the Kyoto mechanisms, a number of concessions had to be made to show a stronger commitment to use the mechanisms in the European trading regime. As a result, two new paragraphs 17 and 18 were added to the preamble of the ET Directive. These paragraphs stress the advantages of linking the Community scheme with schemes in third countries and the importance of the Kyoto project-based mechanisms CDM and JI in increasing the cost-effectiveness of the Community scheme. Importantly, the Council also changed the language of the linking provision (Article 24 in the proposal, Article 25 in the final text) from 'the Community may conclude agreements with third countries' into 'agreements should be concluded with third countries'. The Council also added a third new paragraph to the review provision (previously Article 26, now Article 30), stating that:

> *Linking the project-based mechanisms, including Joint Implementation (JI) and the Clean Development Mechanisms (CDM), with the Community scheme is desirable and important to achieve the goal of both reducing global greenhouse gas emissions and increasing the cost-effective functioning of the Community scheme. Therefore, the emission credits from the project-based mechanisms will be recognized for their use in this scheme subject to provisions adopted by the European Parliament and the Council on a proposal from the Commission, which should apply in parallel with the Community scheme in 2005.*

In second reading the European Parliament insisted that a new sentence be added to this paragraph, stating that 'the use of the mechanisms shall be supplemental to domestic action, in accordance with the relevant provisions of the Kyoto Protocol and the Marrakesh Accords'. Parliament also succeeded in significantly altering the language of preambular paragraph 18.[133] With the Parliament's amendments the question on whether credits from the project-based mechanisms may come into the Community scheme from 2005 or from 2008 onwards was left to be decided in the Directive on the project mechanisms.

CDM and JI in the EU trading regime: opportunities and threats

Before discussing the threats and opportunities for allowing such credits to be used to comply with the obligations under the EU trading regime, it should first be pointed out that an *indirect* link between the ET Directive and these project-based mechanisms already exists. As already discussed above, the ET Directive does not determine the size of the cap for the sectors covered by the Directive. The size of the cap is to be determined by the Member States, depending on their GHG reduction commitments and the allocation of the responsibility to achieve these commitments across the different sources and sinks in their territory, including the sources covered by the ET Directive. During 2008–2012 Member States can use CERs, ERUs, RMUs and AAUs to meet their commitments under the Protocol and the EU joint agreement under Article 4 of the Kyoto Protocol. The Member State could buy such units[134] and provide larger allocations to entities in the trading programme, subject to the allocation rules and the Member State and Community review of proposed allocations for 2008–2012. This will affect the stringency of the cap, in particular the impact of the cap on the amount of domestic reductions that will be achieved under the ET Directive.

A proposal for a Directive on project-based mechanisms is to envisage a *direct* link between project-based mechanisms and the ET Directive. The link would allow participating entities in domestic emissions trading programmes under the ET Directive to use credits generated by the project-based mechanisms for compliance with their obligations under the ET Directive.

There are a number of reasons for allowing entities to directly use project-based mechanisms for compliance with their targets under the ET Directive. The most frequently used reason is that the inclusion of project-based mechanisms can reduce the compliance costs for sectors covered under the ET Directive by geographically broadening the range of opportunities to reduce emissions in another Member State or outside the EU at lower costs. A second reason is that the use of project-based mechanisms can engage sources and sinks not covered by the ET Directive in implementing cost-effective reduction options. Including project-based mechanisms can bring a net benefit to project developers who are not under an obligation to limit their emissions pursuant to the ET Directive, by giving them a financial incentive to reduce their emissions if those emissions can be reduced at a lower cost than the market price for allowances under the ET Directive. The use of the project-based mechanisms within the EU could thus be seen as a piecemeal and ad hoc way to extend of the coverage of the trading regime to gases and sectors not covered under the ET Directive. Including CDM and JI in the ET Directive will also provide an important boost to the use of these instruments. Before the Commission proposed the Linking Directive there was only very little interest from companies in the use of CDM and JI. The combination of emission caps under the ET Directive with the possibility to use CERs and ERUs generated by JI and CDM will create an important incentive for companies to participate in the CDM and JI markets.

There are, however, also a number of important policy reasons against creating a direct link. The first is, that the ET Directive is a domestic implementation measure and should in the first place be used to reduce GHG emissions *inside* the EU, rather than embarking on a path which will make the EU's ability to achieve its emission reduction target dependent on buying sufficient credits from elsewhere. A related concern is that opening the EU trading regime for CDM and JI credits may flood the EU market with these credits and thus avoid the need for intra-EU reductions. While the Kyoto Protocol allows private entity trading, it only does so under the responsibility of Parties and to the extent that Parties wish to allow this. The EU trading regime would be the first and hitherto only regime that would allow the use of these credits at the domestic level, and indeed creates an incentive for doing so by imposing a cap on the sources included in the regime. Since the regime only covers a relatively small number of sources (recent estimates are around 15,000 installations within the EU 25[135]) and will initially concentrate on CO_2 only, the demand for credits under the ET Directive could potentially be overwhelmed by the huge supply of credits available under CDM and JI; therefore, exposing the EU regime to the full global CDM and JI could have negative impacts on its functioning. An additional concern is that using JI to expand the coverage of the EU trading regime within the EU does not actually lead to reductions, as the credits earned by those investments will be used to avoid reductions in the sectors covered by the regime. Instead, this broadening of coverage can provide an impetus for sectors to avoid other policies and measures limiting their emissions or to avoid being brought directly into the EU trading regime. An important argument used by environmental NGOs against 'linking' are the doubts that some of the CDM and JI projects that are currently in the pipeline in reality provide sustainable development benefits or represent real reductions compared to what would have happened otherwise (i.e. these projects are not 'additional'). Using such credits inside the EU trading regime would undermine its environmental credibility.

The European Commission attempted to balance these concerns and opportunities in its proposed Linking Directive. The proposal, however, drew sharp criticism, not only from environmental NGOs,[136] but also from industry.[137] Environmental NGOs, however, found themselves isolated in their opposition to any link with the project-based mechanisms. With a proposal for a Linking Directive already on the table and not only the Commission but also the European Parliament, the Council and industry in favour of establishing a link with the project-based mechanisms, the question was no longer *whether* such a link should be created but *how* such a link should be designed.

A remark should be made about the widely heard criticism, in particular from industry and a number of Member States, that the proposed Linking Directive may restrict access to the Kyoto mechanisms. It can be argued that this criticism is unfounded: the Linking Directive aims to do exactly the opposite. As already discussed above, the Linking Directive, once in force, will for the first time allow legal entities to use the project-based mechanisms to comply with their obligations. Rather than limiting the market, this will create a whole new market for CDM and JI credits and provide a true incentive for the use of those mechanisms, even if the EU decides, for whatever reasons, not to allow all Kyoto units to be used under the ET Directive.

The proposal for a Linking Directive

Soon after the adoption of the EU ETS and the success of the Marrakesh negotiations in November 2001, the Commission placed the elaboration of a proposal for a Directive to link the project-based mechanisms to the ET Directive high on its agenda. In January 2002 the Commission revived ECCP Working Group I. The first meeting of this new working Group I took place on 27 February 2002, and was followed by three further meetings.[138] These meetings were used to exchange ideas among stakeholders on a number of issues related to the possibility of linking project-based mechanisms with the trading regime set up under its ET proposal. Important issues which were discussed included the timing of the inclusion of the various project-based mechanisms, in particular whether any of these mechanisms could be linked before 2008, as well as the question of which types of projects could be linked to the trading regime. In relation to the latter, environmental NGOs argued that should any projects be linked at all, which they should not, these projects should be limited to specific types that conform to high environmental standards.[139] In particular, projects using carbon sinks or nuclear energy should be excluded. Industry continued to take the view that the Kyoto Mechanisms should be incorporated 'as is' in the ET Directive. The conclusions of the group, adopted at its 4th meeting in September 2002, called for 'the early adoption of legislation regarding the recognition of project credits 'as a matter of particular priority' and stated that 'the Commission should aim to make its proposal for a Directive linking JI/CDM credits with the EU emissions trading scheme early in 2003' and that 'the Council and the European Parliament should aim at adopting this legislation so as to allow its implementation as from the commencement date of the EU emissions trading scheme'.[140]

The Commission had planned the adoption of its proposal early in the first half of 2003, but the proposal was delayed in the adoption process within the Commission. One of the reasons for the delay was the disagreement between DG Environment and other Commission services on the need to limit the influx from credits from project-based mechanisms into the EU trading regime. This disagreement will be further discussed below.

The Commission finally issued its proposal on 23 July 2003.[141] Following negotiations, the final version was formally adopted on 27 October 2004. The text of the Directive itself is little over five pages and takes the form of an amendment to the ET Directive, thus fully integrating the use of the CDM and JI into the EU ETS. Although the proposal is very short, the wide spectrum of views on whether and how CDM and JI credits can be used under the EU trading regime caused lively discussions in Council and Parliament. The following paragraphs will provide a background to the five key issues in the negotiations relating to the linking directive:

- Which projects should a link be established with?
- When should a link be established?
- How should this link be established?
- Should there be any quantitative limits on a link?
- How can double counting be avoided?

How to link

There are a number of important differences between the operation of the EU trading regime and the functioning of the Kyoto mechanisms. The most salient differences are that units traded under the EU trading regime are fully 'fungible' (all units are fully interchangeable and thus have the same value) and that Member States are required to accept all EU allowances for compliance, irrespective of the company that the allowance was originally allocated to and the country it was originally allocated by. All EU allowances are furthermore treated the same under the ET Directive's banking rules. As discussed in Part I, under the Kyoto Protocol trading regime there are differences between the various credits (CERs, ERUs, AAUs and RMUs) with regard to their generation, banking and use. Parties under the Kyoto Protocol on the other hand, remain free to choose whether they wish to accept credits from another Party or project.

The question thus arises how the differences between these regimes will be reconciled. Mechanisms for linking the Kyoto and EU regimes are built into the 'entry point' of the Kyoto units into the EU trading regime in the Linking Directive. The Linking Directive provides for a new Article 11(a) to be included in the ET Directive. The Commission's original proposal had allowed for the conversion of CERs and ERUs from CDM and JI projects for use in the Community scheme. This conversion would have been done by a Member State by issuing one new allowance in exchange for one CER or ERU (if it so chose). By giving Member States discretion to convert the Kyoto units into EU allowances, the EU appeared to maintain the full fungibility of units. This would have given business more certainty about the possibilities of using these units.

The final text of the Linking Directive no longer uses the 'conversion' concept. Paragraphs 1 and 2 of Article 11(a) now read that 'Member States may allow operators to use CERs and ERUs from project activities in the Community scheme...'. Although the text still refers to the fact that this 'shall take place through the issue and immediate surrender of one allowance by the Member State in exchange for one CER or ERU held by the operator in the national registry of its Member State', this exchange now takes place in the registry of that Member State after the surrender of the credit, without having an impact on the operator's possibility to use the CER or ERU for compliance purposes.

What projects to link with?

A key issue during negotiations was whether the Linking Directive should allow all project credits generated under the Kyoto Protocol to be included in the EU ETS, or whether these should be limited to a subset of these projects. It should be noted that this question would arise even in the absence of an ETU ETS, as it would be for the EU and Member States to decide whether to give domestic actors access to international project-based mechanisms to offset domestic obligations, and if so, on what terms.

During the international negotiations for the rules for the project-based mechanisms prior to Marrakesh, the EU put forward a number of proposals to guarantee the environmental integrity of these mechanisms. These proposals included a definition of 'supplementarity', the exclusion of sink credits from the CDM, the prohibition of nuclear energy projects and opposition to Article 3.4 forestry credits in the first commitment period.[142] To achieve international agreement at Marrakesh, the EU, like other major players, had to compromise on a number of its positions. It has not been decided whether the resulting international rules will also be applied by the EU when it implements the Marrakesh Accords, or whether the EU will choose to adhere to its pre-Marrakesh positions.

There are a number of 'environmental integrity' arguments for Member States and stakeholders continuing to advocate a selective inclusion of project-based mechanisms. These arguments include questions on:

- whether specific types of projects constitute real emission reductions;
- whether those can be measured or verified;
- the need to pursue real reductions rather than temporary storage; and
- the environmental and social impact of specific project types.

There are a number of approaches to selectively allow the use of JI and CDM credits for compliance with the obligations under the EU trading regime. A first possibility could be to limit the types of instruments with which links may be established. The EU could decide not to link with CDM or JI. Individual EU Member States could decide to adopt a policy of complying with their targets under the EU burden-sharing agreement on the basis of domestic action alone, which in principle amounts to a decision not to use the Kyoto mechanisms. In view of the Council's Common Position requesting a link to be established, such a blanket limitation was not possible. A second possibility was to limit the types of projects. Such restrictions could take the form of a positive list or a negative list. A positive list specifies the projects from which credits can be used. Thus a positive list could specify project types (for example, renewable energy), project sizes (for example, less than 100 kt CO_2) and other characteristics of projects that generate acceptable credits.[143] A negative list accepts all project-based credits except those specifically excluded. A negative list might, for example, specify that credits generated by nuclear, forest management projects and large dams cannot be used for compliance with obligations under the ET Directive.

Concerns related to 'additionality' could be addressed through a more specific definition of this concept in the linking Directive. A more specific definition would, however, be difficult to apply as it would mean that the underlying project of every credit surrendered for compliance with emission limitations under the EU ETS would need an additional separate test of additionality to comply with the EU requirements. This would require the EU to establish administrative and institutional machinery similar to the Executive Board of the CDM, at least in relation to its baseline and methodologies panel. The application of different additionality tests by the EU and internationally by the CDM Executive Board would create additional transaction costs and further fragment the nascent carbon markets. Concerns related to the environmental and social impact of projects could be addressed through the application of existing EU and regional rules on environmental impacts, access to information and access to justice.[144] A key issue here was whether the application of those rules should be limited to projects within the EU or whether these rules, or the concepts underlying them, should also be applied to projects outside the EU.

It was also widely acknowledged that effective enforcement of any type of limited access would have been complicated or even impossible. Any restrictions would need to be implemented on an EU-wide level. The reason for this is that the ET Directive does not allow any restrictions on the transfer of EU allowances. Once a project credit comes into the trading scheme, either the project credit, or an allowance that is 'swapped' for this credit, can freely move around. This means that investors could circumvent any national rules by choosing to bring the project-credit into the trading scheme in the Member State with the least restrictions. This same possibility of circumventing EU-wide restrictions would also be created if the EU trading regime were to be linked to trading regimes outside the EU.

The Linking Directive tackles these issues through a combination of methods. As already seen above, the 'gateway' in Article 11(a) states that Member States 'may' allow the use of CDM and JI credits. Article 11(a) furthermore contains a 'negative list' of types of projects inadmissible for conversion into EU allowances. This list contains nuclear facilities and land use, land-use change and forestry projects ('sinks'). The Directive, however, qualifies the exclusion of nuclear projects to projects that are excluded in accordance with the UNFCCC and Kyoto Protocol and subsequent decisions adopted thereunder. This language was included as certain Member States did not wish to see a categorical exclusion of nuclear projects. The current formulation brings any uncertainties associated with the Kyoto exclusion of nuclear projects into the EU trading regime (see Part I). The exclusion of sink projects is consistent with the EU's opposition to sink projects in the CDM during the international negotiations. When the Commission released its Linking Directive proposal it was technically difficult to include sink projects, as the international rules for the inclusion of sinks in the CDM had not yet been agreed upon. Agreement at COP-9 in Milan in December 2003 on the rules for the inclusion of sink in the CDM[145] means that the Commission and the Member States had to re-evaluate the possible inclusion of LULUCF in the EU trading scheme through the Linking Directive. Technical issues including how the two new types of credits, the tCER and lCER, both of which are of a temporary nature, can be reconciled with the EU trading scheme, which is based on permanent reductions were central to this evaluation. These issues could not be resolved in the context of the agreement on the Linking Directive and LULUCF projects are not included in the EU ETS. The question of the inclusion of sinks will, however, come up again in the revision of the ET Directive in 2006.

Earlier unofficial versions of the Commission's proposal also excluded 'hydroelectric power production incompatible with the criteria and guidelines of the Wold Commission on Dams in its year 2000 Final Report'. This exclusion was removed in the final proposal, but the final text of Article 11(b) requires Member States to respect relevant international criteria and guidelines during project development. The Directive however, adds a new paragraph 2(l) to review the provision set out in Article 30 of the ET Directive stating that the review shall also consider the impact of project mechanisms on host countries, particularly on their development objectives, including whether the future use of credits from JI and CDM large hydroelectric power production projects (exceeding 500 MW).

The Linking Directive furthermore contains a number of guarantees in a new Article 11(b) that is to be included in the ET Directive. Paragraph 1 of this new provision requires that Accession countries take into account the 'acquis communautaire'[146] in the establishment of baselines for JI and CDM projects. This provision was included to avoid accession countries using CDM and JI projects to bring their infrastructure in line with EU requirements. As these countries are obliged to do so under their Accession Treaty, providing ERUs or CERs in return for such projects would clearly not be 'additional' under the rules of the Kyoto Protocol. Paragraph 5 of the new Article 11(b) of the ET Directive will require Member States authorizing private or public entities to participate in CDM and JI projects to ensure such participation is consistent with the UNFCCC and Kyoto Protocol and decisions adopted thereunder:

- real, measurable and long-term benefits related to the mitigation of climate change;

Paragraph 7 will allow the Commission to adopt further guidance on provisions relating to double-counting and participation of legal entities in JI projects under the 'comitology' procedure set out in Article 23(2) of the ET Directive.[148]

When to link

The EU ETS is scheduled to start on 1 January 2005, three years ahead of the Kyoto Protocol's first commitment period. Credits resulting from JI and CDM currently only have value for developed country Parties to comply with their quantified emission limitation and reduction obligations under the Protocol. JI credits can only be issued for emission reductions and carbon sequestration projects from 2008 onwards. CDM credits can, however, be issued for emission reductions or carbon sequestration projects starting from 2000.

Thus the link between the EU ETS and the project-based mechanisms raised the question of when installations covered under the ET Directive should be allowed to use such credits. ERUs from JI projects cannot be used pre-2008, as they will not yet exist. Although a JI-like structure could have been created pre-2008, such a structure would be cumbersome to design and implement for a three-year period only, and its links with actual JI crediting under the Kyoto Protocol would not be guaranteed.

CERs could be used pre-2008. Although under the Kyoto Protocol these can only be used by Parties to fulfil commitments from 2008 onwards, their recognition under the EU ETS pre-2008 would promote investments under the CDM for the 2005–2008 period. As already mentioned above, the European Parliament strongly opposed this option because use of CERs pre-2008 would allow installations to avoid making reductions inside the EU pre-2008, prompting greater domestic reductions in the EU but making achieving the EU's Kyoto target during the Protocol's first commitment period more costly. From the point of view of fulfilling their Kyoto commitments, it may be more desirable for Member States to wait until the first commitment period, as this will reduce intra-EU emissions pre-2008 and ensure an increased availability of CERs during the Protocol's first commitment period, since CERs for reductions pre-2008 will not come onto the EU market until 2008. Industry has, by contrast, already indicated its preference that CDM credits be included in the 2005–2007 period of the EU trading regime, as this lowers costs for industry.[149]

Allowing CDM credits to enter the EU trading regime raised some interesting policy issues. For instance, what would happen to the CERs once they have been surrendered by an installation to compensate for emissions during the 2005–2008 period? Logically they should be cancelled by the Member State authority, as they have already been used to offset greenhouse gas emissions. But since Member States have no emission limitation objective themselves pre-2008, strictly there is no need for doing so, and Member States could decide to hold on to these CERs and use them to offset emissions during the first commitment period, although by doing so they would use one CER to offset two tonnes of GHG emissions. Allowing CDM credits to come into the EU trading regime pre-2008 could also have an impact on the decisions of Member States to allow banking from the pre-2008 to the post-2008 period.

The Linking Directive limits the inclusion of ERUs from JI projects in the EU ETS to post-2008 periods. CERs from eligible CDM projects may be used in the 2005–2007 period, subject to the provisions of paragraphs 2 and 3 of Article 11(a).

How much to link

As already stated above, the discussion on the amount of CDM and JI credits coming into the EU trading regime was one of the reasons for the delay in the adoption of the Linking Directive proposal by the Commission.

Concerns have been expressed that unlimited linking of the EU ETS with the Kyoto Protocol's project-based mechanisms could have a significant downward impact on the price of the allowances traded in the EU trading regime, reducing the incentive for companies to reduce their emissions within the EU.

This could undermine the EU's domestic implementation of the supplementarity obligations in the Kyoto Protocol and the Marrakesh Accords.

One way to address these legitimate policy concerns is through a quantitative limitation on the amount of credits that are introduced into the EU trading regime. In the negotiations on the Kyoto Protocol and the subsequent Marrakesh Accords, the EU tried to introduce a quantitative limit on the use of credits traded under the Kyoto mechanisms as a means to implement the supplementarity principle.[150] Although no quantitative limit was introduced in respect of supplementarity *per se*, the Marrakesh Accords did introduce quantitative restrictions relating to the mechanisms which are explained in Part I:

- a cap on the amount of sink credits for projects other than afforestation and reforestation projects that each Annex I country can issue under Article 3(4) of the Kyoto Protocol;[151]
- a prohibition on the banking of RMUs and limits on the banking of CERs and ERUs into the following commitment period to a maximum 2.5 per cent of a Party's assigned amount for each of these units;[152]
- a cap on the total amount of CERs from land use, land-use change and forestry project activities that each Annex I Party can acquire limited to 1 per cent of its assigned amount.[153]

These specific quantitative restrictions all concern credits that some Parties felt could, because of the uncertainty of the real reductions they represent and in particular the potentially large quantities in which they could be generated, undermine the Kyoto Protocol targets by providing cheap and plentiful reduction opportunities, thus providing a disincentive for domestic action.

The draft Linking Directive proposal that went into inter-service consultation included a paragraph stating that 'Member States may convert CERs and ERUs from project activities for use in the Community-scheme up to 6 per cent of the total quantity of allowances allocated by the Member State' for each of the ET Directive's compliance periods. Industry, as well as a number of Member States and other stakeholders, strongly lobbied other Commission DGs to oppose the inclusion of this paragraph.[154] As a result of this opposition, the proposed cap was replaced by a provision that triggers a review at 6 per cent, on the basis of which a cap may be proposed through a comitology process of 'for example' 8 per cent.[155]

During the negotiations on the Linking Directive proposal a number of Member States, led by the UK, argued for the introduction of an installation-specific cap on the use of project credits. The rationale offered was that such a cap would give greater legal certainty to industry on their use of the Kyoto mechanisms for compliance with their targets under the EU trading scheme. This certainty is not given in the original Commission proposal, which allows for a cap to come in at a later stage and does not define the exact nature that any such cap would take. The final Directive text gives effect to an installation specific cap for the 2008–2012 period which is to be specified by each Member State in its NAP. During 2005–2007, Member States may allow operators to use CERs but the Directive does not refer to a cap for this period, leaving it to each Member State to determine any limits for the use of CERs and ERUs.

Double counting and JI in accession countries

'Double counting' refers to a situation in which CERs or ERUs are issued as a result of reductions that also lead to a reduction in emissions from an installation covered by the ET Directive. Double counting could, for instance, occur if an installation decides to stop its on-site power generation and instead buy its power from an external source. By doing so it would reduce its own GHG emissions and could free up allowances. The operator could decide to buy electricity generated by renewable energy. The operator of the renewable electricity power plant could, however, also attempt to obtain credits for the installation of renewable electricity capacity. By freeing up allowances through moving power-generation off-site and by giving credits for the generation of renewable electricity that replaces this, the reduction in GHG emissions would be credited twice. The issue of double counting is not limited to situations in which the ET Directive and the project-based mechanisms are directly linked. It can also occur without a link between the two regimes, although in the case of such a link the double counting issue could be brought directly into the EU trading regime if the CDM or JI credits are subsequently converted into EU allowances.

The Linking Directive proposal included provisions to prevent double counting by requiring Member States to ensure that no ERUs or CERs were *issued* for reductions of anthropogenic emissions of greenhouse gases or removals by sinks resulting from project activities that reduce or claim to reduce greenhouse gas emissions from installations covered under the ET Directive. Prohibiting the issuance of ERUs and CERs for emissions for installations covered under the ET Directive would, however, have posed a problem for JI projects that are currently being implemented in accession countries. The moment those countries become EU Member States and are subject to the ET Directive, these projects would no longer be able to issue ERUs and will thus cease to exist as JI projects. While JI projects could in principle be converted into emission trades under the ET Directive, such a conversion could create legal and contractual difficulties for projects that are already in the pipeline.

The Commission's proposed approach to double counting came under criticism from Japan, a number of accession countries and a number of Member States. Japan felt that, by practically excluding JI projects in the accession countries, the EU was claiming the JI reduction opportunities in those countries for itself and restricting countries like Japan from accessing this market. Japan's reasoning was strengthened by its argument that early informally circulated drafts of the Registries Regulation prohibited the export of EU allowances outside the EU, thus de facto keeping those reductions within the EU. A number of accession countries were concerned that the Commission's proposal would impede development of JI in these countries whilst a number of Member States pointed out that financial incentives may be taken away for renewable energy.

The final text of the Directive states that CERs and ERUs should not be issued as a result of project activities undertaken within the community that also lead to a reduction in, or limitation of, emissions from installations covered by the ET Directive unless an equal number of allowances is cancelled from the registry of the Member State of the CERs or ERUs origin.

II.9 Conclusion

This part has given a background to a range of issues relevant for understanding the evolution and future development of the EU ETS and the Linking Directive proposal. The adoption of the ET Directive has deepened the process of implementation of climate policies in the Member States, a process which will be vital in determining the instrument's success.

With the adoption of the EU ETS, the EU has started a challenging experiment with a unique new regulatory tool. Developments in Europe are anxiously watched by US industry and academics, many of whom feel that they are losing out on a unique opportunity to participate in this learning experience and shape the future of emissions trading globally.

The implementation of the EU ETS also provides an opportunity for the multilateral process to tackle climate change. A successful EU ETS will be able to demonstrate to both developed and developing countries that GHG emission reduction targets can be reached in a cost-effective manner. The Linking Directive strengthens this message by demonstrating the opportunities and benefits of emission reduction projects in economies in transition and developing countries. A successful EU trading regime will thereby provide an important step forward in achieving the objective of the UNFCCC and the Kyoto Protocol.

Notes

1. Council Decision 2002/358/EC of 25 April 2002 concerning the conclusion, on behalf of the European Community, of the Kyoto Protocol to the United Nations

Framework Convention on Climate Change and the joint fulfilment of commitments thereunder.

2. Carbon dioxide (CO_2), methane (CH_4), nitrous oxide (N_2O), hydrofluorocarbons (HFCs), perfluorocarbons (PFCs) and sulphur hexafluoride (SF_6).

3. Report from the Commission under Council Decision 93/389/EEC as amended by Decision 99/296/EC for a monitoring mechanism of Community greenhouse gas emissions, COM(2003)735 of 28 November 2003. The report gives an overview of EU policies and measures that have recently been adopted or are in the pipeline in table 3 on p16.

4. Ibid.

5. Commission's proposal for a Directive establishing a scheme for greenhouse gas emission allowance trading within the Community and amending Council Directive 96/61/EC, COM(2001)581, 23 October 2001.

6. Directive 2003/87/EC of 13 October 2003 establishing a scheme for greenhouse gas emission allowance trading within the Community and amending Council Directive 96/61/EC, [2003] OJ L275/32.

7. Commission Proposal for a Directive of the European Parliament and of the Council amending the Directive establishing a scheme for greenhouse gas emission allowance trading within the Community, in respect of the Kyoto Protocol's project mechanisms, COM(2003)403 of 23 July 2003. Available through: http://europa.eu.int/comm/environment/climat/home_en.htm.

8. N. 1 above.

9. For a detailed overview of the background to the EU Burden-Sharing Agreement see Ringius (1999). An overview of the negotiations and results of COP-3 can be found in Yamin (1998).

10. See the conclusions of the 1895th Council meeting, Brussels, 18 December 1995.

11. Ringius (1999, p6).

12. Ibid, p24.

13. Ibid, p26.

14. See the conclusions of the 1990th Council meeting, Brussels, 3 March 1997, as well as the conclusions of the 2017th Council meeting, Luxembourg, 19–20 June 1997 and the Conclusions of the 2033rd Council meeting, Luxembourg, 16 October 1997.

15. Council Directive 96/61/EC Concerning Integrated Pollution Prevention and Control, [1996] OJ L257/26.

16. The Directive lists in its Annex II a number of Directives which contain these ELVs and provides the possibility for drawing up new ELVs on the basis of its Article 18.

17. Resolution of the Council and the Representatives of the Governments of the Member States, meeting within the Council of 1 February 1993 on a Community programme of policy and action in relation to the environment and sustainable development – A European Community programme of policy and action in relation to the environment and sustainable development, [1993] OJ C138/1.

18. For an overview of the various new approaches to environmental regulation and how they fit into existing approaches, albeit more from a US approach, see Stewart (2001). See also the overview of literature pro and contra the use of new instruments in footnote 1 of that article.

19. For an analysis of the two major trading schemes in the United States, see Schwarze and Zapfel (2000).

20. Countries that have introduced different forms of taxation include the United Kingdom, Germany, Austria, the Netherlands and Belgium.

21. Proposal for a Council Directive introducing a tax on carbon dioxide emissions and energy, COM(1992)226, 30 June 1992; Amended proposal for a Council Directive introducing a tax on carbon dioxide emissions and energy, COM(1995)172, 10 May 1995; and Proposal for a Council Directive restructuring the Community framework for the taxation of energy products, COM(1997)30, 12 March 1997. After agreement in Council in March 2003: this Directive was finally adopted in October 2003: Council Directive 2003/96/EC of 27 October 2003 restructuring the Community framework for the taxation of energy products and electricity, OJ L283 of 31 October 2003, p51.

22. For an overview of recent agreements see: Barth and Dette (2001).

23. Commission Communication on a Voluntary Agreement with the Automobile Industry Concerning the Reduction of CO_2 emissions, COM(98)495 of 29 July 1998.

24. More information and the text of the various agreements can be found at: http://europa.eu.int/comm/environment/co2/co2_agreements.htm.

25. See http://www.benchmarking-energie.nl.

26. See Part I and Part III of this book for more discussion on trading-based approaches.

27. Such as the integrated environmental permit based on the Dutch Environmental Management Act and the Integrated Pollution Control permit based on the UK's Environment Act.

28. The use of negotiated agreements for the implementation of EC environmental law is, however, limited, and must usually be backed up by a regulatory framework. See Communication from the Commission to the Council and the European Parliament on Environmental Agreements, COM(96)561, 27 November 1996.

29. See the various contributions in the Special Issue on Emissions Trading (2000) *Review of European Community and International Environmental Law*, vol 9, no 3. A regularly updated overview of the various trading initiatives can also be found on the UNCTAD homepage: http://www.unctad.org/ghg/etinfo/etinfo.htm.

30. For a brief overview of the Dutch NO_x trading proposal, see FIELD and IEEP (2002). An up-to-date account of the development of the Dutch NO_x trading regime (in Dutch) can be found on: http://www.emissierechten.nl.

31. Relevant documents for the UK trading regime can be found at the DEFRA UK Emissions Trading Scheme website:
http://www.defra.gov.uk/environment/climatechange/trading/index.htm.

32. Presentation by Mr Brian McLean, Director of the Clean Air Markets Division at the US Environmental Protection Agency at the first meeting of the European Climate Change Programme (ECCP) Working Group 1 on 4 July 2000. See also the summary record of this meeting, available on the Internet at: http://europa.eu.int/comm/environment/climat/wg1_minutes.pdf.

33. See below the section on arguments against emissions trading.

34. See comment by J. Lefevere during the second meeting of ECCP Working Group 1, 19 July 2000, reflected in the summary record of that meeting, published on the Internet at: http://europa.eu.int/comm/environment/climat/eccp.htm. See also Krämer (2001, p30).

35. These 'Hahn/Noll auctions', named after the authors of the paper that first introduced this idea, are, however, rarely used in practice. See also Schwarze and Zapfel (2000, p289, in particular footnote 21).

36. This approach has been used in a number of US trading programmes. See Schwarze and Zapfel (2000, p289).

37. Information on the UK auction and its results can be found on DEFRA's emissions trading website: http://www.defra.gov.uk/environment/climatechange/ trading/index.htm.

38. Information on the Danish trading regime can be found at the website of the Danish Energy Agency: http://www.ens.dk/uk/energy_reform/emissions_trading/index.htm.
39. See note 37 above.
40. See note 37 above.
41. See Climate Action Network Europe's (CAN-E) reaction to the Commission proposal: 'Emission trading in the EU: let's see some targets!', 20 December 2001, published on the Internet at: http://www.climnet.org/EUenergy/ET.html. See also CAN-E in its reaction to the Commission proposal, note 41 above, which describes the Kyoto trading system as 'riddled with flaws'.
42. For a discussion of the ethical aspects of emissions trading in a broader context, including JI and CDM, see also Ott and Sachs (2000).
43. For a discussion of the ethical aspects of emissions trading in a broader context, including JI and CDM, Ott and Sachs (2000). See also Part III, Chapter 4 by Anderson and Bradley in this volume.
44. M. J. Sandel, 'It's immoral to buy the right to pollute', *New York Times*, 15 December 1997.
45. FCCC/CP/2001/13/Add.2, p2, discussed in Part I.
46. See Articles 4.4, 4.8 and 4.9 of the UNFCCC and Article 3.14 of the Kyoto Protocol.
47. Ibid. and M.J. Sandel 'It's immoral to buy the right to pollute', *New York Times*, 15 December 1997.
48. For a detailed discussion of 'hot air' see Part III, Chapter 4 in this volume. Lack of stringent targets and hot air were highlighted by CAN-E in their comments on the Commission's ET proposal and the need to set clear and ambitious targets, note 41 above.
49. This is done in some of the US trading programmes. See Schwarze and Zapfel (2000, p292).
50. For an evolution of the thinking on emissions trading in the EU see Zapfel and Vainio (2002).
51. Communication from the Commission to the Council and the European Parliament, Climate Change – towards an EU Post-Kyoto Strategy, COM(1998)353, 3 June 1998.
52. Communication from the Commission to the Council and the European Parliament, Preparing for Implementation of the Kyoto Protocol, COM(1999)230, 19 May 1999.
53. COM(2000)87.
54. COM(2000)88.
55. Council Directive 96/61/EC Concerning Integrated Pollution Prevention and Control, [1996] OJ L257/26.
56. Decision 93/389/EEC for a monitoring mechanism of Community CO_2 and other greenhouse gas emissions, as amended by Council Decision 1999/296/EC [1999] OJ L117/35, now replaced by Decision 280/2004/EC of the European Parliament and of the Council of 11 February 2002 concerning a mechanism for monitoring Community greenhouse gas emission and for implementing the Kyoto Protocol, [2004] OJ L49/1.
57. Council Conclusions 2278th meeting, Environment, Luxembourg, 22 June 2000.
58. European Parliament resolution of 26 October 2000, [2001] OJ C197/219 and 400.
59. All submissions as well as a summary of the comments made are published on the Internet at: http://europa.eu.int/comm/environment/docum/0087_en.htm. See also Krämer (2001) who discusses the Green Paper and some of the reactions to it.

60. The ECCP WG1 terms of reference, background documents, meeting reports, interim and final report are published on the Internet at: http://europa.eu.int/comm/environment/climat/eccp.htm.

61. See, for instance, the letter by UNICE from 25 June 2001 to James Currie, Director General of DG Environment of the European Commission, which states that 'the Commission had made clear its intention to propose a Community emissions trading scheme towards the end of this year, with further consultation of stakeholders already planned. We regret that the current proposal pre-empts those sound intentions, without taking into account outcomes of the European Climate Change Programme, where major issues were raised of how an EU framework should be linked to separate Member State approaches, and how individual companies should become involved'. Available through: http://www.unice.org.

62. Decision 5/CP.6, The Bonn Agreements on the implementation of the Buenos Aires Plan of Action, FCCC/CP/2001/5, 36–49.

63. See note 5 above.

64. COM(2001)581, 23 October 2001, Council Decision 2002/358/EC concerning the conclusion, on behalf of the European Community, of the Kyoto Protocol to the United Nations Framework Convention on Climate Change and the joint fulfilment of commitments thereunder.

65. Communication from the Commission to the Council and European Parliament on the implementation of the first phase of the ECCP, COM(2001)580, 23 October 2001.

66. An elaborate description of the co-decision procedure and the roles of the various EU institutions therein can be found in Craig and de Búrca (2002).

67. See European Parliament, first reading report on the proposal for a European Parliament and Council Directive establishing a scheme for greenhouse gas emission allowance trading within the Community and amending Council Directive 96/61/EC, 13 September 2002, Rapporteur: Jorge Moreira da Silva, PE 232.374.

68. Climate Action Network Europe's position on the original Commission proposal, 'Emissions trading in the EU: let's see some targets!', 20 December 2001, available through: http://www.climnet.org.

69. See, for instance, the letter by UNICE, a key industry lobby group in Brussels, from 10 December 2001 to the Belgian Council Presidency, which states that 'Most believe that at least this initial phase should be on a voluntary basis, since a prime principle of emissions trading should be to offer motivation and clear market signals to companies', available through http://www.unice.org. Stronger opposition to mandatory trading came from the German industry association BDI, which opposed a mandatory trading regime. See, for instance, the 'Statement of the German Business on the Proposal for a Directive Establishing a Framework for Greenhouse Gas Emissions Trading within the European Community', 21 January 2002, available through http://www.bdi-online.de.

70. See for instance Haites and Hussain (2000).

71. The evaluation report of the first year of the UK emissions trading scheme can be found at: http://www.defra.gov.uk/environment/climatechange/trading/index.htm.

72. See for the Commission's interpretation of the *force majeure* provision Commission Communication of 7 January 2004 on guidance to assist Member States in the implementation of the criteria listed in Annex III to Directive 2003/87/EC establishing a scheme for greenhouse gas emission allowance trading within the Community and amending Council Directive 96/61/EC, and on the

circumstances under which *force majeure* is demonstrated, COM(2003)830, pp23–4, further discussed below.

73. See the 'Statement of the German Business on the Proposal for a Directive Establishing a Framework for Greenhouse Gas Emissions Trading within the European Community' of 21 January 2002, note 69 above.

74. See for the text of the German climate change agreement: http://www.bmu. bund.de/en/1024/js/topics/climateprotection/agreement.

75. Although Article 3.9 of the Kyoto Protocol states that negotiations for the second commitment period must begin by 2005, the Protocol does not specify when these negotiations should end or the length of the second commitment period.

76. These figures are from the Explanatory Memorandum to the Proposal, note 5 above, p10.

77. The final number of installations will only be known once all national allocation plans have been submitted. At the time of writing the UK draft national allocation plan includes about 900 installations, the Dutch national allocation plan up to 300 and the Irish national allocation plan up to 100. See for an overview of the NAP process and a list of national allocation plans with links to the Member State websites containing these plans: http://europa.eu.int/comm/environment/climat/emission_plans.htm.

78. See note 78 above.

79. See note 69 above.

80. See note 67 above, p89.

81. See also EIPA research paper 00/GHA, 'Governance by Committee, the Role of Committees in European Policy-Making and Policy Implementation', Maastricht, May 2000, available through http://www.eipa.nl/Topics/Comitology/comitology.htm.

82. Decision 280/2004/EC of the European Parliament and of the Council of 11 February 2004 concerning a mechanism for monitoring Community greenhouse gas emissions and for implementing the Kyoto Protocol, OJ L49, 19 February 2004, p1.

83. See note 67 above and European Parliament, second reading report on the Council common position for adopting a European Parliament and Council Directive establishing a scheme for greenhouse gas emission allowance trading within the Community and amending Council Directive 96/61/EC, 12 June 2003, Rapporteur: Jorge Moreira da Silva, PE 328.778.

84. Original Commission proposal, note 5 above, p10.

85. Ibid, p3.

86. The application of the IPPC Directive to existing installations has, for instance, been postponed to 2010 for Latvia and Poland and to 2011 for Slovenia, compared to 2007 for existing Member States. For a summary of the various transitional measures see: http://www.europa.eu.int/comm/enlargement/negotiations/chapters/chap22/index.htm.

87. For a more elaborate discussion of the relation between the ET Directive and the IPPC Directive, see below.

88. See note 68 above.

89. See note 67 above.

90. See also the analysis by Krämer (2001).

91. See also the BDI position paper on the original Commission Proposal, note 69 above.

92. The UK government has decided to take its self-proclaimed target of –20 per cent by 2010 rather than its burden-sharing target of –12.5 per cent as the starting point for the allocation under the ET Directive.

93. More information on these programmes can be found on the Internet at: http://www.carboncredits.nl.
94. See for a more elaborate discussion on state aid and the ET Directive: König et al (2003) and Pfromm (2003).
95. Community guidelines on State aid for environmental protection, (2001) OJ C37/3. See for a discussion of these guidelines Fernández Armenteros (2001) and Vedder (2001).
96. See, for instance, the remark by C. Boyd during the first meeting of ECCP Working Group I on 4 July 2000, stressing the need to have input from DG competition in the discussions and asking the Commission to ensure that officials from DG attend the WG I meetings. This remark is reflected in the summary record of that meeting, published on the Internet at: http://europa.eu.int/comm/environment/climat/eccp.htm. The Commission's competition DG was in fact regularly invited by DG Environment officials to attend the ECCP meetings but never turned up.
97. See Commission Decisions on State aid No N 416/2001 – United Kingdom Emission Trading Scheme; N 653/99 – Denmark, CO_2 quotas and State aid; and N 123/2000 – United Kingdom, Climate Change Levy for the free allocation of allowances for companies entering into Climate Change Agreements.
98. Commission Decision on State aid No N 416/2001 – United Kingdom Emission Trading Scheme.
99. Ibid.
100. Commission Decision on State aid No N 653/99 – Denmark, CO_2 quotas and State aid.
101. UNICE, in its comments to the proposal that was leaked in June 2001, for instance stated that 'guidance to member States in Annex III, on criteria for allocation plans, is not nearly clear enough to avoid a danger of single market distortions being caused'. See note 61 above.
102. Commission Communication of 7 January 2004 on guidance to assist Member States in the implementation of the criteria listed in Annex III to Directive 2003/87/EC establishing a scheme for greenhouse gas emission allowance trading within the Community and amending Council Directive 96/61/EC, and on the circumstances under which *force majeure* is demonstrated, COM(2003)830.
103. Ibid, para. 47.
104. See in particular the letter of UNICE to Caroline Jackson, Chairman of the European Parliament's Committee on the Environment, Public Health and Consumer Policy from 2 June 2003, available through http://www.unice.org.
105. See note 68 above.
106. In first reading Parliament proposed that from 2005 to 2007 70 per cent be allocated free of charge with the remaining 30 per cent allocated by means of auction. For the second period covered by the scheme Parliament proposed that all allowances were to be allocated by means of an auction. See note 67 above, p52.
107. See note 56 above.
108. European Commission, DG Environment, Directorate C, 'The EU Emissions Trading Scheme: How to develop a National Allocation Plan', Non-Paper for the 2nd meeting of Working Group 3 of the Monitoring Mechanism Committee, 1 April 2003, available through http://europa.eu.int/comm/environment/climat/emission_plans.htm.
109. Communication from the Commission of 7 January 2004 on guidance to assist member States in the implementation of the criteria listed in Annex III to

Directive 2003/87/EC establishing a scheme for greenhouse gas emission allowance trading within the Community and amending Council Directive 96/61/EC and on the circumstances under which *force majeure* is demonstrated, COM(2003)830.

110. Commission Decision 2004/156/EC of 29 January 2004 establishing guidelines for the monitoring and reporting of greenhouse gas emissions pursuant to Directive 2003/87/EC of the European Parliament and of the Council, OJ L59, 26 February 2004, p1.

111. ECJ 21 September 1989, Case 68/88, *Commission* v *Greece* (community funds).

112. See note 19 above, at 288–9.

113. See 'UNICE Comments on the Proposal for a Framework for EU Emissions Trading' of 25 February 2002, available through: http://www.unice.org.

114. The latest Commission Annual Survey on the implementation and enforcement of Community environmental law starts with the conclusion that: 'The last five years have seen a growing difficulty in the timely and correct implementation as well as proper practical application of EC environmental legislation. This is reflected in the number of complaints received and infringement cases opened by the Commission every year. As in the earlier years, in 2002 the environment sector covered over one third of all infringement cases investigated by the Commission. The Commission brought 65 cases against Member States before the Court of Justice and issued 137 reasoned opinions on the basis of Article 226 of the EC Treaty'. See the Commission's Fourth Annual Survey on the implementation and enforcement of Community environmental law, 2002, 7 July 2003, SEC(2003)804, Commission Staff Working Paper, available at: http://europa.eu.int/comm/environment/law/4th_en.pdf.

115. See also Krämer (2001, pp38–40).

116. Member States only have to apply the national legislation implementing the IPPC Directive to existing installations from October 2007 onwards – see Box III.1 above.

117. European Commission Non-Paper on Synergies between the EC Emissions Trading Proposal and the IPPC Directive, 22 January 2002, published on the Internet at: http://europa.eu.int/comm/environment/climat/non-paper_ippc_and_et.pdf.

118. BREF documents adopted so far can be found at: http://eippcb.jrc.es/pages/FActivities.htm.

119. Commission Decision 2000/479/EC on the implementation of a European pollutant emission register (EPER) according to Article 15 of Council Directive 96/61/EC concerning integrated pollution prevention and control (IPPC), [2000] OJ L192/36. The EPER can be accessed via the Internet at: http://eper.cec.eu.int.

120. European Commission, note 117 above.

121. Commission Decision on State aid No N 416/2001 – United Kingdom Emission Trading Scheme. Since the IPPC Directive does not come into force for new installations until October 2007 there is significantly more leeway.

122. For a discussion on the use of emissions trading within the boundaries of the IPPC Directive.

123. Ibid.

124. European Commission, note 117 above.

125. Ibid.

126. These texts are part of the Marrakesh Accords, adopted in November 2002, and can be found in FCCC/CP/2001/13/Add.2.

127. See notes 1 and 5 above, as well as the Communication from the Commission on the Implementation of the First Phase of the European Climate Change Programme, COM(2001)580, 23 October 2001.
128. See note 68 above, and more recently, CAN Europe press release of 23 July 2003, 'Commission shoots its own emission trading system full of holes', available through: http://www.climnet.org.
129. See note 69 above, and more recently: UNICE letter to Margot Wallström, Commissioner, DG Environment, of 4 July 2003, 'Unice preliminary comments on linking the Kyoto Protocol mechanisms (JI and CDM) with the EU emission trading scheme', available through: http://www.unice.org.
130. Explanatory memorandum, note 5 above, p17.
131. COM (2003)403 of 23 July 2003 see note 7 above.
132. The term 'project mechanisms' was included because it could cover linkages with non-Kyoto-related mechanisms.
133. In the Council's Common Position this paragraph read: 'The recognition of credits from project-based mechanisms for fulfilling obligations under this Directive as from 2005 will increase the cost-effectiveness of achieving reductions of global greenhouse gas emissions and will be provided for by a Directive for linking project-based mechanisms including Joint Implementation (JI) and the Clean Development Mechanism (CDM) within the Community Scheme'. The final text of this paragraph, after the second reading compromise, now reads: 'Project-based mechanisms including Joint Implementation (JI) and the Clean Development Mechanism (CDM) are important to achieve the goals of both reducing global greenhouse gas emissions and increasing the cost-effective functioning of the Community scheme. In accordance with the relevant provisions of the Kyoto Protocol and Marrakesh Accords, the use of the mechanisms should be supplemental to domestic action and will thus constitute a significant element of the effort made.'
134. For instance, through initiatives such at the Dutch ERUPT and CERUPT programmes – see above. For more information on the ERUPT and CERUPT programmes: http://www.carboncredits.nl.
135. See note 77 above.
136. See the CAN Europe press release of 23 July 2003, note 128 above, as well as the letter from CAN Europe to the Commissioner for the Environment Margot Wallström of 28 February 2003, available through http://www.climnet.org.
137. See the UNICE letter to Margot Wallström, Commissioner, DG Environment, of 4 July 2003, note 129 above.
138. The terms of reference, agenda, minutes and conclusions of these ECCP meetings are published on the Internet at: http://europa.eu.int/comm/environment/climat/ji_cdm.htm.
139. See, *inter alia*, CAN Europe's reaction to the ET proposal, note 68 above.
140. Environmental NGOs, in a letter to the Commission, disagreed with the representative nature of these conclusions, objecting in particular to this paragraph. ECCP Working Group on JI/CDM, Conclusions, 15 November 2002, available at: http://europa.eu.int/comm/environment/climat/jicdm/jicdm_final_conclusions.pdf.
141. Note 7 above.
142. Various EU position papers for the international negotiations can be found through: http://europa.eu.int/comm/environment/climat/cop.htm.
143. During the international negotiations the EU in fact advocated such a positive list.

144. These include Council Directive 85/337/EEC of 27 June 1985 on the assessment of the effects of certain public and private projects on the environment, Directive 2003/4/EC of the European Parliament and of the Council of 28 January 2003 on public access to environmental information and Council Directive 96/61/EC of 24 September 1996 concerning integrated pollution prevention and control.

145. Decision -/CP.9 on Modalities and procedures for afforestation and reforestation project activities under the clean development mechanism in the first commitment period of the Kyoto Protocol.

146. The 'acquis communautaire' encompasses the total body of Community law, including legislation and related instruments such as judgments of the European Court of Justice that are in place in the Community.

147. These requirements are included in Article 12(5) of the Kyoto Protocol and in para 43 and the preamble of the CDM text, FCCC/CP/2001/13/Add.2, pp36 and 20.

148. See Box III.6 above.

149. See note 129 above.

150. See also the discussion on international emission trading in Part I .

151. Decision 19/CP.7, FCCC/CP/2001/13/Add.2, p63, para 28.

152. Ibid, p61, paras 15 and 16.

153. Decision 11/CP.7, FCCC/CP/2001/13/Add.1, p61, para 14.

154. See, for instance, UNICE's letter of 4 July 2003, note 129 above.

155. The 6 per cent and 8 per cent figures are based on the study *KPI Technical Report: Impacts of Linking JI and CDM Credits to the European Emission Allowance Trading Scheme (KPI-ETS)*, by Patrick Criqui and Alban Kitous, available through: http://europa.eu.int/comm/environment/climat/emission.htm.

References

Barth, R. and Dette, B. (2001) 'The integration of voluntary agreements into existing legal systems', *Environmental Law Network International Review*, vol 1, pp20–9.

Craig, P. and de Búrca, G. (2002) *EU Law – Text, Cases and Materials*, 3rd edn.

Denne, T. (1999) *Aggregate versus Gas by Gas Models of Greenhouse Gas Emissions Trading*, CCAP, Scoping Paper 6, published on the Internet at: www.field.org.uk/papers/papers.htm.

Fernández Armenteros, M. (2001) 'Overview of the community guidelines on state aid for protecting the environment', *Environmental Law Network International Review*, vol 1, pp36–42.

FIELD (Foundation for International Environmental Law and Development) (2000) *Designing Options for Implementing an Emissions Trading Regime for Greenhouse Gases in the EC*, final report on a study for the DG ENV, European Commission, available at: www.field.org.uk/climate_4.php.

FIELD (Foundation for International Environmental Law and Development) (2001) *Study on the Legal/Policy Framework Needed for Establishment of a Community Greenhouse Gas Emissions Trading Scheme*, report, available at: www.field.org.uk/ PDF/ETreport.PDF.

FIELD and IEEP (2002) *Assessment of the relation between Emissions Trading and EU Legislation, in particular the IPPC Directive*, study conducted for VROM, final report, 31 October 2002, available at: www.field.org.uk/PDF/FINALReport31Oct.pdf.

Haites, E. and Hussain, T. (2000) 'The changing climate for emissions trading in Canada', *Review of European Community and International Environmental Law*, vol 9, no 3, pp264–75.

Hargrave, T. (1999) *Identifying the Proper Incidence of Regulation in a European Union Greenhouse Gas Emissions Allowance Trading System*, CCAP, Scoping Paper 4, published on the Internet at: www.field.org.uk/papers/papers.htm.

IPCC (Intergovernmental Panel on Climate Change) (2001) *Third Assessment Report – Climate Change 2001*, published on the Internet at: www.ipcc.ch.

Johnson, S.M. (2001) 'Economics v. Equity II: the European Experience', *Washington and Lee Law Review*, vol 58, p421.

König, C., Braun, J.-D. and Pfromm, R. (2003) 'Beihilferechtliche Probleme des EG-Emissionsrechtehandels', *ZWeR*, pp152–86.

Könings, M. (2003) 'Emission trading – why state aid is involved: NO_x trading scheme', *Competition Policy Newsletter*, no 3 (autumn), pp77–9; also available at: europa.eu.int/comm/competition/publications/cpn/.

Krämer, L. (2001) 'Grundlagen aus europäischer sicht, Rechtsfragen betreffend den Emissionshandel mit treibhausgasen der Europäischen Gemeinschaft', in H.-W. Rengeling, *Klimaschutz durch Emissionshandel*, Köln: Carl Heymanns Verlag KG 30.

Ott, H. and Sachs, W. (2000) *Ethical Aspects of Emissions Trading*, contribution to the World Council of Churches Consultation on 'Equity and Emissions Trading – Ethical and Theological Dimensions', Saskatoon, Canada, 9–14 May, Wuppertal Papers No 110, September.

Pfromm, R. (2003) 'Die entgeltfreie Allokation von Emissionszertifikaten – eine wettbewerbsrechtliche Sackgasse?', pp537–42.

Ringius, L. (1999) *Differentiation, Leaders and Fairness: Negotiating Climate Commitments in the European Community*, CICERO Report 1997:8, published on the Internet at: www.cicero.uio.no/media/99.pdf.

Sagoff, M. (1999) *Controlling Global Climate: The Debate over Pollution Trading*, Report from the Institute for Philosophy and Public Policy. Published on the Internet at: www.puaf.umd.edu/IPPP/winter99/controlling_global_climate.htm.

Schwarze, R. and Zapfel, P. (2000) 'Sulfur allowance trading and the regional clean air incentives market: a comparative design analysis of two major cap-and-trade permit programs?', *Environmental and Resource Economics*, vol 17, pp279–98.

Scott, J. (1998) *EC Environmental Law*, Longmans European Law Series

Stewart, R. B. (2001) 'A new generation of environmental regulation?', *Capital University Law Review*, vol 29, pp21–182.

Vedder, H. (2001) 'The new community guidelines on state aid for environmental protection – integrating environment and competition', European Competition Law Review, vol 22, no 9, pp365–73.

Yamin, F. (1998) 'The Kyoto Protocol: origins, assessment and future challenges', *Review of European Community and International Environmental Law*, vol 2, pp113–27.

Zapfel, P. and Vainio, M. (2002) *Pathways to European greenhouse gas emissions trading history and misconceptions*, Fondazione Eni Enrico Mattei, Nota di Lavoro 85.2002, available through: www.feem.it/web/activ/_activ.html.

Part III

Development and implementation of the Kyoto mechanisms worldwide

Chapter

1 Emissions trading under the Kyoto Protocol: how far from the ideal?   153
   Richard Baron and Michel Colombier

2 Trading through the flexibility mechanisms: quantifying the size of   166
   the Kyoto markets
   Odile Blanchard

3 Implementation challenges: insights from the EU Emission Allowance   183
   Trading Scheme
   Fiona Mullins

4 Joint Implementation and emissions trading in Central and   200
   Eastern Europe
   Jason Anderson and Rob Bradley

5 Implementing the Clean Development Mechanism and   231
   emissions trading beyond Europe
   Martijn Wilder

6 The Clean Development Mechanism: a tool for promoting long-term   263
   climate protection and sustainable development?
   Mark Kenber

7 Determination of baselines and additionality for the CDM: a crucial   289
   element of credibility of the climate regime
   Axel Michaelowa

8 Creating the foundations for host country participation in the CDM: 305
   experiences and challenges in CDM capacity building
   Axel Michaelowa

Chapter 1

Emissions trading under the Kyoto Protocol: how far from the ideal?

Richard Baron and Michel Colombier

1.1 Introduction and scope

As the negotiations at Kyoto drew to a close, it became clear that a new form of environmental regime was being created: the European Union and its allies had given up their attempt at a list of common and coordinated policies and measures applied to industrialized countries. After intensive communication and the support of many industrial players, economists and a fraction of non-governmental organizations, the United States had convinced other Parties to agree on a set of legally binding objectives and the possibility of trading greenhouse gases (GHG) via a market mechanism, emissions trading. On paper, emissions trading would make the international regime cost-effective: at the margin, no Party would spend more to reduce than another Party. Last but not least, a carbon price would emerge and the market would then reveal one of the most crucial pieces of information for future negotiations: what it really costs to bring GHG emissions down. The Kyoto Protocol would become the perfect illustration of standard environmental economics. Or so the proponents of the tool – including these authors – would argue.

Although the Protocol awaits Russia's ratification to enter into force and emissions trading has not nearly developed into what it is supposed to be – a full-blown cap-and-trade system – we can look in hindsight at how emissions trading was perceived and promoted before Kyoto, and compare this somewhat ideal picture with what current policy developments and various government statements suggest it may eventually become. As developed country Parties are starting to think about future commitments and developing countries are considering adhering to the regime at some later stage, an objective assessment of international emissions trading is necessary to measure its true potential. This is, of course, an arduous task, in the absence of any track record for international emissions trading – project-based mechanisms are another story altogether. What we propose here is, rather, to present some of the problems that the regime may encounter, and hopefully to trigger some thinking about potential solutions and also about what emissions trading is really designed to do: minimize the cost of achieving a pre-agreed environmental target. Our thesis is that discussions on this issue may have suffered from oversimplifications and oversights that raised high expectations about what emissions trading can deliver in terms of economic efficiency and, maybe more importantly, could lead the negotiation of future commitments into a dead end.

Section 1.2 recalls how emissions trading was introduced and promoted in the climate change policy context. Section 1.3 presents our views on the implications of governments' participation in international emissions trading. Section 1.4 discusses the European Union Emission Allowance Trading Scheme (EU ETS), a cap-and-trade regime that is to be embedded into the broader international framework of the Kyoto Protocol Article 17, and discusses potential problems and solutions.

1.2 Emissions trading: focusing on economic efficiency

From the early notion of joint implementation to emissions trading

The United Nations Framework Convention on Climate Change (UNFCCC) included a direct reference to what would later become emissions trading: the possibility for countries to achieve their emission commitments 'jointly': one country's excess emissions would be offset by another's over-achievement. By avoiding excessive abatement costs in the first country, this would lower the overall cost of bringing emissions down. In its more elaborate form, 'joint implementation' as it was known in the Rio treaty would trigger international transfers of avoided emissions (or emission allowances) and the emergence of a full-fledged carbon market, with a price helping investors and governments alike to take action at least cost. This theory was backed up by the development of the USA SO_2 allowances trading programme implemented by the Environmental Protection Agency, which was starting to show how effective emissions trading really was. Strict SO_2 emission caps, primarily on coal-based power generation plants, combined with the possibility of buying emission allowances from another plant if emissions rose higher than the initial allocation, triggered enough allowance transactions to create a credible price for this new 'commodity'. The price, even today, turns out to be much lower than originally anticipated, for a host of reasons: desulphurization technology became cheaper and low-sulphur coal was suddenly available at a much lower cost than its dirtier competitor thanks to a reform of the railway industry. The introduction of ambitious emission caps, a clear set of rules about long-term allocations of allowances and the possibility of selling excess allowances (i.e. to generate a profit from cutting pollution) surely contributed to encourage industry to look for cheaper and cleaner alternatives (Ellerman et al, 1997). Even if they complain about the monitoring costs induced by the system, power producers recognize its economic efficiency and its superiority over command-and-control.

If economic theory allowed defining the right approach to reduce SO_2, there are clear reasons to posit that it should work even better for greenhouse gases. Sources are numerous and so are abatement possibilities: exploiting them under the guidance of a carbon price would work better than seeking a tedious one-by-one approach to every kind of source and gas. Further, unlike SO_2, most greenhouse gases are not local pollutants. While allowing some flexibility to exceed emission limits with the help of emissions trading could be problem for a pollutant like SO_2, it would be neutral in terms of climate change since GHG are well mixed in the atmosphere and affect the Earth's greenhouse effect in the same way regardless of where they are emitted.[1] Further still, it does not matter whether a ton of CO_2 is emitted today or in five years' time. Time flexibility, or 'when flexibility', could therefore be added to what was known as 'where flexibility' – the possibility of reducing emissions somewhere else if it was cheaper to do so, the main feature of emissions trading (Weyant, 1999).

Making the case for international GHG emission trading

A great number of economic assessments of international GHG emissions trading were done as governments sought evidence of the benefits of this rather new instrument. A trading scenario was compared with a reference scenario in which every region or country had to meet its target unilaterally. The purpose of this comparison was to test whether emissions trading could really make a significant dent in the cost of reaching the common emission target. Making reductions more affordable would also be a way to encourage more ambitious reduction objectives.

Modelling does require simplifying some of the policy choices from what is available to governments in the real world. Economists are also reluctant to test instruments that they judge inefficient. If pollution is to be reduced, environmental economics suggests, all sources should pay the same price for their emissions. Any other policy would lead to unnecessary cost. In modelling terms, this means that:

- individual countries or regions achieve domestic reductions through a single policy instrument, a tax on emissions or its equivalent, a domestic emissions trading system. In terms of efficiency, these two instruments are equivalent;
- international supply and demand for allowances is immediately balanced, a single price of carbon emerges from this equilibrium and remains stable throughout the commitment period.

This is standard application of micro-economic theory: an economically efficient outcome requires that, at the margin (to remove the last tonne to achieve compliance), no source pays more than another. These scenarios simply assume that what works for cost-minimising economic agents (power plants in the United States) also applies to countries or regions. To make this work, they must also assume that all sources within the countries pay the same price. In other words, under the Kyoto Protocol, all sources in all Annex B countries would pay a carbon tax set by the international emission trading market; the tax would be known to all, not vary throughout the period and allow all countries to meet their common objective jointly. This rosy picture painted by economic models does not really match the reality of domestic policy decision-making and international relations, as we show in the next section. For a start, market experiments confirmed the intuition that a unique and stable carbon price throughout five years or so is not a probable scenario.[2]

Devil in the details?

Even if we abstract ourselves for a minute from the policy realities and follow the more simplistic policy approach assumed in these scenarios, major elements were omitted that cast a shadow on the attractiveness of emissions trading. Babiker et al (2002) recently showed that not all countries may win from engaging in international emissions trading, a shocking news for UNFCCC delegates that have been presented with economic results showing that all countries would win from emissions trading. Why may this not be the case? Some countries have already fairly high taxes on energy (especially on gasoline and diesel used in transportation) and these taxes introduce economic distortions: their high level causes more harm than good to the economy. Raising them further would mean more economic loss. A country that can meet its own emission goal at €15 per tonne of CO_2 but can sell unused allowances at a profit if it raises its domestic carbon price to €20 (the international carbon price) may in fact lose macro-economically if its existing energy taxes are too high. These authors suggest that some European countries (France, Germany and the United Kingdom) could end up in this awkward position.

A carbon tax added to existing gasoline taxes is unlikely to be the first and only option to regulate emissions from transport. Yet this is what the Kyoto scenarios run by global macro-economic models implied: a blanket increase in fossil-fuel prices based on their carbon content. And for some time everyone believed this would be ideal, if only it were possible. These new results do not mean that emissions trading would systematically hurt countries that sell GHG on the international market, but they confirm what international trade economists had known for some time (but no one had really mentioned in the climate community while discussing Kyoto): not all trade is economically beneficial.

1.3 Article 17 of the Kyoto Protocol: throwing governments into the cost-minimization game

Several features of emissions trading as it is introduced by the Kyoto Protocol's Article 17 differ from the basic elements of a system introduced to regulate pollution among industrial sources. We will not analyse in depth the causes for such differences but simply explain how they may affect the efficiency of the regime compared with an industry-based one.

First and foremost, Parties, i.e. governments, will be bound by the international treaty that defines emissions objectives for 2008–2012. Companies and individuals whose activities release GHG are not recognized as such by the Treaty, even if all recognize their prominent role in bringing emissions down. At the end of the commitment period, governments of countries with emissions above their original targets will need to have acquired sufficient emission allowances from other countries or individual projects to match these excess emissions and restore compliance. It is only then that the overall environmental objective agreed by all Parties will be reached.

The Protocol recognizes the possibility of governments devolving part of their so-called assigned amounts to entities on their territory and allowing them to trade allowances internationally as well. At first sight, this could very much resemble the industry-based regime mentioned above: governments need to allocate their assigned amount to industrial players who will then funnel the cost of carbon throughout the economy. Policy options are available for that purpose, such as an allocation of the assigned amount to all producers and importers of fossil fuel energy in a country.[3] The policy reality is altogether different: emissions trading is well suited for large industrial sources who have control over their own emissions and can invest in the right technology to curb them. Letting fossil-fuel producers regulate the emissions downstream through a price increase is not necessarily the most cost-effective way to reduce CO_2, nor is it a very politically attractive option.[4]

From current experience in the EU, Canada and Japan, large portions of national inventories will remain outside the coverage of domestic or regional emissions trading systems: transportation, the residential sector, small businesses and forestry. Governments are taking measures to curb these activities' emissions, but current GHG inventories reveal that these non-trading sectors have contributed to most of the increase in countries' emissions to date. In contrast, industrial emissions have remained stable or declined in many countries (IEA, 2003). And no government currently projects to require individual car owners to manage their GHG balance on the international carbon market. Rather, governments have introduced or plan to introduce specific policies to curb emissions from sectors that would not be directly connected to the international emissions trading regime. Policies include regulations to reduce the electricity consumption of appliances, voluntary agreements with car manufacturers on CO_2 emissions of future car vintages, fiscal incentives for building insulation and refurbishing, obligations on renewable energy supply, to name a few. Such domestic policies take a number of months or years and lengthy consultations to reach implementation stage. If countries are guided by both the near- and long-term goal of climate change mitigation, they are unlikely to either bring these negotiated measures to a halt or harden them significantly because the international carbon price turns out to be surprisingly low or high. A number of such measures have been designed in the last few years without a specific international carbon price as a guide for action, nor with triggers that would make them adjustable to such a price.[5]

We should also recognize that there are good reasons why governments should not seek to adjust policies to carbon prices.[6] Governments are faced with complex policy choices that involve both the near and longer terms. Affecting long-term transportation needs, for instance, could require new infrastructure for public transport, regulations on telecommuting, or even new tax laws to encourage people to move closer to their workplace. The effects of such decisions on CO_2 emissions over the next five years may be minimal and the transition cost high, not at all in line with the prevailing international carbon price. It may, however, be what is needed to really curb future CO_2 emissions, even if it is in stark contrast with the early modelling results where it was assumed that a unique price signal should guide near-term decisions one period after next. The fact that some governments have already indicated how much they intend to rely on emissions trading to achieve compliance, without knowledge of the future CO_2 price, is evidence that they are not building their compliance strategy on the basis of near-term economic efficiency only. If they were, domestic reductions would be calibrated so that their cost matches the international price, which would then indicate the quantity of allowances to be bought internationally – clearly not the case here.

For these reasons, we can abandon the assumption that international emissions trading will bring compliance at least possible economic cost: rather, Article 17 provides Parties with a tool to achieve compliance with their Kyoto target given their country's economic, social and technological realities (also known as 'national circumstances'). This is different from using the international GHG market as a guide for domestic policy-making.

Further, governments could be inclined to use international emissions trading as a new instrument in their international relations toolkit. A government may give preference to allowances sold by a friendly government, when given a choice. It could also decide to invest primarily in joint implementation or clean development mechanism projects hosted by countries with whom it seeks to strengthen commercial relationships. Obtaining least-cost GHG reductions may be paramount from an economist's perspective, but governments could find value in including other elements in their international GHG transactions.

Another important distinction between an industry-based allowances trading regime such as the USA SO_2 allowances programme and emissions trading under the Kyoto Protocol is the absence of a financial sanction in case of non-compliance – there are indeed very few instances of international agreements in which Parties established the possibility of financial sanctions against each other. In this context, one downside of emissions trading is that governments could sell allowances strictly for a financial purpose and not undertake emission mitigation policies. Rogue players could sell their allowances for a quick profit, pull out of the international agreement and in the end jettison the Kyoto Protocol's environmental goals and architecture. The Marrakesh Accords managed to prevent this risk by requiring Parties with commitments to retain enough allowances to cover their expected emission levels or 90 per cent of their initial holding. Relying on the acquisition of allowances may not be a compelling option for a government, knowing that the sanction could be much less costly than the trading alternative.

Governments will very certainly need to rely on international emissions trading to achieve compliance in the Kyoto commitment period. Measured in terms of net transfers over the commitment period,[7] they would represent the lion's share of transactions (Natsource, 2003). Governments will be major players on the international GHG market, but the market is not likely to guide their domestic policy and deliver cost-minimization at the micro level. The latter goal seems more suited for an industry-based system.

1.4 The EU Emission Allowance Trading Scheme: a step closer to the ideal?

In contrast with an intergovernmental system – in which industry has of course also a role to play – the European Commission has promoted an inter-industry GHG trading regime spanning large industrial sources across Europe, later agreed by the European Parliament and the Council of Ministers, which should start operating in 2005, in the majority of EU-25 countries.

The ETS, as it is known, applies to various industrial activities whose investment and production choices should be strictly motivated by profit maximization and cost minimization. We are therefore closer to the theoretical model best able to bring economic efficiency into industries' actions to reduce their GHG emissions. European decision-makers were motivated by a concern over compliance costs when they introduced the proposal: because industry is more open to international competitiveness, both from inside and outside Europe, it would be best not to burden it with excessive GHG reduction costs, while making sure its emissions are under control. The EU ETS gives each source an access to the same GHG mitigation potential as its competitors. It does not mean that each source will feel it is treated fairly, an issue that is linked to its emission quota allocation, but at least the ETS brings economic efficiency in an international GHG reduction effort that will eventually span 25 countries and thousands of industrial installations.

As governments and industry negotiate over the national allocation plans for 2005–2007, differences in rules governing each country's implementation of the system may create political tensions, for example if one country is particularly generous for new entrants into the regime while another requires that they buy most of the needed GHG allowances from the market. These questions are of course critical, but we think it is also important to tackle two issues in the bigger picture: first, the relation between the EU ETS system and emissions trading under Article 17, as described above. The second issue is the rather near-term quota allocation under the EU system.

One Protocol, two greenhouse gas markets

The EU ETS and emissions trading under the Kyoto Protocol – assuming it enters into force – appear to be disconnected, while one could argue that the same 'commodity' is being traded: avoided greenhouse gas emissions. Installations covered by the system will have access to two streams of allowances for compliance with their objectives:

• the EU allowances, i.e. those distributed to other installations by their respective governments; and

- emission reduction units generated by projects, via joint implementation or the clean development mechanism.

In other words, a power generator in Germany cannot at present use assigned amount units from the Russian government to comply with its objectives under the EU ETS.[8] The German government, however, can do so, under Article 17 of the Kyoto Protocol. Thus, as it stands, the Directive creates an allowance market that is embedded in, yet not really connected to, the full Kyoto regime.[9] Many factors suggest that there will be two separate emissions trading regimes, with two different prices:

- The EU ETS allowance price should depend primarily on the industry's GHG reduction potential and related cost. The potential in sectors outside industry is based on technologies, policies and behaviours that are entirely different from the industry's – an avoided tonne of CO_2 from operating a gas plant instead of a coal plant carries a cost that is different from that of avoiding emissions through building insulation or improved access to public transport.
- Assuming ratification by Russia, there is a very large oversupply of assigned amount units in the current Kyoto Protocol setting compared with a 'full' Kyoto setting that includes the USA.[10] Unless part of the excess available to these countries in transition is made accessible to EU industry via bilateral agreements between the EU, Russia and Ukraine, only governments would be in a position to acquire assigned amount units from these countries, presumably at a rather low cost.
- Project-based units would be accessible to both governments and industrial sources in the EU ETS but are unlikely to be available in quantities large enough to set the price on both markets.

This scenario of an industry-to-industry market (as in the EU ETS) and a government-to-government market existing in parallel could create a policy problem if the two prices differ widely – especially if EU ETS allowances are traded at a much higher price than other Kyoto units. Industry could argue that it puts an unfair burden on it because it is open to competition from firms in countries without GHG commitments, while other sectors enjoying a low cost of compliance are not in that position. The EU Directive does not allow EU governments to buy assigned amount units in order to sell them back into the EU ETS market. The only mechanism that could be used for that purpose is a reserve for new entrants, or a portion of total allowances that a government could set aside and make available for later distribution, but the quantity has to be set in advance. Unless most European governments agree to set aside a portion of their assigned amount for this purpose, it would not be suited to adjust the internal

EU ETS price to the government-to-government price. Absent such mechanism, the perception of an undue pressure on industrial activities could seriously undermine the political viability of the EU ETS.[11]

Of course, Russian and Ukrainian governments could also decide to sell their assigned amount units at the price of EU ETS allowances.[12] The apparent economic efficiency of a single international carbon price would in effect hide a major market distortion, with Russia and Ukraine forming a 'carbon cartel'.

Short-term allocations and industry investment cycles

While the EU ETS is legislatively disconnected from the entry into force of the Kyoto Protocol, it is of course influenced by its design. Its first commitment period (2005–2007) is meant to be a learning phase for a fuller regime that will span the Protocol's five-year commitment period. The European Directive lays out broad criteria that governments should follow to set emission quotas for industry over these two periods, spanning a total of eight years. This fits the broader Kyoto picture, but, while it seeks to provide a signal for investment decisions, it seems out of sync with the lifetime of industrial physical capital stock – power generation being the extreme, with installations operating for longer than 30 years. In contrast, SO_2 allowances were allocated for 30 years to power plants in the USA, creating a form of asset that is managed alongside generation and depollution investments. We must recall that the Kyoto Protocol as it stands does not provide that kind of time horizon, but also that if the EU ETS was meant to operate independently from its UNFCCC parent and focus on industry, a longer-term allocation would have been welcome. It would have removed existing uncertainty over 'what happens next', helping industry focus on the kind of emission reductions that the UNFCCC ultimate goal requires in the longer run. There is otherwise a possibility that investments that are deemed rational under a long-run GHG constraint are not undertaken because they bind the installation to a low emission path and a correspondingly lower allocation when the next allocation plan is negotiated.

Allocating CO_2 emission quotas over 15 years or more would of course raise major technical problems and would be politically daunting: uncertainties over emerging or yet-to-be tested technologies and over future markets for industrial output could turn these allocations into windfall profits or the kiss of death, with disturbing economic and social consequences. At the very least, industry and government alike should be clear about the principles that will guide future allocations and these principles should encourage rather than discourage future reduction efforts. An increasing reliance on auctions rather than 'grandfathering', as suggested by the EU ETS Directive, may be a way forward.

1.5 Conclusion

We have described how the emerging international GHG trading regime is far from the ideal that was promoted in the early days of the Kyoto Protocol negotiation. It seems that the original idea of a grand market where all supply and demand for GHG emission reductions would be confronted to create a single price is out of reach. The initial model which extrapolated an industry-based system into an intergovernmental one was idealistic, as it did not take into account some potential negative economic effects of equalizing the price of carbon across all Parties. But it also lacked realism: not all GHG sources are suited to participate in international emissions trading, and some GHG mitigation policies should anyway be based on longer-term and broader considerations than the prevailing carbon price.

The current situation points to a two-tiered international GHG market: an industry-based market emerging from the implementation of the EU ETS and a broader regime under Kyoto driven by government transactions. The first one brings what is expected from emissions trading systems: the possibility for sources to minimize compliance costs and to rationalize their GHG mitigation options. We have yet to see how the system will evolve and guide industrial choices, given the short-term nature of allocations and the much longer investment cycles of some industrial sectors. The second regime is in fact entirely new: rather than guide governments, i.e. Parties, in their domestic policy choices, it is in fact a mechanism that allows them to commit to precise emission objectives with the legitimate possibility of meeting them through the acquisition of other Parties' unused emission allowances.

The coexistence of these two systems may create problems. Our hope is that future developments will clarify how much we can expect from this international trading regime in the countries' collective effort to mitigate greenhouse gas emissions. It seems crucial to recognize that international emissions trading is only the tip of the iceberg that this ambitious task represents: with the exception of industrial players that are equipped to respond to a market signal and to fully benefit from trading, other activities demand more complex policies. For a host of reasons, the negotiation of mitigation commitments has focused on quantitative targets and flexibility mechanisms. Whether commitments are based on per capita entitlements or historic responsibility (Brazilian proposal), they assume large financial transfers from developed to developing countries on the basis that the latter would undertake efficient mitigation policies, and, again, that a unique carbon price would guide such policies. Yet the analysis in this chapter suggests there are good reasons for questioning the centrality of trading in the future design of the climate regime, even if it is a welcome instrument to facilitate compliance. This should be a useful lesson for developing countries that are starting to think about deepening their future participation in the climate regime.

Notes

1. See OECD (1997) for an early presentation of how emissions trading may be applied to control greenhouse gas emissions.
2. See IEA (2001) for a review of simulations of international GHG emissions trading involving both industry and governments. See also Baron et al (2002).
3. See, for instance, Hargrave (2000).
4. See Harrison and Radov (2002) for a brief discussion of this point.
5. A carbon or GHG tax could play that role. But this would mean that its level would be adjusted according to the international price of carbon, which is not likely to deliver the policy certainty that activities need to make a proper investment in less GHG-intensive technology. Further, market experiments revealed how policy inertia makes it difficult for governments to optimize their compliance strategy from a near-term cost perspective. See, for instance, IEA (2001), Baron et al (2002).
6. We already mentioned the potential negative economic impacts of adjusting fossil fuel taxes to reflect the carbon market price (Babiker et al, 2002).
7. Net transfers are the difference between a country's original assigned amount and its final holding of units when compliance is assessed. This is different from total transactions which represent the sum of all transfers and acquisitions that may take place during the commitment period (more often referred to as the carbon market size), which is much more difficult to forecast.
8. The EU ETS directive allows for linking with similar domestic systems in other Parties with Kyoto commitments. Even under these circumstances, entities would only be allowed to buy assigned amount units from other industrial entities with a commitment in the Russian system, and not from the government, to avoid massive inflow of hot air into the European regime.
9. The existence of a barrier between the two regimes is also recognized by GHG emission brokers, a sector that is generally known for creating bridges across markets that regulators had not foreseen (Von Butler et al, 2003).
10. See Part III, Chapter 2 by Odile Blanchard in this volume for estimates of the effect of the USA's withdrawal on international carbon prices; see also IEA (2001).
11. The trading system for NO_x control in California was brought to a halt when prices peaked. This led to the temporary exclusion of power generators from the system. See SCAQMD (2001) for a description of the NO_x price problem and solutions that were discussed by regulators.
12. A market experiment gathering industry and governments under a scenario of the Protocol's entry into force without the USA showed how Russia could have full control over the price of carbon (Baron et al, 2002).

References

Babiker, M. H., Reilly, J. M. and Viguier, L. L. (2002) *Is International Emissions Trading Always Beneficial?*, Report No 93, December, Massachusetts Institute of Technology, Global Change Program, available at:
http://web.mit.edu/globalchange/www/MITJPSPGC_Rpt93.pdf.

Baron, R., Boemare, C. and Jakobsen, A. (2002) *Trading CO$_2$ and Electricity in the Baltic Sea Region – Report on the Simulation of the Baltic Sea Region Energy Co-operation*, International Energy Agency, Paris, October.

Ellerman, D., Schmalensee, R., Joskow, P. L., Montero, J. P. and Bailley, E. M. (1997): *Emission Trading under the U.S. Acid Rain Program – Evaluation of Compliance Costs and Allowance Market Performance*, MIT Center for Energy and Environmental Policy Research, Cambridge, MA.

Hargrave, T. (2000) *An Upstream/Downstream Hybrid Approach to Greenhouse Gas Emissions Trading*, Center for Clean Air Policy, June, available at: http://www.ccap.org/pdf/Hybrid1.pdf.

Harrison, D. and Radov, D. B., (2002) *Evaluation of Alternative Initial Allocation Mechanisms in a European Union Greenhouse Gas Emissions Allowance Trading Scheme*, Report by National Economic Research Associates prepared for DG Environment, European Commission, March.

IEA (2001) *International Emission Trading: From Concept to Reality*, International Energy Agency, OECD, Paris.

IEA (2003) *CO$_2$ Emissions from Fossil Fuel Combustion, 1970–2001*, International Energy Agency, OECD, Paris.

Natsource (2003) *Governments as participants in international markets for greenhouse gas commodities*, Report for IEA, IETA, EPRI and IDDRI, September, available at http://www.iddri.org/iddri/telecharge/climat/rapport_natsource.pdf.

OECD (1997) *International Greenhouse Gas Emission Trading*, Annex I Expert Group on the UNFCCC, Working Paper No 9, OECD/GD(97)76.

SCAQMD (2001) *White Paper on Stabilization of NO$_x$ RTC Prices*, South Coast Air Quality Management District, 11 January 2001.

Von Butler, B., Pavlovic, I. and Reamer, J. (2003) *Kyoto Protocol and European GHG Trading: Linking the Two Systems*, Evolution Markets Executive Brief, 25 July edition.

Weyant, J. (ed.) (1999) 'The costs of the Kyoto Protocol: a multi-model evaluation'. Special Issue of the *Energy Journal*.

Chapter 2

Trading through the flexibility mechanisms: quantifying the size of the Kyoto markets

Odile Blanchard

Since the adoption of the Kyoto Protocol in 1997, the flexibility mechanisms have moved from formal definitions and scope to practical tools, ready for implementation, with a number of initiatives giving effect to them in advance of the entry into force of the Kyoto Protocol.[1] The focus on implementation has been underpinned by the completion of the detailed rules on JI, CDM and IET adopted by COP-7 in the Bonn Agreement and the Marrakesh Accords, including the rules relating to carbon sinks through land use, land-use change and forestry (LULUCF) activities under Articles 3.3 and 3.4.[2] Rules and modalities for CDM sinks projects were agreed by COP-9 held in Milan in December 2003. All these additional rules have a significant bearing on the degree of reliance Parties might place on the Kyoto mechanisms, the carbon price and hence the amount of trading that might now be expected compared with expectations about the size of the carbon market in 1997.

Potential participants in the carbon markets established by the Kyoto Protocol have also changed over the years. Initially, when the Kyoto Protocol was adopted, all Annex I countries with targets listed in Annex B of the Protocol could be considered candidates to trade part of their assigned amount units and emission reduction units from JI projects; similarly all non-Annex B countries could be considered candidates to host CDM projects and sell certified emission reductions. The rejection of the Kyoto Protocol by the United States in early 2001 followed by Australia's similar decision in June 2002 put those two potential participants out of the trade game. At the time of writing (March 2004) Russia has yet to make a clear decision about whether it intends to ratify the Kyoto Protocol. On the other hand, as discussed elsewhere in this book, the European Union (EU) is moving forward to implement an emissions trading system among the member countries from 1 January 2005 onwards and a number of other Kyoto-consistent schemes are being put in place in other jurisdictions.

The rules agreed by the COPs concerning the flexibility mechanisms and carbon sinks and the changing number of participants in the trading system impact and potentially change the amount of trading and the permit price compared with the agreement reached at Kyoto in 1997. The objective of this chapter is to analyse the impacts on the carbon market through five cases or scenarios that reflect the chronological landmarks in the negotiation process and the building of the Kyoto Protocol flexibility mechanisms.

Section 2.1 addresses methodological issues and presents the general assumptions adopted. Section 2.2 displays the differentiating characteristics of the carbon markets through five scenarios. Section 2.3 scrutinizes the trade features of the carbon markets under the five cases in more detail. The modelling results outlined in this chapter suggest that if countries that have ratified the Kyoto Protocol decide to give legal effect in the face of non-ratification by the United States of America (USA) and Russia, the compliance costs of these pro-Kyoto countries do not fundamentally differ from those estimated assuming participation by the USA and Russia. Additionally, because of the presence of hot air, exclusion of the USA and Russia from the picture means that from an environmental perspective, a decision by pro-Kyoto countries to proceed without these two countries still generates a significant degree of domestic and overseas action to reduce GHGs, boosting carbon markets and still providing many of the environmental gains originally agreed at Kyoto.

2.1 Methodology and assumptions

The results presented in the following sections are based on the POLES[3] model and the ASPEN[4] software, both developed at LEPII-EPE.[5] POLES is a world simulation model for the energy sector (European Commission, 1996). It works in a year-by-year recursive simulation and partial equilibrium framework, with endogenous international energy prices and lagged adjustments of supply and demand by world region. GDP and population are the main exogenous variables. The world is divided into 38 countries or regions. For each country or region, the model articulates the following modules: final energy demand by key sectors, new and renewable energy technologies, conventional energy and electricity transformation system, fossil fuel supply. The main outputs comprise detailed world energy outlooks to 2030 (demand, supply and price projections by global region), marginal greenhouse gas emission abatement costs by country or region, and technology improvement scenarios.

The ASPEN software input data comprise the marginal abatement costs (MACs) assessed by the POLES model for the various countries/regions. The MACs allow simulation of the emission permit supply and demand for any specific market size in 2010. A market equilibrium price (the emission permit price) may thereby emerge. As with other models trade is assumed to perform on a perfectly competitive market.[6]

For the five cases analysed below, 2010 is chosen as the representative year of the first commitment period of the Kyoto Protocol. The emission reduction objectives of Annex I Parties are those set in Annex B of the Kyoto Protocol. The targets of the EU Member States are those set out in the EU Burden-Sharing Agreement.[7] These targets can be achieved through domestic reductions as well as through the three flexibility mechanisms adopted in the Protocol: IET, JI and the CDM. As stipulated in the Marrakesh Accords, the units generated by the three mechanisms are totally fungible: whatever their origin, the units are equal to each other for compliance purposes. Annex B countries import units of emission reduction through the flexibility mechanisms when the marginal cost of their domestic reductions is greater than the international emission permit price (Criqui et al, 1999; Energy Journal, 1999). Table 2.1 presents the marginal abatement cost associated with the Kyoto target for a few countries.

Table 2.1 *Marginal abatement costs of a few countries*

Countries	Kyoto Protocol target relative to base year	Marginal abatement cost to meet target (1995 $/t CO_2e)
USA	−7 %	30
France	0 %	38
Germany	−21 %	11
United Kingdom	−12.5 %	7
Australia + New Zealand	+8 %	16
Japan	−6 %	32
Former Soviet Union	0 %	0
Poland + Hungary	−6 %	0

Notes: The targets of France, Germany and the United Kingdom are those of the burden-sharing agreement (Council Decision 2002/358/EC of 25 April 2002); the base year is 1990, except for Poland (1988) and Hungary (1985–1987).
Sources: UNFCCC for targets and base years, POLES model for MACs.

The marginal abatement costs of the Former Soviet Union (FSU),[8] Poland and Hungary, and more generally of all the Eastern European Economies (EEE)[9] are nil because these regions meet their Kyoto targets without any mitigation action, through what is commonly called 'hot air'. 'Hot air' refers to the excess emission allowances that the FSU and EEE are entitled to in the Kyoto Protocol. Due to the severe economic slump that these countries experienced in the 1990s, their 'business as usual' emission projections in 2010 are around 30 per cent below their Kyoto emission targets according to POLES.

The five cases are based on a set of common assumptions. An accessibility factor is assigned to JI and CDM projects, meaning that only a proportion of the overall economically efficient emission reductions through JI and CDM projects is considered feasible. The accessibility factor reflects the possible institutional pitfalls or technical difficulties in the identification, definition and implementation of the projects (lack of infrastructures or expertise for instance), as well as the difficulty of estimating the project baselines. A 50 per cent accessibility factor is assumed for JI projects in the former Soviet Union, 100 per cent in the other Annex B countries. CDM projects are assigned a 10 per cent accessibility factor in the non-Annex B countries.

2.2 Characteristics of the five cases

The five cases differ in terms of the participating countries or the amount of sink credits that countries are entitled to use towards compliance with their Kyoto commitments. Table 2.2 presents the assumptions associated with each case. The first case, namely the Kyoto Protocol Initial Deal (ID) reflects the main elements that were agreed upon in Kyoto in 1997 at a time when all Annex B parties agreed to take on an emission reduction target. In this scenario, all Annex B parties participate in emissions trading, as all of them initially adopted the Kyoto Protocol. Similarly, all non-Annex B parties are potential hosts for CDM projects. The ID scenario does not allow any sinks to be counted towards compliance in the emission reductions agreed as part of the Protocol. This is because although the Kyoto Protocol included provision for sinks in the articles related to LULUCF, the Protocol did not include actual figures for sinks credits nor generate legal certainty that, in fact, these would be counted towards Article 3 commitments as the accounting method was left to future consideration by the Conferences of the Parties. Excluding sinks from the ID case aims at showing what the negotiations following those in Kyoto brought in economic terms, when sinks allow the different Parties to lower their emission reductions.

The second case, called the Hague Missed Compromise (MC), is based on hypothetical participating countries and decisions, should COP-6 have resulted in an agreement. All parties are assumed to be potential participants in emissions trading, JI or CDM projects (depending on their status as Annex B or non-Annex B countries). The emission credits for sinks are set to 3 per cent of the base-year emissions, referring to what had been proposed in the Hague that could be added to the assigned amount units of the Annex B parties through LULUCF projects.[10]

Table 2.2 *Assumptions of the five cases*

Case	Trading Participants	Accessibility of flexibility mechanisms	Sinks credits
The Kyoto Protocol Initial Deal (ID)	All Annex B (emissions trading and JI projects) All Non-Annex B (CDM projects)	50% of JI projects in the Former Soviet Union 100% of JI projects in other Annex B countries 10% of CDM projects	No
The Hague Missed Compromise (MC)	Same as ID case	Same as ID case	3% of base-year emissions
The Bonn-Marrakesh Accords (BM)	Same as ID case except US	Same as ID case	As defined in the Bonn-Marrakesh Accords
Trading among Committed Countries (CC)	Same as ID case except US, Russia, Australia	Same as ID case	Same as BM case
Trading restricted to Europe (EU)	European countries	Same as ID case	Same as BM case

The third case (BM) sketches the Bonn-Marrakesh Accords. The USA is excluded from the trading system as it had officially rejected the Kyoto Protocol by this time. This case incorporates the modalities related to sinks agreed upon in the Bonn-Marrakesh Accords. Sinks credits are the sum of the following items:

- the maximum accountable credits for forest management under Article 3.4 inscribed in Appendix Z of the Bonn Agreement and modified in the Marrakesh Accords;
- the credits for afforestation, reforestation and deforestation activities under Article 3.3;
- the credits for agriculture management under Article 3.4;
- the maximum importable credits for sinks through CDM projects related to afforestation, reforestation and deforestation (1 per cent of the importing party's base-year emissions).

The latter credits are not accounted for in FSU and EEE. These two regions are assumed not to import sinks credits from non-Annex B countries, as they domestically have excess allowances through hot air relative to their emission reduction targets.

The fourth case (CC) builds on the BM case for the sinks credits but reflects the current political situation in terms of ratifying countries: the USA and Australia have formally rejected the Kyoto Protocol and Russia has not ratified it. The CC case assumes, however, that countries that have ratified the Protocol find ways to give legal effect to Kyoto. This coalition of the ratifying countries, called the Committed Countries (CC), is assumed to continue to engage in IET, JI and CDM projects among each other.

Finally, in the last scenario, called the EU case, trading is restricted to Europe to illustrate the European Union's current initiative of implementing its own emissions trading system (EU ETS). The countries comprise all the European states broken down into Western Europe and Eastern Europe. The POLES model currently does not allow for a complete breakdown of the 25 Member States of the European Union.[11]

2.3 Analysis of the market features of the five cases

Depending on the assumptions adopted, the results differ in terms of internationally traded amounts and emission permit price. This section examines and compares the results of the five cases described above. More emphasis is put on relative figures and changing features from one case to the other than on absolute figures.

In the Initial Deal case (Table 2.3), all countries may trade and no sinks credits are added to the assigned amounts of the countries. Due to the levels of the countries' respective marginal costs, 58 per cent of the emission reductions required are achieved through trading. The international permit price is around US$8 per tonne of CO_2 equivalent.[12] The purchasing regions are the USA, Western Europe (except the UK[13]), Canada-Australia-New Zealand and Japan. They buy credits as soon as the abatement cost of their domestic reductions is higher than the international permit price. The Annex B Parties of FSU and EEE represent the bulk of the selling regions. They sell almost 90 per cent of the traded volume, both through 'hot air' and reductions achieved domestically (either JI projects or domestic measures). Non-Annex B countries sell the remaining credits through CDM projects. All the sellers trade Kyoto units as long as the marginal abatement cost of their reductions is lower than the international permit price.

The ID case highlights the prevailing weight of the USA on the permit demand side (almost two-thirds of the traded volume) and FSU on the supply side (over 70 per cent). It also shows the complementarity of both regions on the permit market. The USA roughly imports as many emission reduction units as the FSU sells hot air permits. Both regions have an interest in trading with each other. These features may justify the stance put by the USA on the flexibility mechanisms in 1997 when negotiating the Kyoto Protocol and the USA's then leniency towards Russia's unambitious emission limitation target, which was tacitly accepted by other Parties as the price of having the USA on board the Protocol.

Table 2.3 Trade characteristics of the Initial Deal

Countries	Permit price at equilibrium ($/t CO2e)	Total abatement cost (M95$)	Purchasers (Mt CO$_2$e)				Sellers (Mt CO$_2$e) Exports			
	8		Required emissions reductions =	Dom. red. to reach target	+ Imports	+ Sinks	CDM	Dom. red. for IET/JI	Traded hot air	Sinks in EEE and FSU
USA		11644	1885	831	1054	–	–	–	–	–
Western Europe		4051	624	227	397	–	–	5	–	–
Canada-Austr.-New-Zeal.		1584	258	118	140	–	–	–	–	–
Japan		1000	159	66	93	–	–	–	–	–
Former Soviet Union		–9353	–	–	–	–	–	200	1029	–
Eastern European Economies		–1813	–	–	–	–	–	89	172	–
Total Annex B		**7111**	**2925**	**1241**	**1684**			**294**	**1201**	
Non-Annex B countries							190			

Note: Totals may not add up due to rounding.
Source: POLES and ASPEN.

Figure 2.1 illustrates the results of Table 2.3. The horizontal axis displays the supply of Kyoto units while the vertical axis shows the permit price. The starting point of the supply curves corresponds to FSU's and EEE's hot air: up to 1201 Mt CO_2e, the FSU and EEE can provide emission reductions at a nil marginal cost. The origin of the demand curve reflects the 2925 Mt CO_2e total reductions required in Annex B countries. The balance between supply and demand is reached when 1684 Mt CO_2e emission credits are traded at a price of \$8/t CO_2e. The gap between the 1684 Mt CO_2 permits traded and the total 2925 Mt CO_2 reductions required represents the 1241 Mt CO_2 domestic reductions achieved by Annex B countries.

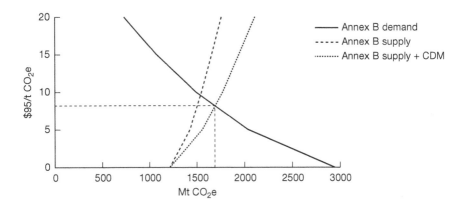

Figure 2.1 *The permit market equilibrium in the Initial Deal.*

Source: POLES and ASPEN.

In the Missed Compromise case (Table 2.4), all countries may trade, and sinks are credited for 3 per cent of the base-year emissions to Annex B countries. Accounting for sinks is tantamount to lowering the other sources of emission reductions to meet the target. Those sinks are assumed costless as Annex B countries are considered to have carried out the eligible-for-credits LULUCF activities on a business-as-usual basis since 1990. Therefore Annex B importing countries are supposed to take advantage of their sink allowances to reach their Kyoto target before turning to domestic reductions or to international carbon markets.

Table 2.4 Trade characteristics of the Missed Compromise

Permit price at equilibrium ($/t CO_2e): 5

Countries	Purchasers (Mt CO_2e)					Sellers (Mt CO_2e) Exports			
	Total abatement cost (M95$)	Required emissions reductions =	Dom. red. to reach target	Imports +	Sinks	CDM	Dom. red. for IET/JI	Traded hot air	Sinks in EEE and FSU
USA	6728	1885	568	1136	182	–	–	–	–
Western Europe	1964	624	133	360	131	–	24	–	–
Canada-Austr.-New-Zeal.	873	258	78	145	35	–	–	–	–
Japan	471	159	45	77	37	–	–	–	–
Former Soviet Union	-5802	–	–	–	–	–	145	1029	123
Eastern European Economies	-1129	–	–	–	–	–	62	172	35
Total Annex B	**3105**	**2925**	**824**	**1717**	**384**	–	**232**	**1201**	**158**
Non-Annex B countries						127			

Note: Totals may not add up due to rounding.
Source: POLES and ASPEN.

Compared to the ID case, the FSU again stands out as the USA's complementary trading partner. But the MC case shows a sharp decline in the permit price to $5/t CO_2e for a similar traded volume: the inclusion of costless sinks lowers the domestic reductions needed to reach the target. Consequently the total abatement costs incurred by Annex B importing countries sharply decline relative to the IC case, as do the revenues for exporting FSU and EEE. The sources of exports differ in some respect. As sinks allocated to the FSU and EEE raise their already beyond-the-target emission permits, sinks are traded along with hot air at nil cost. CDM projects therefore become less competitive and CERs from CDM projects decrease. Had a compromise been reached in The Hague, the sinks credits would have eroded the environmental impact of the initial deal while simultaneously considerably reducing the abatement cost of Annex B importing countries.

The BM case illustrates the consequences of the USA's rejection of the Kyoto Protocol on the emission permit market but includes Russia (Table 2.5 and Figure 2.2). The sinks credits, accounted for as indicated in the Bonn-Marrakesh Accords, differ from those included in the MC case. The USA's withdrawal from the Kyoto negotiation process implies that the supply of credits outweighs the demand for them. The international permit price is therefore nil. On the supply side, the hot air and sinks excess allowances from the FSU and EEE amount to 1342 Mt CO_2e. On the demand side, Annex B importing countries (excluding the USA) must overall cut their emissions by 1041 Mt CO_2e to meet their targets. After subtraction of the sinks credits to which the importing countries are entitled, their remaining required emission reductions amount to 795 Mt CO_2e. The demand for permits is lower than the supply by 547 Mt CO_2e. It is worth noting that excluding the sinks from the analysis leads to the same conclusion. The sinks do not play a major role in this context.

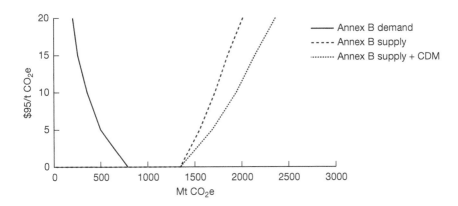

Figure 2.2 *Permit supply and demand in the Bonn-Marrakesh case.*

Source: POLES and ASPEN.

Table 2.5 *Supply and demand characteristics in the Bonn-Marrakesh case*

Countries	Emission reduction target 2010 (Mt CO_2e)	Sinks (Mt CO_2e)	Remaining required emission reductions	Hot air (Mt CO_2e)	Surplus (Hot air + Sinks) (Mt CO_2e)
Western Europe	624	75	549	–	–
Canada-Austr.-New-Zeal.	258	110	148	–	–
Japan	159	60	99	–	–
Former Soviet Union	–	127	–	1029	1156
Eastern European Economies	–	14	–	172	186
Total Annex B	1041	386	795	1201	1342

Note: Totals may not add up due to rounding.
Source: POLES and ASPEN.

The BM case shows that Annex B countries apart from the USA can globally reach and do better than their Kyoto target without any mitigation action. The emission reductions stemming from the economic slump of the FSU and EEE more than offset the projected increased emissions of the other Annex B countries in 2010. In this situation, the model assumes that Annex B countries do not use the CDM projects to meet their targets and thus non-Annex B countries are left aside of mitigation actions. As Kopp puts it: 'A Kyoto Protocol without the US is like musical chairs with one too many chair – there's a lot of marching around but nothing happens' (Kopp, 2001).

In the model, the USA's withdrawal from the Kyoto process has a negative impact on the FSU and EEE as the lack of demand for their excess allowances implies a collapse of the carbon price. In the real world, one way for the FSU and EEE out of this situation is to use their market power on the other Annex B countries by restricting access to their emission surpluses (Manne et al, 2001). If the FSU and EEE can decide to sell only part of their excess allowances to the other Annex B countries in the first commitment period, they will be able to bank the remaining credits for the subsequent commitment periods. The quantity of credits that they will decide to bank will be determined so as to maximize their benefits (Blanchard et al, 2002). According to POLES, when the FSU and EEE exert market power, their benefits are highest if only 15 per cent of their hot air is traded in the first commitment period and 85 per cent is banked for the next periods. The corresponding permit price comes out at \$4/t CO_2e in 2010. When the share of hot air traded by the FSU and EEE increases, their benefits decrease from that point. They are nil when over 55 per cent of their hot air is traded and 45 per cent banked.

If Russia does not ratify the Kyoto Protocol, the BM case is assumed to be out of reach of the international community. Exploring the CC case could therefore be a more accurate representation of present-day realities. The CC case (Table 2.6) builds on the assumption that an agreement among the ratifying countries is reached outside the Kyoto Protocol, allowing emissions trading, JI and CDM projects to proceed among countries that have ratified the Protocol.[14] As the biggest potential traders (the USA and Russia) are outside the market, the traded volume is inevitably far lower than in the ID and MC cases. The remaining countries of the FSU (mainly the Ukraine) and EEE are the major suppliers and bring credits to the market mainly through their hot air.

For the ratifying countries, the total compliance costs of proceeding with Kyoto are lower than in the MC scenario. This means that should committed countries decide that they want to put the Protocol into legal effect without the USA and Russia, they would face compliance costs that would be lower than if the Protocol entered into force with the USA and Russia. More significantly, the CC scenario is environmentally superior to the BM scenario: there is more focus on real GHG reductions, domestically in the EU and in developing countries through the CDM. This is because Committed Countries engage more in real reductions than playing musical chairs with paper trades disconnected from environmental benefits.

Assuming that the remaining FSU-Annex B countries and EEE exert market power in the CC case illustrates what may actually happen if Russia stays out of the Kyoto Protocol. The results do not differ much from those shown in Table 2.6. The EEE and the remaining FSU-Annex B countries would maximize their benefits when trading 55 per cent of their hot air. The permit price would rise to \$6/t CO_2e, therefore allowing more CDM projects and domestic reductions to be achieved.

Finally, the EU leadership case (Table 2.7) highlights the potential volume that can be traded among the European countries. The EU case assumes that all sectors can trade and that emission credits can be acquired from JI and CDM projects.[15] It does not relate to the modalities of the EU ETS directive as not all Member States have defined their national allocation plan yet, nor clearly defined additional policies and measures for the non-trading sectors and corporations to reach the Kyoto reduction targets.

Almost two-thirds of the required reductions stem from trade. Hot air from the Eastern European countries is sold to the Western states and outweighs CDM projects in non-Annex B countries. The permit price (\$5/t CO_2e) is higher than in the CC case because the zero-cost credits from the EEE's hot air represent a lower share of the credits supply.

Table 2.6 Trade characteristics of the Committed Countries case

Permit price at equilibrium ($/t CO$_2$e) 4		Purchasers (Mt CO$_2$e)				Sellers (Mt CO$_2$e) Exports			
Countries	Total abatement cost (M95$)	Required emissions reductions =	Dom. red. to reach target	+ Imports	+ Sinks	CDM	Dom. red. for IET/JI	Traded hot air	Sinks in EEE and FSU
USA	0	–	–	–	–	–	–	–	–
Western Europe	1768	624	120	429	75	–	–	–	–
Canada-New-Zeal.	300	167	34	66	67	–	–	–	–
Japan	295	159	34	65	60	–	–	–	–
Former Soviet Union (excl. Russia)	–859	–	–	–	0	–	31	216	6
Eastern European Economies	–714	–	–	–	–	–	24	172	14
Total Annex B	790	949	188	559	202	–	54	388	20
Non-Annex B countries						97			

Note: Totals may not add up due to rounding.
Source: POLES and ASPEN.

Table 2.7 *Trade characteristics of the European Union leadership case*

Permit price at equilibrium ($/t CO$_2$e) 5	Purchasers (Mt CO$_2$e)					Sellers (Mt CO$_2$e) Exports			
Countries	Total abatement cost (M95$)	Required emissions reductions =	Dom. red. to reach target	+ Imports	+ Sinks	CDM	Dom. red. for IET/JI	Traded hot air	Sinks in EEE and FSU
USA	–	–	–	–	–	–	–	–	–
Western Europe	2314	624	162	387	75	–	3	–	–
Canada-Austr.-New-Zeal.	–	–	–	–	–	–	–	–	–
Japan	–	–	–	–	–	–	–	–	–
Former Soviet Union	–	–	–	–	–	–	–	–	–
Eastern European Economies	–1082	–	–	–	–	–	65	172	14
Total Annex B	**1232**	**624**	**162**	**387**	**75**	–	**68**	**172**	**14**
Non-Annex B countries						133			

Note: Totals may not add up due to rounding.
Source: POLES and ASPEN.

2.4 Conclusion

Referring to the chronological landmarks of the climate negotiations, this chapter has shown that the size of the international emissions permit market and the permit price critically depend on who participates and what units are tradable.

The traded *volume* is highest when all Parties participate either in emissions trading, JI or CDM projects (Initial Deal of the Protocol). It is smallest (but positive) when trade is restricted to the European countries. Worldwide the USA and the former Soviet Union are the biggest players and act as complements. In the initial deal of the Kyoto Protocol, according to the POLES model, the USA represents almost two-thirds of the demand for permits and the former Soviet Union 70 per cent of the total permit supply (mainly through its hot air). Their impact on the market is significant. When the USA withdraws from the Kyoto Protocol process and the other Parties remain in the game, the permit supply of the former Soviet Union outweighs the overall demand on the market. In this case, the Kyoto target is globally met by the Annex B countries (apart from the USA) without any mitigation action, and trade does not take place (unless the former Soviet Union takes proactive action to control the market). The trade benefits that the former Soviet Union would have gained if the USA had stayed in the Kyoto Protocol vanish with the USA's withdrawal. Without obliterating numerous other – more political – considerations, this fact may contribute to explain Russia's lengthy deliberations about whether to ratify the Kyoto Protocol without some assurance that they will, in fact, be able to control the carbon market on terms favourable to their long-term interests.

Limiting trade to the countries that have thus far committed themselves through ratification of the Kyoto Protocol or to the European countries revives the carbon market, as both the USA and Russia are out of the game. The Eastern European countries become the major suppliers of credits, mainly through their hot air.

Hot air clearly impacts very much on the market features. Considering the potential amounts assessed in the POLES model, hot air offers wide opportunities to reduce emissions at no cost. It thus contributes to a low (even nil) permit price. Accounting for sinks also lowers the permit price, as sinks lower the overall required emissions reductions. In this context, credits from CDM projects in non-Annex B countries are hardly competitive. They therefore cover only a small share of the traded volume.

The results presented are inevitably sensitive to the model structure and the assumptions adopted. They particularly rely on the marginal cost calculations and the business-as-usual emission projections. As already underscored, relative figures are more important to analyse than absolute figures. The results relative to the respective roles of the USA and the former Soviet Union in shaping the market are in accordance with other studies (Manne et al, 2001; Buchner et al, 2001; Babiker et al, 2002; den Elzen et al, 2002; Berk and den Elzen, 2004).

Currently, though, the USA and Russia are out of the trading game, and the European Union is about to implement its own emissions trading system. The environmental impact of EU leadership is important politically in keeping the carbon markets 'alive' but in terms of emission reductions the market size is significantly smaller than in the initial deal of the Kyoto Protocol. The interaction between the carbon market established by the EU ETS and trading through IET, JI and CDM will be affected by many other factors beyond the scope of this paper. A key issue is the allocation of allowances by the EU member states as this impacts how much international demand is created for the Kyoto mechanisms.[16] This will become clearer once Member States complete their national allocation plans, which will in future focus on exploring how allocation and other features of the European Union market affect the development of the Kyoto mechanisms worldwide.

Notes

1. See the Introduction by Farhana Yamin, the contributions by Lefevere (Part II) and Wilder (Part III, Chapter 5), and the Conclusion by Erik Haites (Part IV) in this volume.
2. See Part I by Farhana Yamin.
3. POLES stands for Prospective Outlook on Long-term Energy Systems.
4. ASPEN stands for Analyse des Systèmes de Permis d'Emission Négociables, i.e. analysis of tradable emission permit systems.
5. LEPII-EPE (formally IEPE) stands for Laboratoire d'Economie de la Production et de l'Intégration Internationale – Département Energie et Politiques de l'Environnement. It is based in Grenoble, France.
6. A perfectly competitive market is the usual assumption with participation by both the USA and Russia. It should be noted, however, that with hypotheses of Russian ratification and the USA out, monopolistic behaviour by Russia (or, depending on the models, by the former Soviet Union) is a common assumption.
7. See Part II by Jürgen Lefevere for explanation of the EU Burden-Sharing Agreement and the EU's joint fulfilment agreement under Article 4 of the Protocol.
8. The FSU region of the POLES model includes Russia, the Ukraine, Estonia, Lithuania and Latvia.
9. The EEE region of the POLES model includes Bulgaria, the Czech Republic, Hungary, Poland, Romania, Slovakia and Slovenia.
10. See FCCC/CP/2000/5/Add.2, p13, online at: http://unfccc.org/resource/docs/cop6/05a02.pdf.
11. For example, the unusual situation of Malta and Cyprus is not specifically addressed in the model.
12. Prices and costs are expressed in 1995 US$ throughout the paper.
13. The UK is the European seller of the 5 Mt CO_2e credits in Table 2.3.

14. The POLES model considers the FSU-Annex B countries as one region. As POLES only provides results for the whole region, simplified assumptions are adopted in the CC case to single out non-ratifying Russia from the other ratifying FSU-Annex B countries (the Ukraine and the Baltic States). Based on IEA data (IEA, 2002), Russian emissions are estimated to represent respectively 74 per cent and 72 per cent of FSU-Annex B countries in 1990 and 2010. Russian marginal abatement costs in 2010 are therefore assumed to be 72 per cent of the FSU's.

15. Other scenarios relating to trading versus non-trading sectors and to linking the EU ETS directive to JI and CDM are explored in Criqui et al (2003).

16. Allocation issues under the EU ETS are examined in Part II by Lefevere and in the contribution by Mullins in Part III, Chapter 3, in this volume.

References

Babiker, M., Jacoby, H. D., Reilly, J. M. and Reiner, M. D. (2002) *The Evolution of a Climate Regime: Kyoto to Marrakech*, MIT Report No 82, Cambridge, MA, MIT.

Berk, M. and den Elzen, M. (2004) *What If the Russians Don't Ratify?*, RIVM Report 728001028/2004, Bilthoven.

Blanchard, O., Criqui, P. and Kitous, A. (2002) *After The Hague, Bonn and Marrakech: The Future International Market for Emissions Permits and the Issue of Hot Air*, Cahier de recherche de l'IEPE 27 bis, Grenoble, available at: www.upmf-grenoble.fr/iepe/textes/Cahier27Angl.pdf.

Buchner, B., Carraro, C. and Cersosimo, I. (2001) *On the Consequences of the US Withdrawal from the Kyoto/Bonn Protocol*, Fondazione Eni Enrico Mattei (FEEM), Report 102.2001.

Criqui, P., Mima, S. and Viguier, L. (1999) 'Marginal abatement costs of CO_2 emission reductions, geographical flexibility and concrete ceilings: an assessment using the POLES model', *Energy Policy*, vol 27, no 10, pp585–601.

Criqui, P. and Kitous, A. (2003) *Kyoto Protocol Implementation: Impacts of Linking JI and CDM Credits to the European Emission Allowance Trading System*, European Commission DG Environment, Service Contract B4-3040/2001/330760/MAR/E1.

Den Elzen, M., de Moor, A. P. G. (2002) 'Analyzing the Kyoto Protocol under the Marrakesh Accords: economic efficiency and environmental effectiveness', *Ecological Economics*, no 43, pp141–58.

Energy Journal Special Issue (1999) *The Costs of the Kyoto Protocol: A Multi-model Evaluation*.

European Commission (1996) *Poles 2.2*, Joule II Programme, DG XII Science, Research and Development, EUR 17358 EN, Brussels.

European Union (2003) *Directive 2003/87/EC of the European Parliament and of the Council establishing a scheme for greenhouse gas emission allowance trading within the Community and amending Council Directive 96/61/EC*, 13 October, Brussels.

International Energy Agency (IEA) (2002) *World Energy Outlook: 2002*, Organization for Economic Cooperation and Development, Paris.

Kopp, R. (2001) 'A climate accord without the US', *Weathervane*, 14 August, available at: www.weathervane.rff.org/features/feature135.htm.

Manne, A. S. and Richels, G. R. (2001) *US Rejection of the Kyoto Protocol: The Impact on Compliance Costs and CO_2 Emissions*, paper presented at the 6 August 2001 EMF Forum, September, available in *Energy Policy*, vol 32, no 4, pp447–54.

Chapter 3

Implementation challenges: insights from the European Union Emission Allowance Trading Scheme

Fiona Mullins

3.1 Introduction[1]

The EU Emission Allowance Trading Scheme provides the framework for the emissions trading scheme across the EU but leaves decisions on many important implementation details to individual Member States.[2] Part II of this book explains the nature of the EU ETS while this chapter focuses on implementation aspects. To implement the EU ETS, Member States have many components to decide on and activities to carry out, including:

- prepare a National Allocation Plan;
- identify, obtain data from and consult the installations that are covered by the scheme;
- decide on banking, and new entry and closure rules;
- prepare their permitting procedures;
- provide guidance on monitoring and verification; and
- prepare a national allowance registry.

The compressed timescales for the Directive's transposition into national law and its implementation to ensure that trading can commence by 1 January 2005 present a wide range of legal, administrative and technical challenges for Member

States, in particular, relating to the critical issue of allocation of allowances. This chapter discusses these challenges, focusing on how identification of installations, permitting procedures, allocation of allowances, treatment of new sources and sources which close, and registries will be handled under the EU ETS. These 'core' issues have been chosen because they are likely to be of interest to mechanisms experts worldwide. To provide concrete details, this chapter outlines the approaches to allocation within the EU by the UK and by Germany, who are the biggest greenhouse gas emitters in the EU and have the largest number of installations covered by the EU ETS. This chapter provides insights about the way in which institutions and procedures are developing under the National Allocations Plans (NAPs) required by the EU ETS which Member States had to provide to the EU Commission by 31 March 2004. Refinements to the NAPs will continue before the final allocations were decided by 1 October 2004 so this chapter provides a snapshot of a changing process at a particular point in time.

3.2 National Allocation Plans

Introduction

The allocation of allowances to EU ETS installations is a Member State responsibility. Timely completion of the National Allocation Plan is the most crucial and most time-consuming factor of the EU ETS for governments during the initial implementation timeframe prior to the 1 January 2005 start date of the scheme. Each country has to define a transparent and fair distribution of allowances in line with requirements set out in Annex III of the Directive (Annex III criteria). The European Commission provided guidance for Member States on how the Annex III criteria should be interpreted in preparing their NAPs. This guidance is in Appendix 4 to this book.[3]

NAPs were supposed to be submitted to the European Commission by 31 March 2004. The ten countries which are acceding to the EU on 1 May 2004 have to submit their NAP by the date of accession. Many countries have not met the 31 March deadline, including those, such as the UK, which had prepared draft NAPs for public consultation well before the deadline. The European Commission requires at least two months from the date of submission to approve the NAPs and can take three months if it chooses. As submission of some NAPs was delayed the time for Commission acceptance and final national decisions on allocation was less than it should have been. Once approved, the NAPs will form the basis of final allocation decisions which should be made in each country by 1 October 2004.

Overview of allocation methods

The Annex III criteria require Member States to take into account five criteria: their Kyoto commitments; assessments of likely emissions developments; the potential to reduce emissions; consistency with other legislation; and non-discrimination between companies or sectors. A number of other Annex III criteria are optional and focus on them depends on political priorities and industry profiles, such as treatment of new entrants and provision for early action and clean technology.

While they must try to meet the Annex III criteria and each country has different political difficulties to work around, the methods used to devise NAPs are, of necessity, strongly influenced by practical concerns. Governments are using the information that they have available, such as emissions information from other policies and emission projections. To the extent possible, given the short timeframe, governments have also requested (in some cases required by regulation) information such as historical emissions and output from their EU ETS installations and have gathered other relevant data such as sectoral economic forecasts.

Depending on the information that is available, governments have developed their NAPs in different ways using analysis at a number of different levels. Figure 3.1 illustrates four levels that are typically considered.

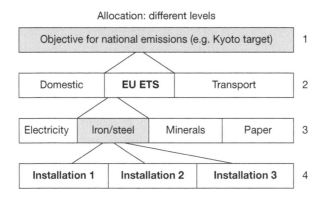

Figure 3.1 *Levels of analysis for National Allocation Plans.*

Source: Presentation to Royal Institute of International Affairs, Fiona Mullins, September 2003.

The first level requires government decisions on the overall objective for national emissions in 2005–2007 in order to be on track for Kyoto compliance in the first Kyoto commitment period 2008–2012. These decisions are influenced by the emission reductions required to meet national Kyoto targets domestically (requiring consideration of 2008–2012 allocation as well as allocation for 2005–2007), whether there is a tougher national objective that a country wishes to meet, whether the country decides to allow for purchase of Kyoto credits to meet its Kyoto target, and emission reductions that are expected from existing climate and energy policies.

The amount of effort that each country needs to make to meet its Kyoto target varies widely. Taking into account projected emissions trends and policies and measures, only three countries appear likely to meet their Kyoto targets through domestic efforts alone, the UK, Germany and Sweden.[4] Huge divergences in emissions trends between EU countries reflect different market structures and different patterns of economic growth. Some countries that face particular problems bringing their emissions into line with Kyoto obligations through domestic efforts have made explicit policy decisions to import Kyoto units. Some countries (such as the Netherlands) have firm policy positions that they will use the Kyoto mechanisms to achieve their Kyoto targets and others (such as Sweden) have in the past made policy statements that they will not.

The second level requires consideration of how much of the national total should be allocated to EU ETS installations in 2005–2007. The decision on what the total constraint should be across all EU ETS installations is a highly political decision in each Member State. The level 2 constraints that governments across the EU choose will define the total pool of emissions that is available to industry. These decisions also define the environmental effectiveness of the EU ETS, the cost of the EU ETS to industry and associated impact on competitiveness, and the cost to industrial and domestic consumers of electricity and energy-intensive products. The impact of the decision on total allocation to all EU ETS installations in a country is therefore the most important and the most controversial part of the allocation process.

Many countries have used their existing energy and CO_2 emission projections to assess the total quantity of emissions that the EU ETS trading sectors and the non-trading sectors such as transport should be allowed to emit. The projections typically take into account climate and energy policies, energy prices and GDP forecasts and so provide a 'business-as-usual' scenario of emissions without the EU ETS but in some countries including the impacts of existing policies and measures. These emission scenarios can then form an authoritative (but not perfect) basis for deciding the extent to which emissions from the EU ETS installations as a whole should be reduced from business-as-usual levels. Some countries are also preparing bottom-up estimates of future emissions from the EU ETS sectors in order to verify the top-down information from model projections.

The third level of analysis requires an assessment of the appropriate share of the EU ETS total for each sector or type of activity. However, countries such as Germany and the Netherlands have not found it necessary to set allocations for various EU ETS sectors. Instead they go straight from the total EU ETS allocation (level 2) to allocation to individual installations (level 4). For other countries, such as the UK and France, sector emission constraints are a key focus for consideration of issues relating to equity and equivalence of effort for EU ETS installations. Projected energy and emission trends and unbiased information on abatement options tends to be better at the sectoral level, which can make assessment easier than with installation-level data. However, energy and CO_2 projections that have been prepared for other purposes than the EU ETS do not tend to match the EU ETS activities precisely. Some countries are able to disaggregate certain EU ETS sectors, such as electricity, from model projections but in many cases the sectoral level data that is available to governments from models or existing policies does not precisely match the emissions from the sectors or activities that are covered by the EU ETS. Because of this, bottom-up data from installations has to be compiled and then reconciled with the sectoral data to the extent possible.

The fourth and final step is to determine the share of the total that each installation will receive. Methodologies for allocation to individual installations determine the share of the total or sectoral constraint that each one gets, not the absolute amount. This is why the total and sectoral constraints have been the main focus of industry and political concern. The typical approach for allocation to individual installations is to use a transparent methodology using available data for the allocation but then adjusting it so that all of the allocations together will equal the total constraint. Many countries are using historic data as the basis for allocation. Fairly recent base-years are being used, from 1998 to 2003, because it is difficult to obtain accurate and consistent emissions data for earlier base-years. Flexibility over the base-years typically is allowed, such as excluding one of the base-years if it was abnormal due to maintenance or low demand or adjusting the numbers to reflect a normal level of business activity.

Once the general method is decided, governments can then add on various other components such as allowing for early action, using benchmarks to reward good performance, or incentives for clean technology. Adjustments can be made for performance of installations against committed voluntary action or the best available technique (BAT), and to take into account projected output in the 2005–2007 period.

These four steps are being used to varying degrees in different Member States for deciding the amount of allowances to be allocated on a grandfathering basis (i.e. for free) to installations.

Member States can decide to auction up to 5 per cent of their allocation in the first EU ETS period and up to 10 per cent in the second. For example, Denmark has opted to auction 5 per cent of its allowances in the first period. Because auctioning is only possible for a small fraction of the total allocation, the grandfathered component will have a far greater influence and is consequently receiving the most attention.

The following sections examine in greater detail how two EU Member States, Germany and the UK, are handling the EU ETS allocation process.

Example: Germany[5]

Germany is the largest CO_2 emitter in the EU so its allocation decision will have a major impact on the EU ETS market. The German government has decided to use its Kyoto target (21 per cent reductions from the 1990 level by 2010 for all six Kyoto gases) as the basis for allocation rather than its more stringent national aim (25 per cent reduction in CO_2 from the 1990 level by 2005).[6] Approximately 2400 German installations will be included in the EU ETS (based on a narrow definition of combustion installations).

There was a contentious debate in Germany over the total allocation to all EU ETS installations. Germany's national emissions have fallen as a result of closure and of efficiency improvements in east German industry. Projections suggest that Germany could achieve a reduction of 32 per cent from 1990 levels by 2010 with existing measures.[7] By the base-period 2000–2003 that the German government is using for its allocation, national CO_2 emissions had fallen by 152 Mt CO_2 or 14.9 per cent from 1990 levels, with the biggest contribution to that reduction from the energy sector and industry. Some stakeholders considered that no further burden should be placed on industry given the reductions already achieved. Others considered that additional reductions of 25 Mt CO_2 from industry per year would be required to reach Germany's national target.

The consensus that emerged was to base the total allocation on the voluntary agreements which form the backbone of Germany's climate change strategy. The voluntary agreements aim for a 45 million tonne reduction from 1998 by 2010. Analysis indicates that the voluntary agreements are consistent with business-as-usual emission trends for 2008–2012 and that it would be economically efficient to reduce industry emissions by far more than this.[8] However, interpretations of the reductions that should be required in line with the voluntary agreements differed enormously: between 488 and 520 million tonnes of CO_2 per year in the 2005 to 2007 period.[9]

The outcome was to allow for 503 million tonnes annually from energy and industry in the period 2005–2007.[10] This can be compared to current emissions of 505 million tonnes of CO_2 annually. The government also indicated that the energy and industry sectors would be allowed 495 million tonnes annually during the second EU ETS period 2008–2012. The EU ETS installations do not precisely match this sectoral breakdown. The government used 2000–2003 emissions data for the 2400 EU ETS installations (which came to 501 Mt CO_2) and applied a reduction factor proportional to that forecast for the whole energy and industry sector for 2005–2007 in order to define the average annual allocation for EU ETS installations as 499 Mt CO_2. This is based on provisional data and so could change in the final allocation on 30 September 2004. The allocation will be in equal annual instalments. All allowances are to be allocated by grandfathering (i.e. free). There will be a small charge for the issue and distribution of certificates and for costs arising in connection with maintaining the registry. Auctioning will not be used by Germany in either the first or second EU ETS periods.

From this debate over the first and second levels of allocation, Germany has gone straight to the fourth level to allocate a share of the total to individual operators. The basic allocation method is to multiply historic emissions in the base-period for each installation by a compliance factor that is calculated precisely so as to adjust the allocations downwards in order to equal the total allocation. There are many variations within this basic method, depending on the specific circumstances of each installation. The base-years that are included in the base-period vary with the date of commissioning of the installation. Some installations which were commissioned after the base-period will use 'announced' or estimated base-data rather than actual emissions in an historic period and will have their allocations adjusted *ex post* once the actual verified baseline data are available. The compliance factor varies from one for process emissions (no reduction) to 0.9765 for energy related emissions (a 3.23 per cent reduction). Thus the adjustment to bring the aggregate of all allocations equal to the total allocation is borne by energy-related emission sources.

There is special provision made for early action with a compliance factor of one for installations which can demonstrate they have achieved certain levels of efficiency per unit of output by specific times. Provision is also made for new combined heat and power (CHP), with annual adjustment to the level of allowances issued free to new CHP based on actual documented net CHP power production in the previous year. Germany has also established a modernization incentive for power generators. From 2008 15 per cent of allowances will be deducted from the baseline emissions recorded for old power stations (such as those powered by lignite). These plants will have the option to transfer their allocation for the old plant to a new plant which would give the new efficient plant a generous allocation. The allocations published in the NAP for each installation do not yet take into account special allocations that companies can apply for. These exceptions and *ex post* adjustment indicate more complex ongoing administration processes for the EU ETS in Germany than in other countries.

Germany has decided to use a 'best available technique' (BAT) benchmark as the basis for allocation to new entrants. New entrants are defined as installations commissioned after 1 January 2005. A reserve of 3 Mt CO_2 per year is set aside for free allocation to new entrants and for increased capacity in new installations. New entrants will have to submit detailed information about their activities and an *ex post* correction will be made to their allocation if their actual activity is higher or lower than that used as the basis for allocation. The new entry reserve includes a special provision of 1.5 Mt CO_2 per year for allocation to fossil fuelled power plants that replace the nuclear power plants that are scheduled for closure during the period.

To be consistent with their new entry rule, Germany will cancel further allocations to any installations that close (with closure defined as activity leading to emissions that are less than 10 per cent of base-period average annual emissions). However, Germany has established a process where closed installations can apply to have their allowances transferred to a new plant within a specified time period (normally three months but up to two years in special cases).

Germany will not allow banking from 2007 to 2008 and will not make use of the opt-out clause.

Example: UK[11]

The UK is the second largest source of CO_2 emissions in the EU with 1500 installations included in the EU ETS and, like Germany has had falling emissions. UK CO_2 emissions were 9 per cent below 1990 levels in 1999, mainly due to restructuring in the energy sector. UK projections are that the UK's emissions will be about 15 per cent below 1990 levels in 2010 if current policies including the 10 per cent renewables target are met, which will be challenging. The UK is therefore comfortably on track to meet its Kyoto target of 12.5 per cent reduction from 1990 (all greenhouse gases).

The UK published its draft NAP in January 2004 and was the first country to do so. However, the UK found that a number of issues raised in the consultation process could not be resolved in time for notification to the Commission by the 31 March deadline. The NAP was notified to the Commission at the end of April 2004, and is still subject to further consultation and revisions before the final allocation is decided (by 1 October 2004).

The decision on the total quantity of allowances to allocate to EU ETS sectors was the major challenge for the UK. As in Germany, the total allocation was decided at political levels based on analysis and advice from government departments. Provisional projections used for the draft NAP showed that with the measures that are currently in place to reduce CO_2, UK emissions will be around 512.4 Mt CO_2 in 2010 (a reduction of 15.3 per cent from 1990 levels). The UK decided to use the EU ETS to ensure that than additional 5.5 Mt CO_2 savings will be achieved from emissions trading. They adjusted their expected 2010 emissions downwards by this amount (to 506.9 Mt CO_2). This gives an expected reduction of 15.2 per cent from 1990 levels in 2010 including the impact of the EU ETS, which will reduce the UK's CO_2 emissions 1.2 per cent further than business-as-usual reductions and overachieve the UK's Kyoto target of 12.5 per cent reduction from 1990 levels (for all greenhouse gases), but is not sufficient to meet the UK's 20 per cent CO_2 national objective. The UK reaffirmed its commitment to its national 20 per cent reduction target in its NAP, which implies that the second-period allocations will be tougher. The 20 per cent objective is well-established in UK policy and is the basis for the Climate Change Agreements (CCA) which will be able to opt out of the first period of the EU ETS on the basis of equivalence of effort. Even greater reductions will be needed in the future to achieve the UK objective to reduce emissions by 60 per cent by 2050 with significant progress by 2020.

In contrast to the German approach, the UK has used a two-stage approach including both level-three and level-four analysis in its NAP calculations. The UK first derived sectoral allocations based on 2005–2007 projections of emissions taking into account the expected abatement from climate change policies and deducting an additional 5.5 Mt CO_2 reductions aimed for from the EU ETS. They then determine the allocation to installations based on defining the share of the sectoral allocation from the share that each installation represented in the five-year base-period from 1998 to 2003. The sectoral constraints are the level at which the UK has addressed variations in likely need for emission allowances and competitiveness concerns rather than trying to address this at the installation level.

The UK worked out constraints for each sector or type of activity using a variety of methods. Projected emissions for the electricity sector and oil and gas were disaggregated from the national energy projection which gives a scenario of emissions from these sectors including expected closures and new entries. The impacts of firm measures, such as the UK ETS, Renewables Obligation and the Climate Change Agreements are taken into account in the projection. Bottom-up sectoral assessments were also prepared using historic data for installations and applying the industry growth rates from the energy projection to reach a 2005–2007 projection. For CCA sectors this was then adjusted in line with the relevant CCA targets.

The general principle for the UK has been to give most industry what it will need in the first EU ETS period. The entire reduction that is attributed to the EU ETS, 5.5 Mt CO_2, is taken from the allowance allocation of the electricity generators. The principle is based on the fact that the UK generators do not face competition from other EU Member States or from other countries and have further low-cost abatement options available to them. This principle has the advantage of being simple and transparent.

Allocations to individual installations will be made based on historic emissions data which must be verified by third-party verifiers before the final allocation decision is made (i.e. 1 October 2004). The UK has taken the share of the total that each installation's emissions represented in the base period (1998–2003) and applied this same share to the sector or activity total that has been decided for 2005–2007 allocations. The base-period is the average of emissions levels in 1998 to 2003 excluding the year in which emissions were lowest. The allocation will be in three equal annual instalments. Use of an historic five-year base-period is the only provision that the UK has made to reward early action.

The UK has made provision in its NAP for a set-aside of allowances to be issued free to new entrants based on a benchmark that will reflect current best technology. Provision is made at the sectoral allocation stage for the reserve based on estimates of expected new entry in each sector for each year of the scheme. The new entrant reserve is calculated separately for each sector and will total about 57 Mt CO_2 or 7.7 per cent of the Phase one allocation. The final numbers will be revised once final verified baseline data are available and final refinements have been made to the energy projection. The reserve quantity is taken off the sector totals before allocating to existing installations.

An additional 4 Mt CO_2 has been taken off the sector totals and added to the new entrant reserve to allow for uncertainties in the projections of new entry. New entrants will receive a free allocation from this new entrant reserve and any remainder that is not allocated to new entrants will be auctioned annually. A portion of this reserve will be ring-fenced for new good-quality combined heat and power in recognition of its special case as an emission-reducing technology that could increase the emissions of an operator compared to base-period levels. The CHP set-aside is based on the likely new CHP installations that are envisaged in the UK's energy modelling. Installations which close will not receive further allocations in future years.

Allocations for individual installations are not listed in the UK's NAP. Final consultations on sectoral splits and growth rates will have to be resolved before allocations to installations can be finalized. Verification of the baseline data could lead to revisions of the base-period data for installations and this will affect the sector totals.

The UK has given industry in the UK ETS and in the Climate Change Agreements (CCAs) the option to opt out. The UK will submit a separate application to the Commission concerning opt-out and has not included its proposal for allowing installations to opt-out of the first phase of the EU ETS in the NAP. Company decisions on whether to use the opt-out provision will depend on the outcome of the CCA target review over the coming months.

The UK will not allow banking of Phase one allowances for use in Phase two. No auctioning is proposed in the UK for the first EU ETS period other than for unused portions of the new entrant reserve.

Major challenges

Overview

The main challenges for EU Member States in their national allocation plans are:

- the decision on the total quantity of CO_2 allowances that can be allocated to EU ETS installations and how much should be expected of non-EU ETS sectors such as transport;
- decisions on the quantity of CO_2 allowances to allocate to each EU ETS sector – treatment of the electricity sector versus process industries is a particular concern;
- lack of data places constraints on the range of feasible allocation approaches at the installation level;
- policy equivalence between voluntary agreements (VAs) and EU ETS allocations;
- decisions on closure and new entry;
- decisions on banking;
- government capacity.

Total cap

A common theme in the allocation processes is assessment of 'what industries will need' in the 2005–2007 period to cover their expected emissions. This reflects the fact that the first EU ETS period is before the Kyoto commitment period when there is less pressure on Annex I countries to take tough measures to limit emissions. Another factor is the extremely tight timeframe, which limits the ability of governments to ensure that they have accurate data and also limits the amount of consultation that is possible with the affected industries. Counterbalancing this is the need for National Allocation Plans (NAPs) to pass the European Commission's assessment of whether they reflect what is needed for the countries to be on track towards meeting their Kyoto Protocol commitments, and of whether they give competitive advantage.

Sectoral approaches

The electricity sector is being treated differently to the primary producers or process industries in many countries. Power generators are approximately 75 per cent of the EU ETS market, and tend to have better historical emissions and output data, homogeneous units of output (kWh), and are expected to be able to pass on the costs of the EU ETS to customers. Impacts on the electricity industry will be very different from the impacts on other industry and this is recognized in the consideration that some countries are giving to using different allocation approaches for different industry sectors. In electricity, studies show that while fossil fuel-intensive plants remain the marginal plant setting wholesale prices, electricity prices will rise as generators pass costs on to consumers, and the generators' profits, particularly for the less fossil fuel-intensive generators, will increase. Primary producers and process industries, on the other hand, typically have diverse products and processes, varying levels of data quality, and compete on an international market which constrains them from passing costs through to consumers. Some are already struggling with low profit margins and low transport costs and exchange rates that favour imports.

Data constraints

Availability of data is the main driver for determining the feasibility of the various allocation methodologies available. Data constraints are leading most countries to opt for basing allocations on historic emissions. Most countries are choosing base-years or base-periods between 1998 and 2002 with flexibility for installations to select a subset of the base-period for their baseline (for example, one of the base-years was abnormal due to maintenance or low demand). Data are available for many installations back to 1998, and this allows some early action to be taken into account. More recent data than 2001 are not available in many countries, although 2003 data can be used in some cases.

Initial estimates were that 5000 installations across the EU and accession countries would be included in the EU ETS. It is now evident that more than 12,000 installations will be covered by the scheme. All countries have found it difficult to identify all of the smaller installations that are involved. Early and comprehensive identification of installations is important as it affects the allocation of allowances, allows time for consultation with all affected installations, and enables the government to communicate key information such as the deadline and requirements for CO_2 permit applications.

The large number of installations results from a provision requiring the aggregation of combustion activities to 20 MW thermal input which brings many previously unregulated installations into the scheme. Identifying the large stationary sources regulated under the IPPC has been relatively straightforward, but identifying the small installations has been difficult, particularly in large countries with many installations such as Germany. Identifying the boundaries of a 'site' and an 'installation' has also caused difficulties. The Commission has produced a non-paper on the definition of installation but this is not a formal requirement so government definitions may vary among Member States.

These data issues have created a great deal of work but will be resolved and should not constitute an ongoing difficulty once the EU ETS is implemented and operating.

Policy equivalence

Voluntary agreements linked to CO_2 or energy targets and to exemptions from CO_2 taxes or future CO_2 policies are particularly important for judging the amount of CO_2 abatement to require from the industries involved. The difficulties of policy interactions can be managed, but resolving them takes up a great deal of government and industry time. The UK appears to be the only country that is seriously considering the opt-out choice in the first EU ETS trading period for industry that is covered by its voluntary agreements. It may be possible to continue the national policies in parallel with the EU ETS in a way that allows voluntary agreements and associated tax rebates to be continued as long as industry felt it useful to do so. However, if industry prefers to leave the voluntary agreements due to the burden of monitoring, reporting and compliance with these, then some governments may allow them to leave the voluntary agreements.

The use of benchmarking is being considered in some cases in order to approximate voluntary agreements that have per unit targets, but problems in obtaining accurate production data are leading to simpler approaches being favoured, possibly with some adjustments to allow for early action in terms of progress against BAT or some other benchmark.

New entry and closure

Closure and new entry rules are delicate political issues. New entrants will face different barriers to entry if, for example, one country makes the new entrants buy allowances while another sets aside allowances and gives them a free allocation. Different treatment of allowances on closure of an installation could create distortions in competitiveness and in industry structure. The difficulty of defining a closed installation could lead to gaming of the closure rules so that a plant is nominally open but emitting very little CO_2. This is an area that would benefit from a harmonized approach across the EU to avoid competitive distortions.

Most countries are making provision to set aside a quantity of allowances for free allocation to new entrants. Since many of the large facilities are planned years in advance it is possible to make an informed judgement of the quantity of allowances that may be needed. It is not yet clear whether any country will require new entrants to buy their allowances. Consistency of treatment of increased capacity at existing installations (for which companies will need to buy allowances) and new entrants (that will either have to buy or be allocated allowances) is one of the issues countries are grappling with. The UK has decided to hold back any further allocation of allowances from installations after they close, but other countries such as the Netherlands have decided to allow installations to retain their allowances on closure in order to encourage the transition away from CO_2-intensive operations.

Banking

The amount of trading that occurs in the early stages of the EU ETS will be affected by national decisions on banking. Full banking is required in the Directive from 2012 onwards but for the first EU ETS period Member States can decide on the extent to which they will allow banking. If some countries allow full banking into the Kyoto period and others do not, industry could sell their unused first-period EU allowances (EUAs) to companies in the countries that allow full banking. This suggests that no country will allow full banking, since governments have to allocate second-period EUAs to its installations to replace any banked units. Both the UK and Germany have announced in draft NAPs that there will be no banking from 2005–2007 into the Kyoto commitment period which sends a strong signal that other countries may follow.

Government capacity

Institutional capacity is strained even in the large EU countries. Typically the Environment Ministries have the lead, but Industry, Energy and Finance Ministries are powerful voices, hold relevant data or modelling capacity, and may be in charge of related legislation. Working groups and inter-ministerial consultations have been set up in most countries to deal with this complexity. Consultation with industry, which is important to get the NAPs right and is required by the Directive, is also time-consuming and further stretches government capacity.

3.3 Permitting procedures

Installations must have a permit to emit CO_2 from 1 January 2005. In addition, only companies with permits can receive an allocation. An existing EU regulation, the Integrated Pollution and Prevention Control Directive, provides the regulatory basis for the EU ETS. This means that agencies that are already providing IPPC permits have much of the capacity needed to issue permits for the EU ETS. However, some countries may decide to establish a new agency for permitting or to place the permitting function within a government agency.

Permitting is the mechanism by which the right to receive an allocation (and the obligation to comply with the EU ETS) is determined. The IPPC permits, which provide the regulatory foundation for the EU ETS scheme, will have to be varied to allow any installation over the threshold size in any of the specified sectors to emit CO_2, and to require them to monitor and report emissions and to surrender allowances equal to emissions at the compliance date (for installations that will require permitting additional to IPCC see Annex 1, EU ETS Directive).

Permitting will also have an important role to play in defining new entrants. Permitting authorities are having to consider the impacts of permit variation rules on incentives for expanding companies to request a permit variation and therefore have part of their operation enter the EU ETS as a new entrant which in many countries will receive a free allocation rather than manage within the installation's original allocation. Permit changes due to a change of operator of an installation might also trigger 'new entrant' status if the permitting function is not rigorous enough.

3.4 Monitoring and verification

The European Commission Decision issued guidelines for monitoring and reporting CO_2 emissions in early 2004.[12] Member States must ensure that CO_2 emissions from installations are monitored in accordance with these guidelines and that each operator of an installation reports the CO_2 emissions from that installation during each calendar year in accordance with the guidelines. In addition, Member States will interpret the Commission's guidance on monitoring and reporting and set their own criteria for verification and for accreditation of verifiers.

These requirements would ideally be identical and they will be fairly similar in accordance with the Commission's guidance. Some countries appear to wish to leave it to verifiers to define their methods in accordance with the EU guidance. Some countries, such as the UK and Germany, are requiring verification of historic emissions data that is used as a basis for allocation, while others will not. These differences do not fundamentally challenge the functioning of the EU ETS. However, they could lead to differences in monitoring and verification costs.

3.5 Registries

EUAs will exist only in electronic format in registries. Transfers of EUAs will be carried out by registries as the final stage of any trade. Registries are very important technical underpinning for the EU ETS. Any failure of a registry could have important commercial implications and an impact on market certainty so have to work well. The registries will also hold information for assessment of compliance (for example, EUA holdings, EUAs surrendered for compliance and emissions). The national EU ETS registries will also serve in the future as national registries for trades that occur under the Kyoto Protocol. From 2008, any EUA transaction by industry will be matched by an AAU transfer by government.

The Commission has prepared an EU Regulation for a 'standardised and secured system of registries in the form of standardised electronic databases containing common data elements to track the issue holding, transfer and cancellation of Allowances, to provide for public access and confidentiality as appropriate and to ensure that there are no transfers incompatible with obligations resulting from the Kyoto Protocol', as required by the Directive.

Each national registry must be a standardized electronic database in line with the EU regulation and has to be in place by 30 September 2004. National registries will have to be compatible but not necessarily the same. The registries will have standard accounts to ensure that national holdings of assigned amounts are recorded and that the commitment period reserve is respected. The registries will also have accounts for each permitted installation in the EU ETS including a holding account showing the EUA holding of that operator for that installation, a compliance account showing EUAs surrendered for compliance, and information on emissions from the installation. The registries will also have trading accounts for holdings of EUAs by entities that are not covered by the EU ETS but wish to trade, including brokers.

3.6 Conclusions

To implement the Directive, Member States must decide how to deal with permitting procedures, the identification of and communication to installations, the transposition of the Directive into national legislation, the preparation of their National Allocation Plan (NAP) and the setting up of national registries. Implementing the EU ETS in the time available requires a major bureaucratic effort, with the allocation process, permitting, legislation and consultation all requiring large amounts of government time. Small government teams typically are struggling to understand all of the issues in the limited time, as well as advise ministers and communicate with industry. Countries, such as Sweden, which have more fully engaged political parties and industry from an early stage have a higher level of awareness but more complex national processes.

The pressures extend to industry as well. Decisions on the allocation of emission allowances to industry are a particularly important component of the implementation process that will determine the environmental effectiveness of the EU ETS, how well the emissions trading market works and the impact on participating industries. However, they have little time in which to prepare, to respond to consultation on the allocation and to present their position to government.

Because of the challenges involved in implementing the EU ETS, it is possible that some countries will not be ready for the 1 January 2005 start date. This could leave some EU ETS participants unable to carry out EUA transfers at the beginning of the scheme. However, as long as all installations have a permit in place for the EU ETS and as long as they at least know their allocations by the start date, trading transactions should be able to take place, albeit that transfers of EUAs would happen once national legislative provisions, allocations of actual EUAs and registries are in place. The latest deadline for trading to comply with the first year of the scheme will be April 2006 as the first reconciliation of emissions and allowances will take place at the end of that month.

Notes

1. Parts of this chapter draw on information prepared for a Royal Institute of International Affairs workshop in September 2003 and the associated report: Fiona Mullins and Jacqueline Karas (2003) *EU Emissions Trading: Challenges and Implications of National Implementation*, November, Royal Institute of International Affairs, available at: www.riia.org.
2. Directive 2003/87/EC of the European Parliament and of the Council of 13 October 2003 establishing a scheme for greenhouse gas emission allowance trading within the Community and amending Council Directive 96/61/EC.
3. Brussels, 7 January 2004, COM(2003)830, final Communication from the Commission on guidance to assist Member States in the implementation of the criteria listed in Annex III to Directive 2003/87/EC establishing a scheme for greenhouse gas emission allowance trading within the Community and amending Council Directive 96/61/EC, and on the circumstances under which *force majeure* is demonstrated.
4. Greenhouse gas emission trends and projections in Europe, Environmental Issue Report No 33, European Environment Agency, 2002, p33.
5. National Allocation Plan for the Federal Republic of Germany 2005–2007, Berlin, 31 March 2004 (translation 7 May 2004).
6. National Climate Protection Programme (CPP) of Germany, October 2000.
7. German Monitoring Report 2001 (*Bericht 2001 der Bundesrepublik Deutschland über ein System zur Beobachtung der Emissionen von CO$_2$ und anderen Treibhausgasen – entsprechend der Ratsentscheidung 1999/296/EG*), reported in the EC Monitoring Report 2002, available at: europa.eu.int/comm/environment/docum/0702_germany.pdf.
8. Paul Klemmer, Bernhard Hillebrand und Michaela Bleuel (2002) *Klimaschutz und Emissionshandel – Probleme und Perspektiven*, RWI-Papiere No 82, Rheinisch-Westfälisches Institut für Wirtschaftsforschung, September.
9. Point carbon 23 March, see: www.pointcarbon.com.
10. Point carbon 30 March 2004, see: www.pointcarbon.com.
11. UK National Allocation Plan 2005 to 2007, available at: www.defra.gov.uk.
12. Decision No 280/2004/EC of the European Parliament and of the Council of 11 February 2004 concerning a mechanism for monitoring Community greenhouse gas emissions and for implementing the Kyoto Protocol.

Chapter 4

Joint Implementation and emissions trading in Central and Eastern Europe

Jason Anderson and Rob Bradley

4.1 Overview

This chapter considers the prospects and progress to date in Central and Eastern European (CEE) countries to participate in JI and IET. It discusses the evolution of JI and the negotiations of targets for the CEEs at Kyoto. Economic and moral arguments favouring domestic action are then explored with a discussion of some practical suggestions for how 'green investment schemes' could be deployed to limit 'hot air' in the Kyoto system, particularly by Russia and Ukraine. The chapter then describes the early experiences of the CEEs under the AIJ pilot phase and the insights gained by many countries through this experiment. The remainder of the chapter explains how prepared CEE countries are for meeting the mechanism participation requirements, illustrating the problems they will face with data on the capacity-related challenges. It then focuses on early 'JI' experiences from participation in carbon funds, principally the Dutch ERUPT programme and the World Bank Prototype Carbon Fund (PCF), before focusing in detail on how the EU ETS will affect the attractiveness of JI.

4.2 Interest in Joint Implementation and emissions trading in EITs

Central and Eastern European countries have always been considered the most likely exporters of carbon credit under international emissions trading.[1] This is because of the high allocation of assigned amount relative to anticipated needs, as agreed under the Kyoto Protocol. There were massive downturns in the levels of greenhouse gas emissions following the disintegration of communism in Eastern Europe, due both to economic chaos and the closure of heavily polluting activities associated with the former regimes, particularly in defence industries. In negotiating the Kyoto Protocol, it was generally considered impolitic to hit the eastern bloc with reduction targets after the momentous and generally welcomed political shifts. In addition, there were already ambitious economic growth predictions emerging from the region showing emissions increases back to near 1990 levels by the first commitment period. As a result, stabilization targets were agreed for most countries.

Almost all countries in the CEE region are predicted to have some excess assigned amount, with predicted inventories in the commitment period below their base year emissions (Table 4.1 and Figure 4.1). The likely exception is Slovenia, which will have to implement successful measures to reach its target; several other countries are projected to be quite close to their targets, including Slovakia, Hungary and the Baltic states. The largest absolute amount of available credit will likely be from Russia, which estimates place in the order of 1000–5000 Mt CO_2 during the commitment period. These figures are often called into question by projections showing Russian economic growth and hence emissions growth rising quickly between now and 2010. There are two reasons for scepticism about these projections, however: they were often being forwarded in an effort to derail Russian ratification of the Protocol, given that emissions trading of excess AA is a major impetus for ratification. Secondly, historical emissions have stubbornly refused to rise in line with projections made in the past; between 1999 and 2001 they continued to be flat and even given some rise in line with growth, the targets seems unlikely to be approached.

In addition to being likely participants in emissions trading, there are 13 countries from the CEE region in Annex I and therefore open for JI: Croatia, the Czech Republic, Estonia, Hungary, Latvia, Lithuania, Poland, Romania, Russia, the Slovak Republic, Slovenia and Ukraine. CEE countries are considered the most likely host countries for Joint Implementation projects. While there is no restriction on JI between any two Annex I parties, the lack of available project finance and the great potential for CO_2 reduction at low marginal cost in Eastern Europe hold special attraction. The potential is especially large in countries not included in the first round of accession to the EU, for reasons discussed below.

Table 4.1 A summary of potentially available AAUs for EITs during the commitment period

Party	Base year information			KP target Art. 3.1	Outlook 2010		Commitment period reserve		Maximal AAUs for transfers
	Year	Base year emissions	KP%		Emissions	% b.y.	90% rule	Latest inventory rule	
Bulgaria	1988	157	0.92	144	99	63	130	99	45
Croatia	1990	n.a.	0.95			n.a.			
Czech Republic	1990	190	0.92	175	135	71	157	135	40
Estonia	1990	32	0.92	29	10	33	26	10	19
Hungary	1985–7	102	0.94	95	79	78	86	79	16
Latvia	1990	31	0.92	29	10	32	26	10	19
Lithuania	1990	52	0.92	48	34	66	43	34	13
Poland	1988	565	0.94	531	411	73	478	411	121
Romania	1989	265	0.92	244	164	62	219	164	80
Russian Fed.	1990	3040	1.00	3040	2815	93	2736	2815	304
Slovakia	1990	70	0.92	64	51	73	51	51	13
Slovenia	1989	19	0.92	18	20	105	16	20	-2
Ukraine	1990	917	1.00	917	769	84	825	769	148
CEE total		5439		5333	4597	85			816

Source: Lecocq and Capoor (2003).

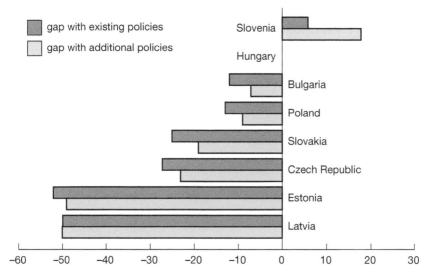

Figure 4.1 *Anticipated 2010 emissions for countries acceding to the EU in 2004 as a percentage of Kyoto target, with current and additional policies implemented*

Note: Only Slovenia seems set for a shortfall.
Source: EEA (2003).

With excess credits available for international emissions trading and energy sector conditions making JI attractive, Central and Eastern Europe would appear to be a key set of countries in the Kyoto Protocol. Certainly this is now more true than ever because of the need for Russian ratification to allow the Protocol to enter into force. But there are a number of complicating economic and political factors that mean use of the mechanisms will not be straightforward for the CEE or potential investors, which will be the focus of this chapter.

4.3 CEE, international emissions trading and 'hot air'

From the beginning, the targets agreed for Eastern Europe in Kyoto were a source of controversy. For all the years of effort leading to the Protocol and the complexity of the negotiations, that so many excess emissions credits were available threatened to undermine the credibility of the accords. NGOs quickly dubbed the phenomenon 'hot air'.

While the United States was still expected to ratify the Kyoto Protocol, hot air, while still denounced by many as a bad outcome of Kyoto, was nevertheless expected to be well absorbed into the overall reduction effort needed. The Bush administration decision to abandon the Kyoto protocol, however, means that hot air potentially becomes a much larger proportion of the overall world reduction effort (Table 4.2). This has two negative outcomes from both buyer and seller perspectives: reducing the level of domestic action and reducing credit prices.

Table 4.2 *Kyoto targets, 1990 emissions, 2010 anticipated emissions (including the effect of policies in place as of 2002) and the difference betweeen 2010 and target emissions*

(in million tonnes (metric) of carbon)	Kyoto limit	1990	2010	Difference
USA	1536	1652	2191	655
Japan	332	353	418	86
EU	1080	1174	1313	233
Other OECD	406	430	546	140
FSU	1385	1413	1064	−321
CEE	429	429	345	−84
Annex B w/o US	3632	3799	3704	72
All Annex B	5168	5451	5895	727

Source: Reilly (2002).

Ensuring domestic action has been a priority of many negotiators and was the premise of the 'supplementarity' provision ultimately agreed in Marrakesh (weak as it was[2]). Meeting more than half (it is commonly felt in the EU) of reduction effort from credits abroad would undermine the credibility of Annex I commitment to actions needed to avoid climate change in the long term. And with supply high and demand low, carbon prices were set to be lower than expected. Given that emissions trading income is one of the primary incentives for ratification of the Kyoto Protocol for countries with hot air, particularly Russia, the incentive is significantly reduced.

The lure of international emissions trading was supposed to be the guarantee of Russian ratification. After all, with perhaps billions of credits to sell over the commitment period, it was reasoned, the revenues would be too attractive to turn away. With the US pullout, however, carbon prices were anticipated to drop significantly, and the drive to ratify would be less pressing. Given the opacity of Russian official thinking, (at the time of writing in March 2004 they still have not ratified) it is impossible to tell the degree to which this factor is hampering their actions. It is argued by some in Russia that what concerns the government more are the eligibility requirements for emissions trading – determining an assigned amount, establishing a registry, and creating an inventory system and submitting annual inventory reports – which will entail definite costs in the short term, while trading revenue is still a long-term prospect and therefore heavily discounted (Tangen et al, 2002). In addition,

Russian official sources tend to play down the potential number of credits, to the point of suggesting there may be a shortfall to reach compliance. Russia's second national communication placed the possible gap at anywhere from −8 per cent to +3 per cent by 2010. Such pronouncements are largely due to ambitious economic growth estimates that are assumed to be paired with emissions rises. Under this manner of thinking, low emissions growth must mean a low economic growth projection, which would be politically unappealing. However, other internal assessments by the Ministry for Economic Development and Trade and Roshydromet are more forthcoming, with estimates from 1500 to 3000 Mt CO_2; external experts consider even these figures may be low (Tangen et al, 2002).

However, Russia stands alone in Eastern Europe in failing to ratify the treaty. Neighbouring Ukraine recently ratified, and has if anything fewer resources to address the challenge of achieving compliance with the rules for mechanisms participation.

4.4 The preference for domestic action

The EU (in its incarnation as the European Community of 15 Member States) has maintained in international negotiations that 'domestic action' is paramount in meeting climate targets. The idea of this is to ensure that most emission cuts (or at least a 'significant' proportion, according to the Bonn Declaration) are made in rich industrialized countries. The reasons behind this preference are worth understanding, because most CEE countries are perceived to lie on the borderline between rich and poor countries, and the relation they play to the EU's idea of domestic action is therefore ambiguous. Nevertheless, in principle the reasons for preferring domestic action are fairly clear and fall into three main categories.

Moral reasons

The world faces a climate change problem principally because of the overuse of fossil fuels by industrialized countries. These countries now enjoy the benefits of their fossil-fuel-based prosperity while the costs and impacts of climate change are borne by developing countries with, on the whole, far lower emissions. Any future in which developing countries will be able to enjoy the fruits of economic development without environmental devastation will require deep cuts in industrialized country emissions. These deep cuts can only be achieved by making fundamental changes to the way energy is produced and used in these countries.

For all these reasons, environmental groups and like-minded governments have generally believed that the overriding priority of any climate regime is for industrialized countries to begin the long process of reducing their own emissions of greenhouse gases. Projects to reduce emissions in developing countries are of course welcome. But if the Kyoto Protocol is not forcing fundamental shifts in the energy sectors of rich countries, it is not doing its job. Only industrialized country emission cuts attack the problem at root, and only thus are rich countries facing up to their responsibilities.

Political reasons

A closely related point concerns the engagement with non-OECD countries. Developing countries (DCs) and countries with economies in transition (EITs) aspire to having economies that resemble those that OECD countries enjoy today. It is unrealistic and indeed unreasonable to assume that they will take on climate targets if industrialized countries decline to do so on grounds of cost.

Prioritizing domestic action therefore is a strong signal to DC and EIT partners that low-emission pathways to prosperity are indeed possible and are being planned and developed. An unwillingness to promise emission cuts in rich countries, on the other hand, would be seen as confirming that such cuts were incompatible with running a rich economy. DC and EIT partners might then conclude, not unreasonably, that their economic aspirations preclude taking on climate commitments. Given that climate change by its nature requires emissions to be limited globally, this message could make the problem impossible to deal with effectively.

Ironically, the global nature of the climate change problem was also used to argue against limiting the use of the flexmex: since an emission cut has the same effect wherever it is made, why not make the cuts where they are cheapest? NGOs argued that this view ignored the political value of rich-country leadership, which means that all emission cuts are not in fact equal.

Technology forcing

Although recent policy discussions have focused on the achievement of the Kyoto Protocol's first commitment period targets at manageable cost, it is widely understood that these targets represent only a first step, and indeed that the emission cuts during this period will be relatively inconsequential compared to the overall challenge of climate change. For environmentalists therefore it is not enough to meet the Kyoto targets: it must be done in a manner which enhances our ability to meet more stringent targets in the future. This means, critically, creating incentives for the development and dissemination of clean and efficient technologies.

The flexible mechanisms of the Kyoto Protocol treat a tonne of CO_2 saving as equal wherever and however it is achieved – and indeed in terms of the direct impact of this saving on the climate this is fair. However, by the same token it does not take into account different levels of technology forcing. Replacing inefficient boilers in the Ukraine with standard German boilers for instance, will give a considerable reduction in CO_2 emissions, but it will not change the range of emission abatement technologies at our disposal. A similar level of emission reduction in Germany would require the invention of new boilers or other innovation in technology or techniques. This would in turn give all countries the potential to make deeper future emission cuts.

The literature on the relationship between environmental policy and technology development is substantial but inconclusive (see Box 4.1), and very little has been done specifically on the effect of domestic action in the Kyoto mechanisms. However, the intuitive importance of technology forcing has remained an important factor for NGOs in favouring domestic action.

BOX 4.1 EVIDENCE FOR CAPS AND TECHNOLOGY FORCING

Relatively little literature has examined the role of supplementarity restrictions on technology and innovation. One paper that has done so (Buonanno et al, 2000) concluded that restrictions on trading do indeed enhance technological change. However, the study also concludes that this effect is counterbalanced by the higher cost of abatement under such restrictions, and thus that they find no rationale for imposing restrictions on trading. It should be noted, however, that a simple economic approach based on marginal abatement costs is unlikely to shed much light on this issue from an NGO perspective. Domestic action in the sense that they advocate it primarily takes the form of policies and measures implemented in industrialized countries, and these policies and measures (energy efficiency programmes, mechanisms for the promotion of renewable energy technologies, removal of perverse subsidies on fossil fuels, etc.) are generally adopted for a range of reasons of which climate protection is only one, and not always the most important. Job creation, security of energy supply, local air quality, reducing congestion and so on are often the primary objectives of such measures, so measuring them in terms of marginal greenhouse gas abatement cost alone is unhelpful. A considerable body of literature looks more generally at the relationship between policy and the development of environmental technology, and tends to support the view that such technology-forcing policies are desirable (for example, Jaffe et al, 2003).

Limiting hot air: green investment schemes

Hot air is clearly unpopular among environmentalists for the reasons listed above, and is also a concern among those fearing it will undermine a well-functioning emissions trading system. But the fear that governments and (if allowed) businesses will take full advantage of trading hot air may not come to pass. Despite the lure of what would probably be very low-cost credits, credit purchasers tend to want to see something real happening in exchange for their money. Businesses may worry about low-cost compliance but they also worry about public image and about creating new market opportunities through engagement in projects. Governments may have limited funding to spend, but voters also like to see public money being put to good effect. Cash transfers in exchange for paper credits do not fit this desire.

Several results follow from this probable reluctance to engage in unlimited hot air trading. First, estimates of traded amounts have to be reduced to account not just for the theoretically most economical course of action (taking account of differing marginal costs of abatement) but also for the limits of political will. National climate plans reflect this: the most recent EU plan for mechanism purchases is that of Belgium, which places international emissions trading as last in a descending list of priority actions: the list begins with domestic action, then EU emissions trading, CDM, JI and finally IET. This last is to be considered only if the rest, which are planned to meet needed reductions, for some reason fall short.

Even countries whose compliance costs are high and which, based on positions during international negotiations, seem to have fewer reservations about the fact of hot air, are expressing doubts about exploiting it to the fullest. Japan and Canada, as members of the Umbrella Group, argued forcefully for full access to mechanisms at low cost. Nevertheless, studies in these countries show an unwillingness to simply 'write a cheque' in exchange for credit. In a survey among Canadian businesses, government officials and NGOs there was a surprising amount of consensus about the desire to see money spent in the name of climate protection achieving some visible impact rather than just compliance on paper.

Potentially of equal influence on the ultimate use of hot air is heterogeneity in seller-side internal priorities. The presumption that trading would be maximized to swell the coffers of whichever ministry controls the sales or the national treasury is facing internal debate. AAU sales under IET could proceed several ways: revenue maximization paces sales such that withholding credits until future commitment periods is a rational course of action; anticipated future emissions growth causes potential sellers to withhold sales for future compliance; internal debate over the use of IET revenues induces control over the volume of sales and destination of the revenue. This last option has been the focus of extensive discussion, particularly in relation to Russia.

In Russia, the struggle over control of climate policy has been a central feature of its chaotic efforts which extend to potential use of the mechanisms. With the prospect of large amounts of income from trading in the offing, the stakes are higher. Rather than seeing trading income go to state coffers, ministries whose portfolios are related to the climate policy, like energy, see IET as a way of earning income that may be used for priorities within the sector. The same reasoning that makes JI attractive in Eastern Europe – low marginal abatement costs coupled with reducing possibilities to raise needed financing to act on the opportunities – means that emissions trading revenue could be put to positive use in the sector. This has led to the development of concepts to recycle AAU income into energy projects, and in so doing to 'green' hot air.

The concept that has been the subject of the most discussion is the 'Green Investment Scheme' (GIS), which was the subject of discussion among experts and academics beginning in 1999 and was first officially presented by the Russian delegation at COP-6 in the Hague in 2000 (Tangen et al, 2002). As proposed, the scheme could be applied in any one of several ways. Emissions trading revenue could be used to do anything from 'soft' activities like capacity building to 'hard' activities like CO_2-reduction projects with additionality requirements. The funds could be managed by the appropriate ministry, or could be dedicated to a fund; the oversight could be Russian, or with international participation. With more pliant rules about project selection and the possibility to address technical assistance and capacity building, gaps could be filled between the kinds of projects and activities that follow formal JI procedures (OECD, 2003).

Already, Japan, Canada and the EU have (unofficially) shown interest in the concept, though there is a continuum of expectations about the choice of soft to hard options. Japan, where abatement costs are among the highest in Annex I, has expressed interest in the soft paths – essentially assuaging concerns about hot air in exchange for at least some positive benefit coming from IET revenue. The EU is at the other end of the spectrum, being concerned about having some measurable CO_2 reductions emerging from funded projects. Canada is somewhere in the middle, supporting the idea of projects but without the need for strict carbon accounting, and with a willingness to examine projects outside of the energy sector (Tangen et al, 2002). Whether support of the scheme would involve direct commitments to purchase AAUs has not been openly discussed, but the implication is that the barrier to purchases may be lowered significantly in buyer countries. Environmental NGOs have expressed guarded support, still emphasizing domestic action, but seeing a GIS as a way of addressing the problematic presence of hot air while giving an incentive for Russian ratification. All CEE countries anticipating a surplus of AAUs in the commitment period could be candidates for such a scheme, but discussion has been limited.

At the moment such concepts are academic, given Russia's non-ratification. With the focus placed on achieving entry into force of the Kyoto

Protocol and on European emissions trading, the subject of hot air trading has faded from view. However, if Russia ratifies, then if anything the need for such a scheme may be higher than ever. The more concrete climate policies become, the more it is emerging that the 'last resort' of IET may to some degree be a certainty.[3]

4.5 AIJ in CEE: early experience with projects

The concept of joint implementation is found within the Convention (Article 2(a)), which allows that parties 'may implement ... policies and measures jointly with other parties ...' At COP-1 it was decided that it was premature to launch a JI system (which at the time meant projects either with an Annex I or non-Annex I host country – there was no CDM until Kyoto), so a pilot phase was launched wherein no credit toward compliance with the (as yet unquantified) convention commitments could be earned. 'Activities Implemented Jointly' resulted in 185 projects, of which 94 were in Central and Eastern Europe (Table 4.3) (Van der Gaast, 2002). This large proportion was largely influenced by the stance taken against JI by many developing countries prior to COP-3, including China and India. Under AIJ, the projects are often an extension of programmes already underway, though the USA, the Netherlands, Japan and Australia established official JI offices. Projects were often clustered in areas where hosts are traditionally active, for example Nordic countries in the Baltic states.

Table 4.3 *AIJ projects in EITs, by country*

AIJ projects in CEE countries	
Latvia	25
Estonia	21
Russian Federation	11
Lithuania	9
Poland	6
Romania	6
Slovakia	5
Czech Republic	4
Hungary	3
Ukraine	2
Bulgaria	1
Croatia	1

Source: Van der Gaast (2002).

While no credit changed hands, indicative reduction amounts were recorded (Table 4.4) from various project types (Table 4.5). Already striking is the diversity of crediting amounts among different project types: although Latvia hosts the most projects, with 25, they account for only 1.5 per cent of the total emissions reductions. Russia, conversely, hosts only seven projects which account for nearly 66 per cent of the reductions under AIJ. The former were primarily small efficiency improvements while the latter included a very large project, an alliance between Germany's Ruhrgas and Russia's Gazprom to reduce losses in natural gas pipelines. Single large projects with non-CO_2 gases continue to be a feature of the project mechanisms today (*CDM Watch*, 2004).

Table 4.4 *Projects and reductions estimated under AIJ*

	Emission reductions		Number of projects	
	CO_2 (tCO_2)	Per cent	Number	Per cent
Bulgaria	0	0	0	0
Croatia	50	0.1	1	1.3
Czech Republic	16,049	23.5	4	5.2
Estonia	1,004	1.5	21	27.3
Hungary	459	0.7	3	3.9
Latvia	1,001	1.5	25	32.5
Lithuania	755	1.1	9	11.7
Poland	2,609	3.8	2	2.6
Romania	1,232	1.8	2	2.6
Russia	44, 992	65.8	7	9.1
Slovak Republic	207	0.3	3	3.9
Slovenia	0	0	0	0
Ukraine	0	0	0	0
Total	68,358	100	74	100

Source: Fankhauser and Lavric (2003).

Table 4.5 *Project types in AIJ*

Project type	Number	Reductions (Mt CO_2)
Afforestation	1	0.3
Forest preservation	2	10.7
Fugitive gas capture	3	31.2
Fuel switch	8	8.8
Renewable energy	28	2.1
Energy efficiency	42	7.1

Source: Michaelowa (2001).

AIJ's aims were the following (Michaelowa, 2002):

- support host country national development priorities;
- reduce compliance costs for investors;
- projects to cover all sources and sinks;
- secure host country government approval;
- real, measurable and long-term environmental benefits that are additional due to the AIJ activity;
- financing must be additional to GEF and ODA funding;
- there will be no credits toward UNFCCC obligations as a result of the projects.

From the start AIJ had difficulties that highlight some of the perils of project activities, particularly when they occur in a political context. The very first AIJ project was a fuel-switch from coal to gas in a municipal CHP plant in the Czech Republic. The mayor of the town in which the plant was located signed an approval letter with investors from the United States. However, the Ministry of Environment was not informed and withdrew approval. Only after long subsequent negotiations was the project allowed to go forward. The confusion and debate over authority led to development of a centralized approval procedure that kept any further projects from taking place for several years (Michaelowa, 2002).

Despite requirements to consider them carefully, with no compliance targets in place for Annex I countries and no credit changing hands, the details of baselines and additionality were not central to project selection. More than half of the projects analysed in Michaelowa (2002) were found not to actually have a baseline study. Monitoring data show that the performance was not all that was expected, with projects overall achieving in the order of 30 per cent fewer CO_2 reductions than was feasible.

Despite hosting the largest amount of reductions under informal accounting for AIJ, Russian experience with AIJ was not particularly encouraging as a model for JI. Nine projects were officially registered, of which six were cancelled or delayed and only three completed (Tangen et al, 2003). Problems under AIJ in Russia that have implications for JI if not solved were identified as falling into three categories – institutional, implementation and funding. The most important failures were: the lack of a strong national focal point symptomatic of the unclear institutional structure of the Russian climate change administration; a constantly shifting legislative, regulatory and taxation system; lack of a memorandum of understanding (MoU), given the impossibility of identifying a ministry with clear responsibility for them in Russia; difficulty finding a local-level partner committed to completing the project; a lack of local technical expertise; problems with local infrastructure and material quality; high transaction costs due to the unpredictability of the economy; and lack of funding, including lack of local co-funding. Overall, with low energy prices discouraging action and an unfavourable investment climate generally in Russia, it has been very difficult to tap into the high theoretical potential for projects.

Despite the difficulties, AIJ has served as a learning experience, which was ultimately the goal of the effort. Several lessons consistently emerge from analyses of AIJ, most of which have to do with national priority setting and institutional capacity. A study of experiences in Bulgaria, the Czech Republic, Estonia, Poland and Slovenia identified the following host country actions for JI to be successful (REC/WRI, 2000):

1. Align and integrate JI into national priorities and a long-term national development vision.
2. Specify clear legal authority and competencies for project selection and approval.
3. Establish actionable project selection criteria.
4. Set up systems to ensure transparency of decisions, public oversight of project performance and information dissemination to business.

In addition, from the standpoint of the improvements to the design of JI as a mechanism, baselines, additionality and credit sharing were the most important issues to be addressed.

4.6 Mechanism participation requirements and CEEs

While there are theoretically opportunities for JI and IET in Central and Eastern Europe, the ability to participate in these mechanisms is determined by compliance with a number of requirements laid out in the Marrakesh Accords.

In order to transfer or acquire ERUs under JI, a Party must be a Party to the Protocol, know its assigned amount, have a system for estimating sources and sinks, have a registry and be submitting annual reports. If these requirements are met, then a host Party can verify the reductions from a project as being additional to business as usual, and can issue the ERUs to the investing party in an amount that is decided through negotiation. This is Track 1 JI. Because both parties are accounting for all of their emissions, the effectiveness of JI in contributing to overall emissions reductions will be measurable, unlike CDM accounting, which relies on comparisons to counterfactual baselines that are inherently unprovable. Because all of the ERUs an investor receives are reduced from the AA of the host, it behoves the host to be sure the reductions are really taking place or else they are stuck with the responsibility for tonnes that never got reduced. Conservative crediting, or a requirement to leave a portion of the credit in the host country, however, can both reduce risk and give extra benefit to the host country.

However, if a host Party does not meet one or more of the above requirements, there is an option for Track 2 JI. This allows projects to continue under supervision of the Annex 6 supervisory committee, which is expected to work very much like the CDM. Just as in the CDM, if hosts do not have inventories and monitoring mechanisms in place, then the impact of miscalculating reductions will not necessarily be dealt with at the end of the day through national accounting. So one has to be certain the reductions are completely attributable to the project and are well measured. This requires some more robust rules.

Although Track 2 projects can take place and accumulate credits without compliance with the eligibility requirements, they can only be issued and transferred by the host once they are in compliance. It is this fact that led many JI supporters to say the extra rules for Track 2 were unnecessary. Nevertheless, the security that emissions are properly accounted for is necessary at the time the project takes place, because the mechanisms for doing so will only come later with later compliance.

Unlike CDM, while JI is allowed to start from 2000, it can only be credited from 2008. Nevertheless, JI projects are taking place, where any credits earned prior to 2008 will be transferred as AAUs in the future. These are all operating under the assumption of needing to meet Track 2 rules, given the uncertainty about meeting Track 1 and the desire to earn credits as early as possible and not waiting for compliance to begin project activities.

Meeting eligibility requirements is no mean feat. Even Western European countries with significant resources will find it a challenge. Analysts generally feel it would be difficult for Russia and Ukraine to achieve compliance prior to the commitment period. Of the other CEE countries those most likely to meet requirements by 2005 include the Czech Republic, Hungary, Poland and Slovakia (Lecocq and Capoor, 2003). Only the Czech Republic and Hungary's registries are under development, with the others lagging behind (Table 4.6).

The risk of non-compliance scuttling current projects is a real one that project developers are well aware of. This risk is not without compensating factors, however: in the Netherlands, for example, domestic measures beyond those already earmarked for action may cost €100/tonne, whereas projects are now being developed for an average of €5/tonne: that some may not bear fruit is considered a risk worth taking (Mulder, 2004).

Table 4.6 *Status of JI administratively in CEE countries*

	JI policy	National registry	Provisional procedues	Dedicated JI office	JI staff
Bulgaria	Yes	No	Yes	Yes	1.5
Croatia	Draft	No	No	No	5
Czech Republic	N/a	Under preparation	Yes	Yes	N/a
Estonia	Draft	No	No	No	1
Hungary	Draft	Under preparation	Under preparation	N/a	2
Latvia	Draft	No	Being drafted	No	5
Lithuania	Under discussion	No	No	No	1
Poland	N/a	No	Yes	Yes	N/a
Romania	Draft	No	Yes	Yes	4
Russia	No	No	No	Being reformed	2
Slovak Republic	Yes	No	Yes	No	2
Slovenia	Proposal	Proposal	No	No	7
Ukraine	No	No	No	No	0

Source: Fankhauser and Lavric (2003).

4.7 Early 'JI' experiences

As of the beginning of 2004 there are 17 projects in the CEE hoping to earn credit, along with several more in some stage of planning (JIQ, 2003). The first two main buyers were the Dutch government and the Prototype Carbon Fund (PCF) (on behalf of its government and industry participants), with new participants including Finland, Italy, Belgium, Austria, Sweden and Denmark. Host countries are in various states of readiness (Table 4.6).

Enthusiasm for participation in JI has varied among potential host countries. Accession countries often hoped to use JI to achieve compliance with the *acquis communautaire*, but questions over baselines have been raised, calling the use of JI for compliance into question. Furthermore, accession countries will be participants in the EU emissions trading system, which covers most of the facilities that might be involved in JI projects. To avoid double counting, JI may not be possible in these facilities. These factors lead some to conclude that JI will be essentially non-existent in accession countries after any cut-off date that may be specified as part of the European emissions trading system. In addition, countries like Hungary initially felt they were going to be too close to their targets to risk exporting ERUs. Those farther from accession and with a larger cushion of excess assigned amount, like Romania and Bulgaria, were more engaged. In almost all cases, the institutional capacity necessary was underestimated and take-up has been slower than originally anticipated as a result.

ERUPT

The first programme designed to earn credit from JI activities (as opposed to merely gaining experience, as under AIJ) was the Dutch ERU Procurement Tender, or ERUPT.[4] Experience with AIJ convinced the Dutch that there were interesting project opportunities, but that financing the whole project, as opposed to just buying reductions, was more expensive than simply doing projects in the Netherlands. The Dutch lay claim to the concept of separating financing for the reduction amount from financing projects entirely, laying the basis for a market.

The first ERUPT tender was in 2000. Currently a fourth round is open, and a fifth is likely. The aim is a total of 34 MT ERUs, which is one-third of the amount sought by the national government from the mechanisms. The first tender showed good response, mostly from Dutch companies, but there was a clear lack of knowledge in the market. They had difficulties closing deals at the local level, such as the needed power purchase agreements. Several projects were rescinded after deadline extensions. The trend through the ERUPT tenders has been from Dutch to international and now more local companies. Given the difficulty of doing business in most JI host companies, the local knowledge is a boon.

Each of the first three tenders elicited some 30 proposals, of which around five were contracted. In the fourth round, there have been 45 proposals, with improved geographical scope, even extending to projects hosted by New Zealand, Germany and Canada. There is a broad spread of proposed project types as well, from fuel switching, renewables, N_2O reduction, coal-mine methane, end-use efficiency and new natural gas infrastructure for households.

The Dutch programme, while on paper a procurement effort that shares the main elements with the way a government buys buses and paper clips, clearly has a more political aspect as well. The approach is top-down, through memoranda of understanding with host governments leading to capacity building and projects, and bottom-up, where interest in project finance from the Dutch in a country, it is hoped, will drive governments to engage in JI.

The first ERUPT experiences with Romania and Bulgaria were top-down initiatives, with MoUs and clear engagement from government offices with the needed authority to approve projects. More recent engagement with Poland and Russia has been, by necessity, more bottom up. Both countries have had difficulty establishing and maintaining interest in JI projects and the jurisdiction for authorization. In the case of Russia this has much to do with the lack of a decision to ratify the Kyoto Protocol: a decision to authorize a JI project would be a signal of presumed ratification that they are unwilling to make now at national level, though regions are willing to consider MoUs. In Poland's case there have a variety of bureaucratic changes that have until recently negatively impacted JI.

Engagement with Russia has been an ongoing feature of the Dutch programme, despite the lack of projects. At present, seven Russian projects have been submitted to the fourth ERUPT tender. It is hoped that this private sector interest will be another inducement to ratification – one that is internal and market-driven rather than coming in the form of constant exhortation from outside governments and interest groups.

In addition to its own tenders, the Dutch are participants in the PCF, and have entered into arrangements with the European Bank for Reconstruction and Development (EBRD) to maintain a new JI fund in which it will invest some €32 million. The fund is intended to focus on energy efficiency and renewable energy projects.

PCF

The Dutch procurement approach is designed to spur market interest – the competitive call for tender yields many more proposals than will actually be chosen. The PCF approach is primarily through involvement in projects from the ground up. As a result the PCF has been far more involved in the details of project development, baselines and crediting calculation than the Dutch.[5]

PCF invests on behalf of its funders, which include the governments of Canada, Finland, Japan (via its Bank for International Cooperation), the Netherlands, Norway, and Sweden, and the private investors BP, Chubu Electric Power, Chugoku Electric Power, Deutsche Bank, Electrabel, Fortum, Gaz de France, Kyushu Electric Power, Mitsubishi, Mitsui, Norsk Hydro, Rabobank, RWE, Shikoku Electric Power, Statoil, Tokyo Electric Power and Tohoku Electric Power. A total of $247 million in projects are in the pipeline, with some 14 per cent in Eastern Europe (PCF, 2003).

The PCF has completed credit purchase agreements for one JI project in each of five CEE countries, including Bulgaria, Hungary, Latvia, Moldova and Romania (JIQ, 2003). In 2003 host country agreements were signed with the Czech Republic, Bulgaria, Romania and Poland (PCF, 2003). Development of Polish projects comes despite previous negative experience: an earlier agreed set of PCF projects was cancelled at the last minute, delaying any re-engagement for quite some time.

Others

Finland has agreements for five JI projects in Estonia, including three district heating, one wind farm and one small hydro plant. *Italy* plans to focus on energy-efficiency projects in Russia through a new tender opening in 2004. *Austria* opened a tender in 2002 that will close in September 2004, and focuses on new projects in heat and power, renewable energy, energy efficiency and demand-side management. A total of €300 million is earmarked for both CDM and JI, and memoranda of understanding have been sighed with the Czech Republic, Slovakia, Hungary, Bulgaria, Romania and Lithuania; MoUs with Estonia and Ukraine are under negotiation (JIQ, 2003). *Belgium* also recently announced €10 million will be made available for CDM and JI tenders (Point Carbon, 2004b).

Table 4.7 *The scope for low-cost JI*

	Tonnes of carbon emissions per GDP in 2010 (US$)	Scope for low-cost JI (ranking)
Bulgaria	1328	2
Croatia	176	12
Czech Republic	380	7
Estonia	360	8
Hungary	205	11
Latvia	296	10
Lithuania	393	6
Poland	402	5
Romania	683	4
Russian Fed.	1164	3
Slovakia	337	9
Slovenia	120	13
Ukraine	2530	1

Source: Fankhauser and Lavric (2003).

4.8 The future potential of JI

A simple indicator of the scope for low-cost JI (Fankhauser and Lavric, 2003) places Ukraine, Bulgaria and Russia out front (Table 4.7). However, as AIJ and early JI experience has shown, institutional capacity, business climate and political backing are, if anything, more important factors. Taking these into account, the EBRD has a three-fold ranking (Table 4.8).

Table 4.8 *Ranking of conditions for JI on three different scales*

Rank	Scope for JI	JI capacity	Business environment
1	Ukraine	Czech Republic	Estonia
2	Bulgaria	Hungary	Hungary
3	Russia	Slovak Republic	Czech Republic
4	Romania	Poland	Slovak Republic
5	Poland	Romania	Lithuania
6	Lithuania	Estonia	Slovenia
7	Czech Republic	Bulgaria	Poland
8	Estonia	Latvia	Latvia
9	Slovak Republic	Lithuania	Croatia
10	Latvia	Russia	Bulgaria
11	Hungary	Croatia	Romania
12	Croatia	Slovenia	Russia
13	Slovenia	Ukraine	Ukraine

Source: Fankhauser and Lavric (2003).

What emerges from the summary score is a story about the importance of institutional factors. Ukraine, while clearly the most energy intensive of all the countries, and having the most low-cost options, is ranked 13th on the two other scales, making the overall prospects for JI less positive. However, should there be some effort on remedying these conditions in Ukraine and Russia, the potential for projects is enormous. At the other end of the scale, Croatia and Slovenia have limited prospects for projects and so have not spent much effort facilitating them. The Czech Republic, Estonia and Hungary are among the countries with the most successful market reforms. Business conditions are amenable to projects, but more efficient energy systems and higher growth using modern technology mean that potential is lower. This translates to greater opportunities in the short term, with conditions thereafter becoming more similar to Western Europe. The remaining countries have median strengths and weaknesses on the three scales that may or may not turn into positive futures for JI.

Although the number of programmes seeking to buy credit via JI is limited, as reported above, a larger number of countries have signed memoranda of understanding (Table 4.9), indicating the possibility of growth in the future. However, the currently low volume of projects being generated to meet this interest may result in stronger competition among buyers chasing a limited number of prospects (Eik and Buen, 2004).

Table 4.9 *Status of Memoranda of Understanding between Buyers and Sellers*

Buyers: Sellers:	Aus	Can	Den	Fin	Fr	Ger	It	Jap	Neth	PCF	Swe	Switz
Bulgaria	s		s				s		s	s		s
Croatia							s		s			
Czech Republic	s		s		s	s		u	s	s		
Estonia			u	s							u	
Hungary	s		u	s			u					
Latvia	s		u	s								
Lithuania			u	s							u	
N. Zealand									p			
Poland		s	s	s			s		p			
Romania	s		s				s		s	s	s	s
Russia		u	u	u			u	u	u			
Slovak Republic	s		s						s		u	
Slovenia							s					
Ukraine			s	s	u				u			

Key: s – signed MoUs; u – MoUs under discussion; p – project-level MoUs.
Source: Point Carbon (2004a).

The attractiveness of JI vs. CDM

During the years between Kyoto and Marrakesh one of the main features of negotiations on the mechanisms were competing concerns between EITs and the G77 that either the CDM or JI would be advantaged by the rules and swallow all available project capital. JI had the presumed advantage of negotiable credit transfers, with no complicated counterfactual baselines needed, while CDM would both be allowed to start prior to 2008 and occur in host countries where, it was assumed, costs were extremely low. Neither assumption about either mechanism has proved entirely true, as Track I JI is not yet feasible and CDM projects are having more difficulty getting off the ground than many predicted.

The stalemate between the two mechanisms may continue, as JI will be limited by the *acquis* and the European emissions trading system, while CDM has trouble attracting buyers willing to pay enough to incentivize projects that are additional to business as usual (and therefore qualify for validation), and institutional challenges remain as difficult in many countries as they are in JI.

What happens in the event of Russian non-ratification?

The failure of the Kyoto Protocol to enter into force would be a severe setback for international climate policy. But for CEE countries and their use of JI and EIT the impacts would be profound; it could also have serious repercussions for the European emissions trading system, depending on the level of commitment Europe shows to its own mechanism.

The Kyoto mechanisms and the assigned amounts that led to the availability of excess credits are both artefacts of the Kyoto process. Absent the Kyoto Protocol, hot air would disappear, and with it any potential revenue anticipated by CEE countries. The potential for JI-type projects would not decrease, however, because of the lower marginal costs for mitigation than in the rest of Annex I. Thus, any non-Kyoto world that allows crediting may find opportunities in CEE countries.

Without Kyoto, however, the question is whether or not there would be the necessary demand for credits. By far the most important demand driver is the European emissions trading system, should it be linked to the Kyoto mechanisms as currently proposed. No provision is made for future projects, or even those already underway, to be able to convert credit for use in meeting obligations under the EU ETS. This therefore is a subject of intense interest to current investors and may eventually yield a crediting system if the non-ratification becomes inevitable.

JI, baselines and the *acquis*

At the moment, JI is being developed much as CDM projects are – while most accession countries may be eligible for Track 1 at some point, the desire to get projects up and running prior to certainty about eligibility means that the working presumption is to prepare everything under Track 2. This has implications for baselines and crediting that make the decision to do JI more complex.

A buyer will naturally try to minimize the risk that a project will fail to deliver contracted tonnes. This leads to choosing projects that are lower risk and on a more stable financial footing. However, if too stable and certain a project is chosen, then the additional benefit of outside funding becomes questionable. Under Track II, as under CDM, the additionality determination could be crucial to acceptance of a project by the supervisory committee (once it is set up).

Financing can be a chicken and egg problem: the buyer is looking for security, such as proof of secured financing for the balance of the project not covered by ERU financing. But if ERUs are integral to the project being carried out, then other financial institutions may want to see proof of a contract to supply ERUs before approving funding. ERU buyers have by necessity become more involved in financing as a result. In the case of ERUPT, the Dutch procurement agency Senter signs a contract for ERUs prior to proof of financing, but it must be secured, and construction begun, within half a year. Once construction starts, more money is released – half is transferred before commissioning, the other half is paid on delivery (Senter, 2004).

The past two years have seen considerable debate about the nature of additionality in the CDM, which has an impact on JI decisions. Proving that the counterfactual baseline would not take place without the CDM or JI is a difficult task, but an inherent part of the logic behind the mechanisms. In the case of CDM, developers are becoming more cautious about the types of projects chosen and have to give the proof of additionality considerable thought. At low carbon prices, the task is not easy. But in the case of JI, most potential host countries are also anticipated to have excess assigned amount. So they may be frustrated by JI and choose simply to do emissions trading. While this does not necessarily assist in improving their energy sector, funding from IET could be recycled in a green investment scheme that has (potentially) some of the benefit of JI without the rules.

The picture is complicated much more in the accession countries by the *acquis communautaire*. There is a clear split in opinion as to whether actions undertaken to meet the *acquis*, thereby reducing emissions, should be taken account of in the baseline for JI, or whether JI itself can be used to achieve such reductions. A similar debate on 'regulatory additionality' has been taking place in the context of CDM. It has been suggested, for example, that the lack of compliance with existing legislation should be taken to be the baseline for projects rather than the legislation itself. But in other cases, project proposals make clear that the moment any legislation is put in place, the baseline will shift to accommodate it. The matter is up for discussion at the 14th meeting of the Executive Board. But what this means for JI is unclear. In so far as CDM is a model for JI, particularly Track 2, it may be influential. But the EU is also in the position of enforcing its own rules, and may simply decide that compliance with the *acquis* is the baseline.

Still, the definition of a baseline will not be simple: the *acquis* is made up of all of the relevant directives and regulations, which are themselves not always particularly clear. Important elements include the Integrated Pollution Prevention and Control Directive (IPPC), the Large Combustion Power Plant Directive, the Landfill Directive, the Building Energy Efficiency Directive, the Renewable Energy Directive and many others. The IPPC may have the most direct impact on power plants. But the requirement to choose 'best available technologies' is not a particularly clear one. The result of complying is likely to be a reduction in energy intensity and CO_2 emissions, but whether that can be translated directly to the complicated calculations necessary to establish project-level baselines is doubtful.

While it is commonly assumed that accession will result in a falling baseline, there is an important exception – the decommissioning of nuclear reactors that do not reach Western European standards. Bulgaria will close all four units of the Kozloduy nuclear plant by 2010; the Slovak Republic will decommission the Bohunice plant prior to the commitment period, and the Ignalina plant in Lithuania may be closed by 2005. These facilities will probably be replaced by thermal power, resulting in higher emissions.

4.9 European emissions trading in Central and Eastern Europe

CEE National Allocation Plans: the story so far

At time of writing, only one of the draft CEE country National Allocation Plans (NAPs) has been formally presented – that of Latvia. Given the complex negotiations that are expected to take place between the Member States, old and new, and between them and the European Commission, detailed comment at such an early stage serves little point. Nevertheless, there are a number of common themes that have emerged as of concern to CEE accession countries.

The NAPs are among the first pieces of EU legislation to be implemented by the new Member States *as* new Member States, rather than a forced adoption of the *acquis communautaire*. Thus it is an important procedural learning process, as well as a complex technical and political one as elsewhere in the EU.

Key CEE country concerns in the NAP process

While lip service was paid during the development of the ETS to the needs and circumstances of the accession states, in practice the system has been designed with only the EU-15 firmly in mind. This has left some important question marks over the implementation of the system and the development of the NAPs by these countries.

For instance, Annex 3 of the Directive makes the Kyoto target (or burden-sharing target as defined under Decision 2002/358/EC) the basis for the setting of the allocation. For the EU-15 this would be expected to produce a target that constrains emissions in the covered sectors. However, in the new Member States this constraint does not apply, as these countries have targets that they can hardly fail to meet regardless of whether they constrain industry emissions or not.

Furthermore, climate change is a low political priority in the new Member States. It is true that greenhouse gas emissions are associated with other issues that are seen as important (such as conventional air pollutants from coal power generation). But climate change as such is rarely cited as a major concern by either governments or the general public.

During the Kyoto negotiations in 1997, the new Member States were grouped with the other economies in transition (EITs) undergoing adjustment from command economies – those of the former Soviet Union. Thus they were given stabilization targets that, given the fact that their emissions had fallen up to 40 per cent from 1990 due to economic restructuring, amounted to implicit bribes to agree the Protocol. The main interest of the CEE countries was the potential for the Kyoto mechanisms to finance new investment in their economies, which were characterized by high emissions and low energy efficiency. In this context many in the CEE countries view the EU ETS as simply a more straightforward mechanism for achieving the same end.

But the EU as a whole tends to view the accession process as also implying the adoption of its values. CEE countries are in effect being asked to prepare themselves for further emission cuts in the name of climate protection and EU 'solidarity'. This is slowly changing the tenor of the debate around the NAPs.

Keeping 'hot air' out of the system

Most Central and Eastern European countries have Kyoto targets that are considerably higher than their projected emissions during the Kyoto commitment period: the difference generally being referred to as 'hot air'. The impact of this hot air on the Kyoto system is potentially large: in the case of the new EU Member States the estimated volume of hot air (to within an order of magnitude) could compensate for the required reduction effort from the EU-15. Potentially therefore the existence of this hot air could compromise all incentives for the EU-15 to make further emission reductions.

Dealing with this hot air is therefore an important priority in terms of Kyoto compliance for those that want to ensure emission abatement action within the EU-15. AAU trading under the Protocol potentially makes it available, and will no doubt be used to a certain extent. But the EU-15 countries are under public and NGO pressure to avoid such trading as far as possible, in particular due to the hot air problem. Introduction of hot air into the EU ETS through the allocation of allowances, however, would allow the use of this free resource under the guise of what is seen as a 'domestic' policy. Considerable effort therefore has gone into trying to ensure that hot air cannot be allocated to entities covered by the ETS.

The Directive attempts to make allocation of hot air to legal entities impossible. Annex 3 states:

> ... *The total quantity of allowances to be allocated shall not be more than is likely to be needed for the strict application of the criteria of this Annex. Prior to 2008, the quantity shall be consistent with a path towards achieving or over-achieving each Member State's target under Decision 2002/358/EC [the burden-sharing agreement] and the Kyoto Protocol.*

In principle this is pretty clear, and Annex 3 of the Directive further stresses that no sector should be given undue advantage relative to others. The Commission has also consistently maintained that State Aid rules would make over-allocation (which in effect amounts to a direct subsidy) illegal.

However, some CEE governments have argued that things are not so clear-cut. Annex 3 also contains several provisions which give a potential fig leaf for over-allocation:

- NAPs states may award credit for early action (i.e. for emission cuts made before the baseline year). This can include the use of efficiency benchmarks.
- They may take into account clean and energy-efficient technologies (a nod towards companies that have invested in technologies such as cogeneration).
- 'Account should be taken of unavoidable increases in emissions resulting from new legislative requirements.'
- Member states must take account of new market entrants.

The possibility of taking early action into account is a particular vulnerability. The precipitate decline in emissions from 1990 in CEE countries was almost entirely in the industry sector, most of which is covered by the ETS. Thus it is entirely possible to present the bulk of the hot air as 'early action'.

The potential for more emission reductions without hot air

Despite this, CEE countries are not expected to exploit these opportunities to allocate hot air to their legal entities. There are several reasons for this.

One is that, just emerging from their accession negotiations, they are less used to defying the Commission and testing the limits of the law than the EU-15, who have long experience in this area. Another is that despite their protestations there is a general realization that there is a resistance in the EU to using hot air more than necessary and that, as they take their place as equal members of the Union, there is little sympathy for the idea that their over-allocation of AAUs is necessary compensation for taking on an environmental commitment.

But most importantly, over-allocation could ruin the ETS as a mechanism for generating more investment in cleaner technologies. And in this case the biggest losers would be the CEE countries themselves. For despite the advances that have been made over the past decade, the CEE region is still full of opportunities to make genuine emission cuts. Ageing power generation infrastructure and industry sectors still heavily dependent on communist-era plant offer real opportunities for improving energy efficiency in many CEE countries. Interest in Joint Implementation reflects these opportunities, but in practice the lower transaction costs of the ETS are expected to make the latter the primary driver for investment in more efficient infrastructure.

Public participation

The sensitivity of the NAPs led to the inclusion in the Directive of a range of safeguards for public scrutiny. The Directive includes this requirement directly within the Article describing the National Allocation Plans.

> *Article 10(i)*
> 1. *For the three-year period beginning 1 January 2005, each Member State shall decide upon the total quantity of allowances it will allocate for that period and the allocation of those allowances to the operator of each installation. This decision shall be taken at least three months before the beginning of the period and be based on its national allocation plan developed pursuant to Article 9 and in accordance with Article 10,* taking due account of comments from the public *[emphasis added].*

> *Annex 3*
> *The plan shall include provisions for comments to be expressed by the public, and contain information on the arrangements by which due account will be taken of these comments before* a decision on the allocation of allowances is taken *[emphasis added].*

Article 3(i)
... 'the public' means one or more persons and, in accordance with national legislation or practice, associations, organisations or groups of persons ...

These requirements, while their general intent is clear, leave wide scope for interpretation in any Member State. However, for Member States without a tradition of public consultation (which certainly includes most or all of the new Member States) they are particularly tricky. One accession state government was approached by NGOs interested in the NAP and told them promptly to mind their own business. Only after continued capacity building from the European Commission and indeed from international NGOs did they come to accept that civil society groups were legitimate participants in the process.

The converse is true for much of industry. As experience has shown throughout Europe, where industry has been publicly owned there is a strong tendency to see government and corporate interests as aligned, and this tendency persists for long after privatization has theoretically separated these interests. The reasons are obvious ones of familiarity and mutual trust: many of those running the newly privatized industry are former colleagues of those that remained in ministries. A general fear of excessive foreign control of industry also prompts many governments to treat their 'national champions' with kid gloves.

The results are sometimes obvious: in the case of the Czech Republic's NAP, for instance, there is no fund of allowances for new entrants, despite the fact that most if not all the EU-15 countries have prepared such a fund. If viewed solely from the point of view of a country's interest in attracting new investment this seems irrational. However, from the point of view of incumbent industry it has the clear attraction that potential competitors are discouraged while existing installations can enjoy a higher allocation (since new entrant funds must be formed by removing allowances from existing installations). Thus the NAP bears the clear fingerprints of excessive industry influence.

Joint Implementation and the EU ETS

To date there has been more talk than action in JI, with only 17 projects contracted in the new Member States, and a few more in the pipeline. Some have identified the emerging EU ETS as a primary factor in dissuading potential JI investors, and indeed this may have added to the uncertainty in the market.

While this may be so, the ETS is just one of many factors causing this uncertainty. By far the greatest is Russia's delay in ratifying the Protocol. Until the Protocol enters into force ERUs do not legally have any status, which undermines the viability of JI projects. Although the EU has intimated that it will recognize ERUs even in the absence of entry into force this commitment is still not convincing enough to project developers; and even if it were it leaves the elephant-sized question of how much JI Russia itself will absorb. Russia's lack of EU-style *acquis communautaire* to confuse the baselines, as well as its huge size and potential for energy-efficiency improvements (Russia has been referred to as 'the Saudi Arabia of energy efficiency'), may make it a magnet for JI investors and undermine JI in other countries – although, as discussed above, Russia has its own problems of political and security risk that might outweigh its advantages.

So JI has had quite enough reasons for delay even without the ETS. In fact, some JI projects seem to have been brought forward more quickly due to the Commission's proposal to exclude projects from the sectors covered by the EU ETS in EU countries: an exclusion was given for projects completed before 31 December 2004. But this presumably accelerated projects already in the pipeline rather than generating new ones. The development process is too complicated to move very quickly – something project investors have come to discover in the past four years.

Still, it remains true that in CEE countries JI and the ETS will effectively be in competition. It is possible that high demand from non-EU parties such as Canada and Japan will give ERUs a price premium over EU allowances. But it seems more likely at present that lower transaction costs in the EU ETS will make JI increasingly marginal in these sectors in the new Member States.

JI, the *acquis communautaire* and double counting

The principle difficulty in defining JI projects is establishing the baseline, i.e. what emissions would have been in the absence of the JI project. This is an important exercise not only in calculating the correct volume of ERUs to be awarded to the project, but for the broader impact it can have on emission abatement policy.

It is well understood that CDM projects can have a perverse incentive effect on the host country, discouraging it from establishing policies and measures to cut emissions. Why phase out certain fluorinated gas compounds, for instance, if one can wait and have each individual project that cuts emissions of those compounds credited under the CDM? In theory, this effect should not apply for JI host countries, since they have an absolute emission ceiling and should benefit in equal measure by reducing emissions through JI or through domestic policy. However, in practice the existence of hot air in most EITs and the reluctance of many countries to buy AAUs give a strong incentive to host countries to maximize JI credit volume, and thus a disincentive to apply policies and measures.

For new members of the EU, the problem of whether to adopt policies and measures does not really arise. The *acquis communautaire*, as the body of EU legislation is called, includes a comprehensive set of such measures. However, the implementation of this legislation – famously 80,000 pages strong – is a long business, and in many instances requires extensive upgrading of infrastructure. Should JI play a role in this process?

For the European Commission, the answer is clearly that it should not. The law is the law, and compliance with it is not the kind of 'additional' measure that a Joint Implementation project should be. Thus its proposal for linking CDM and JI to the ETS states that:

> *Member States shall take all necessary measures to ensure that base-lines for project activities, as defined by subsequent decisions adopted under the Kyoto Protocol, undertaken in countries having signed a Treaty of Accession with the Union fully comply with the* acquis communautaire, *including the temporary derogations set out in the Treaty of Accession. (Article 11(ter)(1))* [6]

Since the ETS itself and the targets set under the NAPs are part of that *acquis*, this implies that JI projects have no role in the ETS sectors, and the Commission's proposal did indeed make this explicit exclusion. The primary concern of the Commission has been to avoid 'double counting', or the awarding of two sets of credits (ERUs and freed-up allowances) for a single emission reduction, and thus inflating the overall emission cap.

However, CEE countries have been alarmed at the proposal, seeing the JI as a promising mechanism that might still generate useful investment flows. Hungary in particular has proposed methodologies that aim to guard against double counting while retaining an incentive for JI. The Japanese government has also taken an extremely active interest in the proposed Directive for the same reason.

Although at time of writing the final negotiations for this Directive have not been completed, it is clear that this exclusion will not apply as the Commission's proposal intended. Instead, projects directly reducing the emissions from ETS sectors will have to retire one allowance for each ERU issued. The provisions for avoiding double counting of emissions reduced *indirectly* in the ETS sectors by a JI project (if, for instance, the JI project in a non-ETS sector reduces electricity demand, thus freeing up allowances in the power generating sector) are at present unclear.

Other climate policy and JI baselines

In addition to the ETS, the *acquis* contains a host of legislation that will affect JI baselines. The Directive on renewable energy[7] for electricity production, for instance, requires that the eight CEE new Member States increase their aggregate share of renewable energy in their electricity mix from 5.4 per cent in

1997 to 11.1 per cent in 2010 – a substantial increase in every country – and put in place financial incentives and other measures to ensure that this target is met. In principle this means that any new renewable energy capacity should only be considered 'additional' if the country has already met or exceeded its target. However, in practice it is far from impossible to make a case that an individual project would not have gone ahead in the absence of the JI project. This is often because the CEE country policy is not in fact sufficient to meet the relevant target; but then again this is also the case for almost all the EU-15. The difference between legal theory and actual implementation, which is often wide in the EU, makes judging baselines on the basis of the *acquis* tricky.

Notes

1. See Blanchard, Part III, Chapter 2 in this volume for expected volumes of trades.
2. Supplementarity, which defines the amount of reduction that must take place within a Party's borders, was argued to be set at anywhere from two-thirds by the African Group, to roughly half by the EU, to unlimited by the Umbrella Group. The final language speaks only of a 'significant' portion being domestic.
3. Domestic action in the EU, for example, will be in large measure driven by European Emissions Trading, where National Allocation Plans submitted to date (March 2004) have been unambitious. CDM credit supply has also developed more slowly than many anticipated. As a result, meeting Kyoto targets may inevitably entail use of IET.
4. Information here is primarily from www.carboncredits.nl and from personal communication with Senter, the Dutch procurement agency. See also Wilder, Part III, Chapter 5 in this volume.
5. More details of the PCF and the World Banks other carbon funds are provided by Wilder, Part III, Chapter 5 in this volume.
6. COM(2003)403 final Proposal for a Directive of the European Parliament and of the Council amending the Directive establishing a scheme for greenhouse gas emission allowance trading within the Community, in respect of the Kyoto Protocol's project mechanisms.
7. Directive 2001/77/EC of the European Parliament and of the Council of 27 September 2001 on the Promotion of electricity produced from renewable energy sources in the internal electricity market, *Official Journal of the European Communities*, 27 October 2001, L283/33.

References

Blyth, W. and Baron, R. (2003) *Green Investment Schemes: Issues and Options*, OECD/IEA document, COM/ENV/EPOC/IEA/SLT(2003)9.

Buonanno, P., Carraro, C., Castelnuovo, E. and Galeotti, M. (2000) *Emission Trading Restrictions with Endogenous Technological Change*, available from: www.papers.ssrn.com/sol3/papers.cfm?abstract_id=235093.

CDM Watch (2004) CDM Status Note, March, available at: www.cdmwatch.org.

EEA (2003) *Greenhouse Gas Emission Trends and Projections in Europe 2003*, EEA Environmental Issue Report No 36, available at: www.eea.org.

Eik, A. and Buen, J. (2004) 'Viewpoint: when will all the ERUs come?', *Carbon Market Europe*, 5 March, available at: www.pointcarbon.com.

Fankhauser, S. and Lavric, L. (2003) *The Investment Climate for Climate Investment: Joint Implementation in Transition Countries*, EBRD Working Paper No 77.

Jaffe, A., Newell, R. G. and Stavins, R. N. (2003) *Technology Policy for Energy and the Environment*, available at: www.nber.org/~confer/2003/ipes03/jaffee.pdf.

JIQ (2003) *Joint Implementation Quarterly*, vol 8, no 4, Foundation JIN, Paterwsolde, The Netherlands.

Lecocq, F. and Capoor, K. (2003) 'State and trends of the carbon market 2003', *PCFplus Research*, available at: www.prototypecarbonfund.org.

Michaelowa, A. (2001) 'The role of Joint Implementation and greenhouse gas emissions trading for project finance', in Birgit Heinze, Gernot Bäurle and Gisela Stolpe (eds), *Financial Instruments for Nature Conservation in Central and Eastern Europe*, Bonn, BfN-Skripten 50, pp63–70.

Michaelowa, A. (2002) 'The AIJ pilot phase as laboratory for CDM and JI', *International Journal Global Environmental Issues*, vol 2, nos 3/4.

Mulder, G., Senter, The Netherlands, personal communication, 26 February 2004.

PCF (2003) Annual Report, available at: www.prototypecarbonfund.org.

Point Carbon (2004a) 'Large interest for ERUs', *Carbon Market Europe*, 5 March, available at: www.pointcarbon.com.

Point Carbon (2004b) 'Breaking news', 26 March, available at: www.pointcarbon.com.

REC/WRI (2000) *Capacity for Climate Protection in Central and Eastern Europe: Activities Implemented Jointly (AIJ)*, available at: www.rec.org.

Reilly, J. (2002) *MIT EPPA Model Projections and the U.S. Administration's Proposal*, MIT EPPA Technical Note 3, March.

Tangen, K., Korppoo, A., Berdin, V., Sugiyama, T., Drexhage, J., Egenhofer, C., Pluzhnikov, O., Grubb, M., Legge, T., Moe, A., Stern, J. and Yamaguchi, K. (2002) 'A Russian Green Investment Scheme: securing environmental benefits from international emissions trading', *Climate Strategies*.

Van der Gaast, W. P. (2002) *The Scope for Joint Implementation in the EU Candidate Countries*, Foundation Joint Implementation Network, Paterswolde, The Netherlands.

Chapter 5

Implementing the Clean Development Mechanism and emissions trading beyond Europe

Martijn Wilder

5.1 Introduction

In the establishment of its regional emissions trading scheme (the EU ETS), the European Union has proposed a Linking Directive which will enable the inclusion of credits from Clean Development Mechanism (CDM) and Joint Implementation (JI) projects to meet commitments under the scheme. The EU ETS represents the first major regulatory step to incorporate the Kyoto Protocol's flexible mechanisms into a domestic emissions trading framework. However, the ability to utilize credits from such projects will be dependent on the host country of a CDM or JI project having in place an appropriate legal framework to implement the project in accordance with the Kyoto Protocol, the Marrakesh Accords and the international climate change rules as they are developed and refined. In the case of CDM projects, this requires the establishment of a Designated National Authority (DNA) by the host country government, which will be responsible for authorizing participation in CDM projects and confirming that they meet the sustainable development criteria for that country.[1]

Over the last three to four years significant developments have occurred in this regard, particularly in Asia and Latin America. This chapter focuses on the extent to which the three flexible mechanisms under the Kyoto Protocol have been implemented outside Europe, focusing predominantly on JI and

CDM. Countries which have set in place an effective framework to implement the Kyoto Protocol mechanisms will be well placed to trade credits into the EU ETS once projects become operational and the Linking Directive proposal has been adopted.[2] The chapter therefore provides an overview of mechanisms developments in Australia, New Zealand, the United States and Canada. It also examines mechanisms related developments in Argentina, Brazil, China, India, Indonesia, Malaysia, the Philippines and Thailand. Finally, it describes developments relating to the World Bank Prototype Carbon Fund and other carbon funds as collectively all these sources of demand and supply will influence the development of Kyoto carbon markets.

5.2 Emissions trading (Article 17)

Introduction

As at the date of writing, the EU ETS is the major Kyoto Protocol-based emissions trading system in existence at a regional and national level. Nonetheless, a number of significant developments have also occurred outside of Europe by developed and developing countries, both with reference to and outside of the Kyoto Protocol framework. This chapter provides an overview of the approach to national emissions trading schemes in a number of Annex I and developing countries as of March 2004. It covers the following Annex I countries: Australia, New Zealand, the US, Canada and Japan, and the following non-Annex I countries: China, Thailand, India, Indonesia, the Philippines, Argentina and Brazil. Developments in these countries are proceeding ahead of the entry into force of the Protocol and are likely to gain momentum if it enters into force.

Annex I Parties and emissions trading

Australia

It is current Australian federal government policy not to ratify the Kyoto Protocol without the participation of the United States and the inclusion of developing country commitments. However, the government has nonetheless committed Australia to meeting its emission reduction commitment in the Protocol and has stated that it is within reach of meeting this target during the first commitment period (2008–2012). Significant work has been undertaken by both the federal and state governments on the possible establishment of a national emissions trading scheme, although the implementation of such a scheme is presently unlikely given the current government's position.

Although there is not currently a national emissions trading scheme based on the Kyoto Protocol in Australia, there is an environmental trading scheme operating on a federal level through Australia's Mandatory Renewable Energy Target (MRET), which endeavours to increase the use of renewable energy sources (RES) in Australia by imposing obligations on electricity retailers to surrender a certain number of 'Renewable Energy Certificates' (RECs) created from electricity generated from RES each year. MRET was introduced by the federal government via the Renewable Energy (Electricity) Act 2002 (REEA) and creates a trading market for statute-based RECs as the compliance mechanism for electricity retailers and certain wholesale energy purchasers which have binding obligations under the scheme. REEA established a target to increase the contribution of the RES in Australia's electricity mix to 9500 GWh by 2010. To date, the MRET has been successful in stimulating renewable energy production and use.

The majority of Australian States and Territories, which have their own parliaments and legislation but do not have the power to ratify United Nations treaties, support ratification of the Protocol and the involvement of Australian industry in national and international emissions trading.

In December 2002, the State Parliament of New South Wales (Australia's most populous state and the largest economy) passed legislation which amended the Electricity Supply Act 1995 (NSW) to place an obligation upon electricity retailers and certain entities including large electricity users to meet mandatory targets to reduce the emission of greenhouse gases from electricity production and use. The legislation and related guidelines, rules and regulations made mandatory a scheme that had previously been functioning on a voluntary basis but had failed to achieve its targeted goal. The emissions trading scheme is known as the NSW Greenhouse Gas Abatement Scheme (the 'NSW Scheme') and commenced operation on 1 January 2003. The NSW Scheme, which is linked to the Sydney Futures Exchange and the London International Petroleum Exchange, is based on Kyoto Protocol concepts and accounting principles and incorporates a conservative interpretation of Article 3.3 of the Kyoto Protocol, enabling credits to be created from carbon sinks.

The goals of the NSW Scheme (to reduce greenhouse emissions from the electricity industry and encourage abatement activity) are met through:

- establishing state greenhouse gas benchmarks and individual greenhouse gas benchmarks for certain participants in the electricity industry and large users of electricity;
- imposing a penalty on relevant entities which fail to meet greenhouse gas benchmarks in any year;
- recognizing activities that reduce or promote the reduction of greenhouse gas emissions; and
- enabling trading in, and use of, certificates created as a result of those activities for the purpose of meeting greenhouse gas benchmarks.

The New South Wales Scheme enables accredited entities to create and trade Abatement Certificates from emission reduction activities within the state or within the national electricity market. At present NSW Abatement Certificates are the only Australian legislative right which could potentially be compatible with Protocol rights, as they are created using monitoring methodologies based on those required by the Marrakesh Accords. It is not possible to trade Abatement Certificates within any other national or international carbon trading regime, although there has been some discussion within the European Union on the possibility of linking the EU emissions trading scheme with Australian and US state schemes.

The NSW push on emissions trading is one of several initiatives to strengthen the sustainability agenda of the state and ultimately to position the state to benefit from the Kyoto Protocol mechanisms. These include establishing a NSW Greenhouse Office and a new Department of Energy, Utilities and Sustainability.

Other state governments of Australia have also implemented greenhouse policies and are currently considering their options in relation to other measures such as emissions trading schemes. The premiers of various Australian states are looking at developing an interlinking trading regime and also at talking to the US about potential coordination between state-based schemes. State and Territory Governments have recently established a working group to develop a multi-jurisdictional emissions trading scheme. In developing an agreed model for this scheme, the working group aims to identify possible options for a scheme (including the potential extension of the NSW Scheme or the adoption of a cap-and-trade system), identify key market participants, assess whether pre-existing schemes can be successfully integrated (including the Commonwealth MRET scheme), focus on the use of Kyoto-eligible greenhouse gas reduction activities and assess the economic, social and environmental impacts of each option within each jurisdiction.

New Zealand

New Zealand formally ratified the Kyoto Protocol on 19 December 2002 and has been an active advocate of the framework as a first step in addressing climate change on a global level. In the first Protocol commitment period, New Zealand is obliged to reduce GHG emissions to 1990 levels. NZ has an advantageous position as, through the increase in forestry development post-1990, it will be a net seller on the carbon markets. The government is looking beyond the first commitment period (CP1), however, and is setting policies to progressively expose the economy to the international price of emissions. The policy package was (with a few changes) approved by cabinet in October 2002.

Climate change policy in New Zealand is managed by the Climate Change Office within the Ministry for the Environment. New Zealand has announced that it will introduce an emissions charge based on the carbon content of fuels in 2007, to coincide with the first Kyoto Protocol commitment period. This charge will be pegged to the international price of carbon, but will be capped at a maximum of NZ$25 per tonne of carbon dioxide equivalent.

The charge will apply to:

- emissions from energy supply and use;
- process emissions; and
- fugitive energy emissions.

Design of the emissions charge is ongoing amid much concern from industry as to how it will integrate with the market mechanisms for electricity pricing. Similarly, the mechanisms for revenue recycling are still to be determined.

The NZ government's position on emissions trading is still 'to retain the option to introduce private sector emissions trading if conditions permit'. Policy implementation with potential linkage to emissions trading has progressed rapidly in two areas:

1. Negotiated Greenhouse Agreements ('NGAs'); and
2. Project Mechanisms.

An NGA is a contractually binding agreement between a firm (or sector) and the New Zealand government. The core of such an agreement is a firm's commitment to be on a pathway to world's best practice in emissions management, in return for a full or partial exemption from the proposed emissions charge.

The NZ government's Projects Mechanism was formally announced in April 2003, with the objective of providing an incentive to New Zealand industry to generate greenhouse gas emission reductions that go beyond business-as-usual. Implementation of the Projects Mechanism occurred in September 2003 with a formal tender round for 4 million emissions units. Tenders closed in October and the first agreements were signed in December. The emissions units on offer are assigned amount units (AAUs), although where requirements under Article 6 of the Kyoto Protocol (JI) are met, transfer of emission reduction units (ERUs) may be requested.

The first tender round attracted 45 bids from a range of large and small organizations. The government stated that this was to be an exploratory round, the learning points of which will be captured in any future rounds. One area that will likely receive focus is the assessment of economic additionality in the absence of standardized tender parameters such as fuel pricing. The emission units awarded by the government are expected to be internationally tradable when the Kyoto Protocol comes into force. The international market will set the price for future emission units. Project owners will be free to trade their units as they wish and in fact several successful project tenderers are already contracting to sell the units provided by the New Zealand government to the Dutch government's ERUPT programme. Emission units will be transferred to project owners annually according to the verified emissions reductions from the project in that year. The awarded units are for reductions that will be delivered during the first commitment period of the Kyoto Protocol.

United States

Although the United States is the leading contributor to global warming on the planet, (accounting for 25 per cent of the world's greenhouse gas emissions), the Bush government has to date resisted undertaking any mandatory actions to reduce greenhouse gas emissions at the federal level. Notwithstanding, there has still been some climate change initiatives at the federal and state government levels, as well as under voluntary partnerships between and among private organizations and non-government organizations to limit greenhouse gas emissions within the United States.[3] These activities have included the introduction of various legislative proposals in the US Congress and the administration's voluntary greenhouse gas programme, as well as a variety of voluntary private-sector initiatives. Emissions trading market activity in the US remains largely dormant, however, reflecting the lack of any mandatory emission reduction requirements in the US and only the initial phases of voluntary reduction and trading schemes.

Different states within the US have established varying GHG emissions reduction (and trading) regimes. For example, Oregon has passed legislative measures that establish carbon dioxide emission standards for utilities and allow credits from CO_2 offsets. New Jersey has adopted new rules to set up an emissions trading programme for the generation and banking of greenhouse gas credits – all six gases under the Kyoto Protocol will be covered. These new rules are in line with New Jersey's administrative order (under the state's Open Market Emissions Trading Registry) to voluntarily reduce greenhouse gas emissions by 3.5 per cent below 1990 levels by 2005. Numerous other GHG reduction initiatives exist in other US states: Arizona has adopted an Environmental Portfolio Standard (EPS) requiring that utilities derive minimum percentages of retail energy sales from RES; California and Connecticut maintain several programmes including a Climate Change Taskforce, GHG Standards for Motor Vehicles, a Climate Action Registry and Renewable Portfolio Standards; Hawaii, Minnesota, Mexico, Nevada, New York, Pennsylvania and Texas all (among other things) maintain Renewable Portfolio Standards; Idaho, Illinois and Nebraska maintain Carbon

Sequestration Advisory Committees and Massachusetts has emission control plans for several power plants.

In the absence of federal regulation, the northeastern States are formulating a strategy to reduce carbon dioxide emissions on a regional basis.[4] The 'Regional Greenhouse Gas Initiative' (RGGI) seeks to apply a cap-and-trade programme to reduce CO_2 emissions from power plants in the participating states 'while maintaining energy affordability and reliability and accommodating, to the extent feasible, the diversity in policies and programs in individual states.'[5] The RGGI aims to offset climate change, impose a carbon dioxide cap on area power plants and open the door to reductions from other industrial sectors by April 2005. Initial goals are to reduce CO_2 emissions from power plants to 5 per cent below 1990 levels by the year 2010, and 10 per cent by the year 2020. State leaders began meeting for the RGGI in September 2003 in an attempt to reach an implementation agreement on standards and protocols by April 2005. Their aim is to develop state-based protocols that would allow international trading. For example, Illinois,[6] New Hampshire[7] and Massachusetts[8] are reviewing legislation that would place caps or targets on emissions of nitrogen oxides, sulphur dioxide, and carbon dioxide (as well as mercury in Illinois and New Hampshire) on fossil-fuelled power plants that meet certain criteria. These programmes would include the possibility of meeting the carbon dioxide requirements through the purchase of credits from certified greenhouse gas offset projects and/or trading of emissions reductions.

Notwithstanding the federal government's position on Kyoto, the United States has been working on developing an appropriate emissions trading scheme since the late 1990s. In June 2001, the US Congressional Budget Office ('CBO') released an analysis on four proposals for the design of a domestic cap-and-trade greenhouse gas emissions trading programme. The criteria for evaluation included: ease of implementation, carbon-target certainty, incremental cost-certainty (i.e. limits on the cost the US economy would bear), cost-effectiveness and distributional effects. The four proposals analysed were:

1. *Upstream Option 1*: based on the proposal made by Resources for the Future and Americans for Equitable Climate Solutions. Upstream fossil-fuel suppliers would be required to purchase allowances (permits), the cost of which would be capped at $25 per allowance. Revenue from the auctions would be distributed equally to US residents and stakeholders adversely affected by the policy due to supply chain effects (for example, increase in electricity prices or products that are energy-intensive).
2. *Upstream Option 2*: similar to Option 1 with two exceptions – the price of allowances would not be capped and revenue generated through auction of allowances would be used to reduce corporate income taxes.

3. *Downstream Option 1*: based on the model proposed by the Progressive Policy Institute. Emissions would be capped at current levels and lowered by 1 per cent each year. Under this option, large carbon dioxide emitters would be required to hold allowances, which would be grandfathered to them based on the current year's emissions estimates and would decline by 1 per cent each subsequent year.
4. *Downstream Option 2*: similar to proposals introduced to the 106th Congress.[9] A cap on emissions would be restricted to units of the fossil-fuel-fired electricity-generating sector above a certain size. Each unit would receive an annual allocation of permits on the basis of their expected annual production multiplied by a generation performance standard.

Although the CBO did not directly endorse any specific option, the analysis concluded that the upstream programmes hold several advantages over downstream programmes, including ease of implementation and incentives to achieve the most cost-efficient emissions reductions for the economy as a whole. The CBO stated that the key design issues where trade-offs would have to be made involve decisions on how to allocate the permits (for example, if auctioned, where does the revenue go?) and whether to cap the price of the permits.

In January 2003, Senators Lieberman and McCain proposed a new national emissions trading bill to the US Senate Commerce, Science and Transportation Committee (the McCain Bill).[10] Though never passed, the McCain Bill proposed several novel emissions trading measures. Through a system of trading emissions allowances under the US EPA administration, the Bill required electricity utilities, industrial plants, transportation and large commercial facilities to reduce their GHG emissions to 2000 levels by 2010 and to 1990 levels by 2016. The residential sector would not be covered while agricultural and forestry entities could opt in. Before 2010, an entity could satisfy up to 15 per cent of its target by international trading, dropping to 10 per cent for 2010–16. The penalty for non-compliance would be a fine per ton of three times the market value of the greenhouse gas emission reductions.

The Chicago Climate Exchange (CCX) is the first US voluntary pilot programme for trading all six GHG emissions. It is a self-regulating, peer-reviewed marketplace with 21 members, including Dupont, International Paper, Rolls Royce, American Electric Power, ST Microelectronics, Motorola, IBM and the City of Chicago. Although the exchange is voluntary, it requires a legally binding commitment by the corporations, municipalities and institutions involved. The exchange trades credits from projects in the United States, Mexico, Canada and Brazil. The combined emissions of the current members is approximately 275 M tons of GHG annually or about 5 per cent of US CO_2 emissions in the year 2000. The members of the CCX have each made a voluntary, legally binding commitment to reduce their greenhouse gas emissions by an increasing amount up to 4 per cent below the average of their 1998–2001 baseline each year between 2002 and 2006, which is the final year of the pilot programme.

In addition to achieving its ultimate objective of reducing greenhouse gas emissions, the CCX also endeavours to build institutions and skills needed to cost-effectively manage emissions and undertake projects.

The CCX creates tradable instruments known as Exchange Allowances (XAs) and Exchange Offsets (XOs). Exchange Allowances function in a similar manner to AAUs and RMUs under the Kyoto Protocol and are allocated to members in accordance with their baseline and emission reduction schedule, and also on the basis of sequestration in sinks and demand-side abatement in electricity use. Exchange Offsets are issued from mitigation projects and registered by members.

The CCX also creates a functioning registry with the ability to identify each XA and XO with an annual vintage. Both types of instruments are equivalent for compliance purposes and may be banked for use in later years, subject to the exchange rules. Two trading platforms have been established (the CCX Trading Platform and the Clearing and Settlement Platform) to execute trades and process transaction information.

Canada

Canada ratified the Kyoto Protocol in December 2002, and has committed to reaching a 6 per cent reduction from 1990 GHG emission levels as its Kyoto Protocol first commitment period target. The Canadian federal government proposed a 'Climate Change Plan' designed to achieve the 240 megatonnes of reductions against business-as-usual levels required to achieve Canada's Kyoto cap. Emissions trading, both through the use of the Kyoto mechanisms and internally in a proposed domestic GHG emission reduction trading arrangement, took a prominent role as a way of achieving reductions on a cost-effective basis.

Of the overall 240 megatonnes of additional reductions anticipated to be required during each of the first commitment period years, the Canadian government acknowledged that fully 25 per cent or 60 megatonnes would be achieved through a number of unspecified programmes yet to be devised. The government committed to buying no less than 10 megatonnes per year under the Kyoto flexible mechanisms, primarily through CDM projects. Measures already in place in Canada prior to 2003, as well as the country's right to achieve the ability to count additional land use and land use management reductions, would achieve a further 80 megatonnes.

Canadian companies have been prominently involved from an early stage in trading international and domestic GHG emission reductions on a voluntary basis. Through voluntary mechanisms like the Pilot Emission Reductions

Trading project (PERT) and the Greenhouse Gas Emission Reduction Trading project (GERT), companies like Ontario Power Generation (OPG), Suncor and TransAlta built significant experience with emission reduction trading. With the advent of the Canadian government's Climate Change Plan, voluntary trading has become far less prominent as the focus now is on preparing for a legislated system. Moreover, significant doubts regarding credit for early action and baseline protection have surfaced and these have served to deter voluntary, pre-legislation reduction activities. Details of the Canadian approach to a domestic emissions trading are described in the Conclusion by Haites in this volume.

While the Canadian government was developing its Climate Change Plan, the Province of Alberta (home to Canada's petroleum industry and the site of major planned expansions in GHG-emitting oil sands processing facilities) has moved to deal with climate change on a basis which it sees as more compatible with the interest of industries and consumers in its province and elsewhere in Canada. Alberta has a statute being considered by its legislature that would require reductions in GHG emissions by 2020 to 50 per cent or less of 1990 levels but importantly this cap is measured relative to economic activity and thus is an 'intensity' target; the parameter that would be measured and controlled under the proposed Alberta statute is GHG emissions relative to Alberta's Gross Domestic Product (GDP).

Japan

Japan ratified the Kyoto Protocol in June 2002. While there is ongoing discussion regarding the introduction of emissions trading in Japan, as yet no formal policies exist for emissions reductions except the amendment of the Energy Savings Law, which regulates the energy efficiency of electric equipment, cars, large factories and buildings. Although the Law has been contributing to the improvement of energy efficiency, GHG emissions are nonetheless still increasing. Japan's Energy Policy was also amended in 2002. This amendment increased the share of natural gas as well as imposing a tax on coal used for certain purposes. A law for the promotion of RES was passed in April 2003 (which obliges electricity generators to use RES).

Several Japanese companies have recently commenced (or are in the process of engaging in) voluntary emissions trading programmes. Mitsui, a Japanese trading company, has begun to develop a carbon emissions trading market and has invested US$6m in the World Bank's Prototype Carbon Fund (PCF). Kansai Electric, also in the top 50 Japanese companies, has similarly invested in Natsource, a CO_2 broker. Mitsubishi has purchased UK emissions trading allowances from Shell and trading house Sumitomo Corp. has obtained approval from the Japanese government to start GHG reduction emissions trading later this year under the framework of the Kyoto Protocol. Specifically, the company aims to reduce some 5 million tons of carbon dioxide annually at an Indian factory producing chlorofluorocarbon, in a joint project with a British chemical company. It plans to sell emission reductions earned through the project to Japanese firms such as electronic power companies.

Emissions trading in Japan is discussed in further detail in the concluding chapter by Haites within this volume.

Developing countries and emissions trading

Many developing countries are already taking action that is significantly reducing their greenhouse gas emissions growth. However, these efforts are driven not so much by climate policy but rather by imperatives for development and poverty alleviation, local environmental protection and energy security. Developing nations offer large opportunities for further emissions mitigation, but competing demands for resources may hamper progress. Developing countries can use policies to leverage human capacity, investment and technology to capture large-scale mitigation opportunities, while simultaneously augmenting their development goals.

The experiences of developing countries in efforts to reduce GHG emissions have implications for future policy at multiple levels – for national efforts within developing countries, for the evolving international climate framework and for other bilateral or multilateral efforts aimed at encouraging emission reduction in developing countries. One broad lesson, given the diversity of drivers and co-benefits, is the need at both the national and international levels for flexible policy approaches promoting and crediting a broad range of emission reduction and sequestration activities. Other policy priorities include continuing to promote market reforms, such as more realistic energy pricing, that can accelerate economic growth while reducing emissions growth.

China

China has been experimenting with pilot emissions trading schemes since 1991 and is viewed by the Asia Development Bank as a pioneer in emissions trading within Asia. The country is trailing Total Emissions Control (TEC) combined with emissions trading to reduce sulphur dioxide (SO_2) emissions. Pilot emissions trading projects undertaken since 1998 have been largely successful and can provide a model for large-scale trading. A September 2000 Amendment to the Air Pollution Prevention and Control Law provided a legal foundation for TEC, and TEC policies gained additional political support when they were set out formally for the first time in the Tenth Five-Year Plan (2001–2005). That plan advocates a 10 per cent reduction in SO_2 from year 2000 levels. Additionally, the plan calls for reductions of 20 per cent from year 2000 levels in two highlighted 'control zones' in eastern and southern China.

China's TEC and emissions trading policies are modelled on the 'cap-and-trade' system used in the United States to control acid rain. Companies that produce less than the allocated pollution can bank units for later use or sell them to firms that have breached the cap. The purchasing firm can use the credits to reduce its aggregate output on paper. The penalty for an industrial source producing sulphur dioxide is five yuan per tonne.

TEC policies, in contrast to concentration-based standards that focus on local air quality, are designed to address issues of acid rain and transboundary emissions across wider regional areas. Emissions trading uses market-based mechanisms to encourage emissions reductions at the lowest possible economic cost.

To accomplish emissions trading, environmental authorities first cap pollution from factories and power plants. If an enterprise emits at levels below the cap, it may accumulate credits or permits toward future emissions, or trade with other emitters who are unable to meet the cap. Thus sellers of credits are compensated for environmental protection efforts and purchasers have an expanded emission quota.

Using market-based mechanisms as instruments for environmental protection is relatively new to China. However, the government has embraced pilot projects in emissions trading, and TEC and emissions trading currently enjoy significant political support from senior State Council leaders as well as the former President and Premier. Recently, Xie Zhenhua, Minister of the State Environmental Protection Administration (SEPA), highlighted Total Emission Control as the key to successful emissions trading and spoke of the need to treat emissions as a 'resource'. He noted that 'emissions trading is widely accepted because it solves more environmental problems with fewer costs,' adding that China will implement an emissions trading policy after current pilot programmes are successfully concluded. Xie also lauded China–US cooperation in the design and implementation of emissions trading pilot projects.

In March 2003, SEPA announced that four heavily industrialized provinces (Shandong, Shanxi, Henan and Jiangsu) and three cities (Shanghai, Tianjin and Liuzhou) would pioneer China's first cross-provincial border emissions trading scheme in SO_2. This scheme represents the nation's first attempt at using economics to curb acid rain; the scheme has since expanded and now includes cross-provincial border trade. Guangdong, Hong Kong and Macao will go on the programme if the initial batch of four provinces and three cities continues to perform satisfactorily.

Foreign governments, including the United States, have stepped up to help China in meeting its ambitious sulphur dioxide emissions reduction goals. In 1999 the administrators of the US EPA and SEPA signed a memorandum of agreement establishing a bilateral project on emissions trading and control of acid rain. The EPA has since been involved in several projects in China, including:

- a feasibility study and training on SO_2 emissions trading project;
- an air quality monitoring demonstration project; and
- an ambient monitoring project.

In addition to the United States, other countries have also provided technical advice on emissions trading and other SO_2 reduction programmes. Japan is active in the area of acid rain management and provides training for managers of power plants and large industries in the two control zones. It has also provided low-interest loans for desulphurization equipment and pilot projects in Shandong, Shanxi and Inner Mongolia through its Green Aid Plan.

The development of the Chinese trading scheme will provide it with valuable experience in trading and could at a later stage be expanded to include other greenhouse gases. The development of the TEC illustrates that the US experience with emissions trading is being absorbed by other countries as they design environmental regulations. The future challenge for both the US and China will be whether these frameworks can integrate greenhouse trading while meeting integral environmental and health related objectives.

Brazil

Brazil's annual GHG emissions are 91 million tons, or 10 per cent lower than they would be if not for aggressive biofuels and energy efficiency programmes aimed at reducing energy imports and diversifying energy supplies. A tax incentive for buyers of cars with low-powered engines, adopted to make transportation more affordable for the middle class, accounted for nearly 2 million tons of carbon abatement in the year 2000. If alcohol fuels, renewable electricity, cogeneration and energy efficiency are encouraged in the future, carbon emissions growth could be further cut by an estimated 45 million tons a year by 2020. Deforestation, however, produces almost twice as much carbon dioxide as the energy sector. Government policy, with few exceptions, indirectly encourages emissions growth in the forestry sector.

India

India's growth in energy-related carbon dioxide emissions was reduced over the last decade through economic restructuring, enforcement of existing clean air laws by the nation's highest court, and renewable energy programmes. In 2000, energy policy initiatives reduced carbon emissions by 18 million tons – over 5 per cent of India's gross carbon emissions. About 120 million tons of additional carbon mitigation could be achieved over the next decade at a cost ranging from $0 to $15 per ton. Major opportunities include improved efficiency in both energy supply and demand, fuel switching from coal to gas, power transmission improvements and afforestation.

5.3 Joint Implementation (Article 6)

Introduction

Outside of Europe, those Annex I countries between whom Joint Implementation Projects are likely to take place (provided they have ratified the Protocol or ultimately do so) include New Zealand, Japan, Canada, the US and Australia. There is also the potential for countries with economies in transition (such as the former Soviet states) to participate in JI projects and emissions trading under the Kyoto Protocol. However, with the failure of the United States to participate in the Kyoto Protocol, the likely revenues from international emissions trading for these countries may be limited, at least during the first commitment period.

New Zealand

The New Zealand government has implemented a variety of legislation and policies to reduce New Zealand's greenhouse gas emissions. However, whereas Australia has chosen to meet its Protocol target outside of the Protocol framework and trading regime, New Zealand has embraced the Protocol as the first step in reducing emissions on a global scale and its domestic laws and policies have been firmly based within the framework established by the Protocol rules.

While the New Zealand government has announced that a domestic emissions trading programme is its preferred policy measure for meeting its Kyoto commitments, the potential design of such a system has not yet been determined. Under the Kyoto Protocol, New Zealand has agreed to stabilize its greenhouse gas emissions to 1990 levels by 2012. National estimates indicate that greenhouse gas reductions in the order of 20 per cent, compared to a business-as-usual scenario, will be required to achieve this target. The government has also indicated that the country will endeavour to meet any future targets in subsequent commitment periods after 2012. New Zealand industry and stakeholders have been actively involved in the development of an effective national climate change policy and will be integral to meeting the Protocol target to which New Zealand is committed.

As discussed above in relation to emissions trading, the measures implemented through this climate change policy include an emissions charge, Negotiated Greenhouse Agreements and the assignment of AAUs to project tenderers. Additional policies include:

- the decision of the New Zealand government to retain all Protocol rights and obligations arising from New Zealand's sink assets; and
- the provision of overseas development aid to Pacific Islands to develop adaptation options for nation-states which are particularly vulnerable to the effects of climate change.

In addition, the New Zealand government has also sought the cooperation of countries which are not parties to the Protocol, including Australia and the United States, to enter into bilateral partnerships to enhance knowledge of climate change and work to reduce emissions.

New Zealand has been a leading source of expertise in the international negotiation on the design of registries. It is unsurprising therefore that it has been able to take the lead in creating a legislative framework for the development of a national registry as provided for in the Marrakesh Accords. Once the registry begins operation (expected to occur in 2005), this will place New Zealand in a position to undertake emissions trading and CDM and JI projects.

The major New Zealand industrial sectors (agriculture, forestry, and industry and manufacturing) have begun to develop adaptation strategies for the impact which New Zealand's climate change operations will have on their business. Companies have also begun to explore the opportunities created from the ability to reduce greenhouse gas emissions and in particular several New Zealand renewable energy companies have already entered into contracts with the government that will assign them AAUs which they are able to sell to purchasers in the carbon market to provide additional finance to projects which would otherwise be economically unfeasible. New Zealand government and industry has therefore begun to adapt to the changes that will be necessary for New Zealand's participation in the Protocol and to position itself to benefit from the opportunities that may arise from international emissions trading and domestic policies.

Japan

The development of an emissions trading scheme in Japan is at an early stage and is discussed by Haites in the Conclusion to this volume. However, there has been significant Japanese interest in both the CDM and JI mechanisms. In the case of JI there have been some suggestions that Japan may limit JI projects in Japan so as to prevent the flow of AAUs from Japan who needs to retain reductions to meet its Kyoto target. However, Japanese companies have been very active in pursuing JI projects outside of Japan and also in undertaking CDM projects. In addition, a number of private Japanese financial institutions are in the process of establishing CDM investment funds along the lines of the World Bank's Prototype Carbon Fund.

So far as project-based mechanisms are concerned, the Japanese Ministry of the Environment's support to operationalize CDM and JI projects consists of the following measures:

- *Project finding and information service:*
  - *Conduct of CDM/JI Feasibility Studies* – feasibility studies are carried out to ascertain promising CDM/JI projects (as proposed by private

companies). Such studies include on-site examination, emission reduction calculation and drafting of relevant accounting documentation.

- *Establishment of Domestic Kyoto Mechanism Support Centre* – this centre gathers information on potential CDM/JI projects and provides it to private companies. The Centre also disseminates information on relevant CDM rules, the drafting of CDM accounting documents and situations of host countries, etc.

- *Provision of financial support:*
  - *Provision of subsidies for CDM/JI projects* – subsidies are provided to feasible and economically attractive CDM/JI projects. These subsidies cover one-third of the cost of construction of required facilities (for example, facilities to convert waste into fuel, wind turbines, etc.). The government then acquires one-third of the generated emission reduction credits created from the subsidized projects.
  - *Establishment of the Kyoto Mechanism Fund* – the Development Bank of Japan has established the Kyoto Mechanism Fund, which will invest in CDM/JI projects and acquire credits. Private companies are expected to participate in the Fund.

- *Preparation for the system to utilize CDM/JI:*
  - *Capacity-building program for Operational Entities (OEs)* – this programme aims to foster the ability of OEs. The MOE selects several draft project design documents (PDDs) and then commissions several applicant entities (AEs) to validate the PDDs on a trial basis.

5.4 The Clean Development Mechanism (Article 12)

There has been significant effort made by a number of European countries (in particular the Netherlands and Denmark) to purchase emission reductions (which ultimately should become CERs) from CDM projects around the world despite the fact that the Kyoto Protocol has not yet entered into force. Increasingly other countries are now consolidating funds to follow this approach, thereby further increasing the demand to find CDM projects from which CERs can be purchased and used within the EU ETS pursuant to the Linking Directive.

The ability to utilize CERs from non-Annex I country CDM projects requires the establishment of appropriate legal frameworks in the host countries to enable such projects to be assessed and approved. Article 12 of the Kyoto Protocol, as further supplemented by the modalities and procedures for CDM Projects in the Marrakesh Accords and subsequent decisions of the CDM Executive Board, sets out the international legal rules applying to CDM projects (the CDM Rules).

Central to the CDM Rules are the 'Participation Requirements' for Parties to the Protocol which require the establishment in host countries of effective institutional and legal frameworks for approving such projects.[11] Specifically the CDM Rules require host countries to:

- ratify the Kyoto Protocol; and
- designate a national authority (DNA) for CDM projects.

No further guidance is given in the CDM Rules as to what the requirements for establishing a DNA are or what the role or function of the DNA is to be. This is a matter for each individual host country and different countries have pursued different approaches as detailed further below. Nonetheless, it is clear that:

- CDM projects must be approved by the host country of a project. It is anticipated this will occur through the DNA, with a practice having been established of issuing 'Host Country Project Approval Letters';
- CDM projects must assist in achieving sustainable development in the host country, as determined by the host county; and
- where environmental impacts of the project are considered significant by the host country, an environmental impact assessment, undertaken in accordance with the procedures of the host country, will be required.

To facilitate the establishment of DNAs there has been significant capacity building in countries including Indonesia, Malaysia and Mozambique by organizations including both UNEP and the World Bank, as well as countries such as Canada, Denmark, the Netherlands and Germany. The result is that the implementation of the CDM in non-Annex I host countries has been quite substantial, although the level of legal detail varies greatly. In addition there is also a range of levels of understanding and implementation of the necessary regulatory frameworks among host country stakeholders.

The following sections examine CDM developments in Asia and Latin America, focusing on countries that have taken the lead in implementing the CDM and are potentially significant suppliers of CERs on the Kyoto carbon markets.

Asia and the CDM

The development of the CDM in Asia, particularly South-East Asia, has been slower than in other regions (such as Latin America) due to a number of factors including continued scepticism over the value of the mechanism for developing host countries.[12] Even where accepted, emissions trading schemes in Asia are mainly subordinate to taxation measures (such as the coal tax in

Japan) or mandatory measures, such as the requirement that retailers sell a certain amount of renewable and non-carbon power. In Japan, the mandatory approach was reinforced in April 2003 when a bill requiring the incumbent electricity utilities to use new energy sources to supply 12.2 TWh a year from March 2010 became law. Japan is one of the few countries in the region with evolved market-based plans although other countries in the region have benefited from cross-border deals.

In contrast, regions such as South America have aggressively sought CDM investment and will probably account for the majority of CERs generated for use by Annex I countries. Notwithstanding this early lead, a number of CDM projects are being developed in South-East Asian countries and a number of capacity-building and project identification programmes are ongoing which are likely to increase the capacity of Asian countries to supply CERs.

China

China has been a strong advocate of the CDM and has been active in the policy debate, leading delegations and undertaking capacity-building efforts to promote and develop CDM projects in China. China also currently has an alternate representative on the CDM Executive Board.

The principal Chinese national authorities for climate change and CDM activities include the Office of the National Coordination Committee for Climate Change and three associated Climate Change Working Groups. The National Coordination Committee for Climate Change consists of 15 member organizations, with the National Development and Reform Commission as the Chair and the Ministry of Foreign Affairs, Ministry of Science and Technology, State Environmental Protection Administration and the China Meteorological Administration as the vice chairs. The National Coordination Committee for Climate Change's *Interim Measures for CDM Project Activities* was revised in 2004. This publication is divided into four sections: general rules, management and implementing agencies, operation procedures and others.

It seems that the final CDM governance structure in China will consist of a National CDM Board made up of seven government ministries and co-chaired by the National Development and Reform Commission (NDRC) and the Ministry of Science and Technology (MOST). A Management Center will also exist to assist the National CDM Board. Projects can be submitted to any member of the National CDM Board or to the CDM Centre. A project approved by the National CDM Board will then be submitted to the Ministers or Vice-Minister of NDRC, MOST and the Ministry of Foreign Affairs. If approved by all three Ministers or Vice-Ministers, NDRC will issue the host country approval letter required for registration as a CDM project.

Initially, China had made statements to the effect that it would only support CDM projects that were 100 per cent foreign owned. However, this stance has shifted so that the government appears to be more open to CDM projects being undertaken as joint ventures between China and other countries. China has tended to oppose the idea of unilateral CDM Projects (i.e. projects where there is no foreign investment and the project is undertaken entirely within the developing host country), as it views the CDM as an opportunity to benefit from technology transfer from Annex I countries in the CDM context. A firm policy on this issue is yet to be resolved. Nonetheless, China continues to actively promote the CDM and there is much action within the country with foreign delegations seeking to develop projects or forward purchase CERs from Chinese projects.

China is in the process of finalizing its DNA for approving CDM projects and is still to make clear its final national policy on CDM. Draft governance structure, policies and procedures are in place for provisional application while waiting for final approval. At this stage, China's State Development Planning Commission (SDPC) represents the country's DNA for CDM projects. The SDPC will establish the National CDM Project Management Centre and will designate a director (in accordance with the entrustment or recommendations of the National CDM Board). The SDPC will also approve (jointly, with the Ministry of Science and Technology and the Ministry of Foreign Affairs) CDM projects according to the conclusions made by the National CDM Board.

CDM projects in China include cooperation with the Dutch Certified Emission Reduction Procurement Tender (CERUPT), the World Bank's PCF and other organizations. Current Asia Development Bank (ADB) CDM projects in China include:

- the Gansu Province Clean Energy Project (Xiaogushan Hydropower Station) (in association with the World Bank);
- the Shanxi Coal Mines Methane Demonstration Project; and
- the Liaoning Environment Improvement Project.

Most of China's current proposed CDM projects have not yet been implemented and are only at the initial stages of development. Principal obstacles to the development of these projects include the following:

- As the Kyoto Protocol is still not yet in force, the Chinese government is reluctant to commit substantial funds, resulting in a limited market for CDM projects and CERs.
- Given the complexity of international laws and regulations concerning CDM projects, project proponents and governments experience difficulty in comprehending CDM modalities, procedures and rules (for example, the definition of 'additionality').

- Many enterprises regard implementing CDM projects as being limited to seeking approval through the normal foreign investment process, largely ignoring CDM-specific implementation rules.

Malaysia

Malaysia ratified the United Nations Framework Convention on Climate Change (UNFCCC) in 1994 and the Kyoto Protocol in September 2002. With the capacity-building support of the Danish government, Malaysia has established its CDM Executive Board – the approved Designated National Authority (DNA) – which will have responsibility for authorizing potential CDM projects on behalf of the Malaysian government and for establishing 'sustainable development' requirements. Malaysia has created an institutional structure for obtaining CDM project approval from the DNA, and three applications are currently awaiting approval.

The Malaysian DNA will be the Conservation and Environmental Management Division of the Ministry of Science, Technology and Environment (MOSTE). Within the MOSTE, the National Steering Committee on Climate Change (comprised of relevant government departments) oversees the formulation of Malaysia's policy and response to climate change. A National Committee on CDM has also been created to approve CDM projects. The Secretariat for both of these bodies is the Conservation and Environmental Management Division (CEMD) of MOSTE.

The Malaysian CDM approval process divides applications from carbon sinks forestry projects from other projects (primarily energy projects). A Technical Committee and Secretariat have been established for these types of projects to undertake technical evaluation of (CDM) project proposals and provide the necessary resources for the energy and forestry sectors in relation to CDM projects. Official preference is for energy projects with a high technology transfer component.

The approval process for CDM projects in Malaysia is set out in Figure 5.1.

Thailand

Thailand initially opposed the CDM and then limited it to projects that received Cabinet approval. However, Thailand has increasingly become involved in carbon credit trading but will be selective about the offers it accepts from developed countries. Thailand has good potential to initiate CDM projects from processes such as biomass and biogas. Specific CDM activities in Thailand include the following:

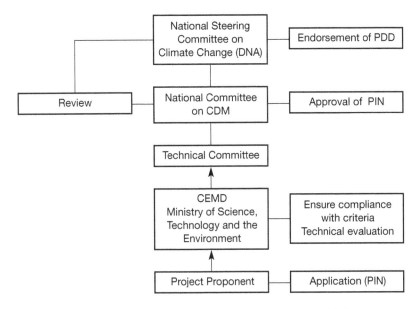

Figure 5.1 *Approval process for CDM projects in Malaysia.*

- Japan's New Energy Development Organization is conducting a range of feasibility studies for CDM projects which include some in Thailand.
- Thailand is part of the 'Developing National Capacity to Implement Industrial CDM Projects in ASEAN' programme of UNIDO. No individual projects have yet been identified, but specific industries and processes have been prioritized.
- The Asian Development Bank's Asia Least-Cost Greenhouse Gas Abatement (ALGAS) Project prepared country studies that identified GHG mitigation potential including some projects which are potential CDM projects in Thailand.

India

The Indian government is in the process of establishing a DNA to approve proposed CDM projects. The Ministry of Environment and Forests will act as the DNA for India. The Ministry will require participants to submit a presentation on the nature of the project and has provided a level of detail as to the information the DNA will require from participants before it will approve the project.

The Ministry has prescribed strict criteria for additionality, sustainable development, baselines, and financial and technological requirements which must be complied with before projects are approved. The Ministry has also stated that the project proposal should clearly indicate risks associated with it including apportionment of risks and liabilities, insurance and guarantees. The proposal should also clearly describe the credentials of the project participants.

Current proposed CDM projects in India include the following (Annex I Party partners are identified in brackets):

- integrated demand-side management with the Andhra Pradesh State Electricity Board (Norway);
- hybrid renewable energy project in Rajasthan (Australia); and
- DESI Power – biomass generation (the Netherlands).

Under the CERUPT scheme, five Indian biomass and wind projects have been approved by the Dutch government to supply CERs.

There is indication that Indian companies have shown interest in the Carbon Ring Consortium (as established by NM Rothschild and E3 International) that is intended to represent an investment vehicle in the Asia Pacific region for carbon credit trading. Carbon credits will be bought from projects that achieve a reduction in GHGs and traded with companies or countries that have not reached their quota of emission reduction. Clean energy projects such as natural gas and wind turbine generation make the country a large beneficiary in the carbon credit trading scheme. Recent reports suggest that in July 2003 the Indian government permitted 12 Indian firms to engage in carbon trading.

Indonesia

Indonesia has proposed the establishment of a National Committee on Climate Change (NCCC) to fulfil the role of DNA for Indonesia. The NCCC will act as a 'one-stop shop' for investment in CDM projects, and will comprise an Emission Offset Board and an Emission Offset Clearinghouse, together with input from ministries of government sectors with an interest in climate change (such as energy, forestry, industry), an Expert Group and a Stakeholder Forum encompassing local government, private sector representatives and NGOs. The NCCC will be involved in the CDM project to approve projects and then later to register CERs created by those projects.

The roles of the various entities in the NCCC can be summarized as follows:

- *Project proponents* – the individual project proponents will develop and submit the project proposal, submit environmental and sustainability impact assessment documents and implement the project itself.
- *Emission Offset Board* – the Board will ensure stakeholder participation in the approval process, hold any public hearings necessary on the impact assessment documents and have the responsibility of ultimately approving the project proposal and recommending its registration with the CDM Executive Board.
- *Emission Offset Clearinghouse* – the Clearinghouse will assist project proponents in 'match-making' between investors and the host, facilitate

stakeholder consultations, assist the board with appraising project proposals and provide necessary information for appraisal, the impact assessment process and all necessary evaluations.

The NCCC is in the process of developing more detailed rules and guidelines, including those on sustainability impact assessment, public participation in the process, standardized forms and membership guidelines.

Specific CDM activities in Indonesia include the following:

- The Asian Development Bank (ADB) and Indonesian government are developing a pilot carbon sequestration project and submitted a paper for COP-9 in December 2003 on sinks.
- The ADB is administering the US$5m Canadian Cooperation Fund on Climate Change, which may lead to CDM projects being developed. Priority will be given to China and India in the field of greenhouse gas reduction activities, Indonesia in the field of carbon sequestration and Pacific Island countries in the field of adaptation.
- UNCTAD is conducting a study on the potential for CDM projects in the Malaysian and Indonesian rubber sectors. The study will look at the potential for using genetic engineering to enhance the potential of rubber plantations as carbon sinks, and the use of more efficient energy technologies in the rubber production process.
- The World Bank, German and Australian governments jointly funded National Strategy Studies (NSS) on climate change for Indonesia. Included in the study was identification of potential CDM projects.
- Japan's New Energy Development Organization (NEDO) is conducting a range of feasibility studies which include some in Indonesia.
- Indonesia is part of the 'Developing National Capacity to Implement Industrial Clean Development Mechanism Projects in ASEAN' programme of UNIDO. No individual CDM projects have yet been identified, but specific industries and processes have been prioritized.
- The Pembina Institute in Canada and Tata Energy Research Institute (TERI) in India are exploring the application of the CDM in Asia. This project is funded by the Canadian International Development Agency and involves assessment of CDM opportunities in India, Bangladesh, Indonesia and China and CDM project development and workshops in Canada and Asia.
- Indonesia is a partner in the South-South-North programme through the Jakarta-based research institute Pelangi.
- The German government is funding Pelangi to assist in the establishment of the Indonesian Designated National Authority (DNA) for the CDM.

The Philippines

The Philippines ratified the UNFCCC in August 1994. The Senate Committee on Foreign Relations sponsored ratification of the Kyoto Protocol on the floor of the Senate in June 2003; interpellation and voting occurred in August 2003. A two-thirds majority in the Senate is required to achieve ratification.

The Philippines' Inter-Agency Committee on Climate Change (IACC) was established in May 1991 and is responsible for coordinating the country's climate change-related activities. It also prepares the country's climate change policies and the Philippines' position in the UNFCCC negotiations.

Capacity development for CDM projects in the Philippines aims to assist in establishing greenhouse gas emission reduction projects that are consistent with sustainable development goals, particularly projects in the energy sector. Specific objectives include generating a broad understanding of the opportunities afforded by the CDM and developing the necessary institutional and human capabilities that allow them to formulate and implement projects under the CDM.

The Philippines government has established the Philippine CDM Operational Framework. The principal objective of this framework is to enable the country to participate in the emerging CDM market in a competitive manner. Specific objectives include:

- creation of a DNA;
- establishing the framework within which the DNA will operate; and
- assessing, developing and implementing capacity-building programmes for key stakeholders to facilitate the effective operation of the DNA.

Specific CDM activities in the Philippines include the following:

- The United Nations Environment Programme (UNEP) has launched the project 'Capacity Development for the Clean Development Mechanism' with financial support from the Dutch government. It will operate in 12 developing countries, one of which is the Philippines.
- The Philippines is part of the 'Developing National Capacity to Implement Industrial Clean Development Mechanism Projects in ASEAN' programme of UNIDO. No individual projects have yet been identified, but specific industries and processes have been prioritized.
- The Asian Development Bank's Asia Least-Cost Greenhouse Gas Abatement (ALGAS) Project prepared country studies that identified GHG mitigation potential, including some projects which are potential CDM projects.

Latin America

Many Latin America countries have been active in the climate change negotiations with Brazil playing a leading role in the design of the CDM.[13] Thus there is much interest in this region in attracting CDM projects.

Argentina

Argentina signed the Kyoto Protocol on 16 March 1998 and ratified it on 13 July 2001. Argentina was the first non-Annex I country to voluntarily announce its intention of taking on emission reduction commitments. At COP-5 in Bonn, Argentina announced its target of a 2–10 per cent emission reduction from the projected emission levels during the first commitment period, determined by the average GDP growth during those years.

In July 1998, an independent Argentine Office of Joint Implementation (OAIC) was created by presidential decree under the Ministry of Social Development's Secretariat of Environment and Sustainable Development. Since that time, the OAIC has been replaced by the *Oficina Argentina del Mecanismo para un Desarollo Limpio* (OAMDL). The OAMDL has been invested with functions which include:

- identifying priority sectors to implement CDM project activities;
- establishing methodologies and procedures for project identification, formulation and evaluation;
- management and national approval of CDM projects;
- identifying sources and facilitating contact between financial resources, buyers and investors of project development; and
- the general management of CDM projects.

In terms of sustainable development criteria, OAMDL has put forward a set of suggested indicators as follows:

- employment generation;
- regional economic development;
- national technology development and hi-tech transfer; and
- social development, poverty mitigation, impacts reduction and co-benefits which flow from these goals.

OAMDL is also involved in the generation of carbon credits from bio-diesel projects in Argentina. This could assist in satisfying some of the international demand for CERs from Annex I Parties to the Kyoto Protocol. As Argentina is one of the world's largest producers and exporters of oilseeds such as soybean and sunflower, it has enormous potential for bio-diesel production. The country could well emerge as one of the first nations 'reaping fuel' at the same time as it contributes to limiting global warming.

Unfortunately, Argentina's current socio-economic crisis is hindering investment decisions and is the main barrier to overcome. The CDM is seen as offering a triggering incentive to encourage producers and investors to develop project activities compatible with additionality requirements. However, as the recently inaugurated administration has promised to eliminate the Ministry of Natural Resources, the fate of the DNA is therefore unknown. Argentina has national technical and policy capacity; however, the country requires financial assistance to create and support more positions in the climate change arena.

With respect to flexibility mechanisms, Argentina generally follows the positions of the G-77 and China. However, a recent announcement declares Argentina's interest in accessing and using *all* mechanisms (both JI and CDM) of the Protocol, a position not shared by any other non-Annex I country but one which is consistent with Argentina's long-standing interest in taking on targets rather than just participating in the CDM. This is evidenced by Argentina's long-standing request for a 'third option' within the Kyoto Protocol, which would not lead to Argentina becoming an Annex I Party but would also not result in Argentina having no quantified reduction obligations as is currently the case for non-Annex I Parties. The new government of Argentina will decide with what vigour the country pursues the third option within the UNFCCC climate change negotiations.

Brazil

Brazil was the first country to sign to the UNFCCC in June 1992, and the Brazilian National Congress ratified it in February 1994. The Convention entered into force for Brazil in May 1994, 90 days after its ratification by the National Congress.

Brazil also played a seminal role in the development of the CDM. In climate negotiations before COP-3 held in Kyoto in 1997, the Brazilian government proposed that, if a developed country exceeded its GHG emissions requirements, an economic penalty would be assessed, and this would be collected in a Clean Development Fund. Monies from the Fund would be directed towards developing countries, who would then use these funds for mitigation projects designed to prevent or mitigate global warming. During COP-3, the proposal evolved into the current CDM.

The United Nations Industrial Development Organization's (UNIDO) *CDM Investment Guide for Brazil* aims at providing CDM project proponents in the country, and CDM investors interested in CDM opportunities in the country, with reliable, recent sources of information regarding CDM opportunities in the energy and industrial sectors of Brazil. The Report was commenced in February 2002 and completed in February 2003.

Principal findings of the Report include the following:

• Global climate change and greenhouse gas emissions are perceived as being very important issues for Brazil, and the country is conducting a variety of efforts in the area of climate change and CDM (including the development of criteria and indicators for appraising CDM projects in the country). Brazil already has an institutional structure that is well placed to deal with CDM projects, including the Interministerial Commission on Global Climate Change, created for the express purpose of coordinating actions of the government in this area, with the authority to verify whether CDM project activities conform with the sustainable development objectives of the country and, if so, provide a formal approval of the CDM project activities in Brazil.

- A large potential for CDM projects exists in the energy sector in the country, in the areas of fuel substitution and energy efficiency, as the 1980s and 1990s were years marked by a growing share of fossil fuel use in Brazil's energy sector. CDM project activities are seen as a way of reversing this trend.

- A large potential for CDM projects also exists in the industrial sector in Brazil in the areas of process change, energy efficiency and fuel substitution, as basic materials industrial subsectors are responsible for an important share of total energy use – and total carbon emissions – by industry in Brazil. CDM project opportunities are widely available in these industrial subsectors, not only because they have, on average, a high specific energy consumption associated with the deployment of best production technologies worldwide, but also because process heat and direct heat based on the use of carbon-intensive, fossil fuels are an important share of end-uses in the industry in the country, and these demands can be easily met (in most cases) by the use of renewable energy sources that are widely available within Brazil.

Various CDM project development activities and types are currently underway in the country at different stages (under consideration, in preparation or ready for submission as a CDM project) which may then be ready for validation provided they are registered by 31 December 2005. These project activities are concentrated in three main mechanisms for carbon abatement: use of renewable energy sources, co-generation or energy-efficiency measures to displace energy obtained from carbon-intensive fossil fuels, and because expertise resides within a small group of governmental and university officials, capacity building in CDM projects is urgently required.

Africa

The development of CDM projects in Africa has been far slower than in Asia or Latin America in part because of the perceived sovereign risk issues that many Annex 1 country investors hold. Nonetheless, the development of national DNAs and the interest in projects remains. In summary, key developments in Africa have been:

- the inflow of international assistance funding to assist in the establishment of DNAs especially in South Africa and Mozambique and with other countries such as Ghana and Zambia to be assisted in the future;
- the undertaking of a number of CDM capacity-building projects in West and Southern Africa under the European Union's Synergy programme and through UNEP;
- the pursuit of initial CDM projects through the Prototype Carbon Fund (such as in South Africa and Ghana) and the pursuit of unilateral CDM projects by a number of multinationals in South Africa.

Canada

Canada's CDM (and JI) office was established in 1998 within the Climate Change and Energy Division of the Department of Foreign Affairs and International Trade to enhance Canada's capacity to take advantage of the opportunities offered by the CDM and JI. It is the federal government's focal point for CDM and JI activities and is responsible for the approval of Canadian participation in these mechanisms. The office facilitates Canadian entities' participation in the CDM and JI and assists them in obtaining emission reduction credits from these projects.

Examples of activities aimed at building a Canadian capacity to develop, register and implement CDM and JI emission reduction or sinks projects have included:

- bilateral agreements with host country governments;
- financing and technical support;
- market identification studies;
- workshops and roundtables with industry, government and stakeholder representatives, domestically and internationally;
- matching project developers with project investors and buyers;
- enhancement of host countries' capacity to assess CDM and JI activities; and
- analytical work in support of CDM and JI policy development and implementation.

Financing support is aimed at reducing the transaction costs for Canadian firms undertaking CDM and JI activities. Canadian entities can apply for financial assistance for the following activities:

- feasibility studies;
- baseline development;
- monitoring plans;

- risk assessments; and
- environmental impact assessments (only as relevant to the CDM or JI project).

5.5 International development agencies

The World Bank, UNEP, the European Union (through its Synergy programme) and many countries in Europe such as Denmark, Norway, the Netherlands and Germany have either directly or through international agencies or their own aid agencies been very active in providing funding to assist in CDM capacity building. For example, Denmark has been providing assistance in South Africa and Malaysia to assist with the establishment of the DNA, Germany and the Netherlands have provided funding for similar programmes in Indonesia and the Netherlands have provided UNEP with funding to run a major international CDM capacity-building programme. It is interesting to note that in some cases this has also been followed up by those same governments entering into MoUs with the countries they are assisting or early contracts with local project developers to secure CERs.

5.6 Other emissions trading activities

Although at the date of writing there are limited operating mandatory emissions trading schemes at a national or international level, nonetheless significant early action has occurred in the carbon market through speculative trades of project-based carbon rights in anticipation that such rights may eventually be recognized under the Kyoto Protocol. Set out below is a discussion of several of the major purchasers in the early carbon market.

The World Bank's Prototype Carbon Fund

Recognizing that climate change will have the most impact on its borrowing client countries, on 20 July 1999 the Executive Directors of the World Bank approved the establishment of the Prototype Carbon Fund (PCF). The PCF, with the operational objective of combating climate change, aspires to promote the Bank's tenet of sustainable development, demonstrate the possibilities of public/private partnerships and offer a 'learning-by-doing' opportunity to its stakeholders. The PCF will pilot production of emission reductions within the framework of JI and the CDM, and will invest contributions made by companies and governments in projects designed to produce emission reductions fully consistent with the emerging framework for JI and the CDM. Contributors, or 'Participants', in the PCF, will receive a pro-rata share of the emission reductions, verified and certified in accordance with agreements reached with the respective countries 'hosting' the projects. As a pilot activity, the PCF does not endeavour to compete in the emission reductions market; it is restricted to US$180 million and is scheduled to terminate in 2012.

The Netherlands Clean Development Facility

The World Bank announced an agreement with the Netherlands in May 2002 establishing a facility to purchase greenhouse gas emission reduction credits. Through CERUPT (the Certified Emission Reduction Unit Procurement Tender), the Netherlands wants to implement CDM by providing funds for the acquisition of Certified Emission Reductions (CERs). CERs from these projects may be issued to the account of the investor country. In order to realize this, host country and investor country both approve the project as a CDM project. The Facility's initial target was to purchase 16 million tons of carbon dioxide equivalent (Mt CO_2e) in the first two years of the agreement. The agreement has now been extended, with a firm commitment to purchase an additional 5 Mt CO_2e by mid-2005, and permits a further purchase of up to approximately 11 million tons of carbon dioxide equivalent. The Netherlands CDM Facility provides an excellent opportunity for many more developing countries to gain invaluable experience by undertaking their first commercial transactions for the purchase of emission reduction credits under the CDM and competing in the emerging global carbon market.

The Community Development Carbon Fund (CDCF)

The CDCF provides carbon finance to small-scale projects in the poorer rural areas of the developing world. The Fund, which was designed in cooperation with the International Emissions Trading Association and the United Nations Framework Convention on Climate Change and became operational in July 2003, is a public/private initiative that has a target size of US$100 million and is still open to subscriptions. The CDCF supports projects that combine community development attributes with emission reductions to create 'development plus carbon', and will use financial innovation to improve the lives of the poor.

The BioCarbon Fund

The World Bank intends to mobilize a new fund to demonstrate projects that sequester or conserve carbon in forest and agro-ecosystems. The Fund, a public/private initiative administered by the World Bank, will aim to deliver cost-effective emission reductions, while promoting biodiversity conservation and poverty alleviation. The BioCarbon Fund was formally approved by the World Bank Executive Board of Directors and officially opened for Participant contributions on 26 November 2003. It started operations on 14 May 2004 with a capital of $10 million. The target size of the Fund is US$100 million.

The Italian Carbon Fund

In 2003, the World Bank entered into an agreement with the Ministry for the Environment and Territory of Italy to create a fund to purchase greenhouse gas emission reductions from projects in developing countries and countries with economies in transition that may be recognized under such mechanisms as the CDM and JI. The Fund is open to the participation of Italian private and public sector entities.

5.7 Conclusion

As can be seen from the above analysis and overview of developments in many jurisdictions, despite the fact that the Kyoto Protocol has not yet entered into force, there has been significant global implementation of its flexible mechanisms. Even in those Annex I countries which have not ratified the Protocol, measures have been taken to address climate change. In addition, several countries with no quantified commitments under the Kyoto Protocol, such as China, are also considering the possibility of implementing national emissions trading schemes. Ultimately the speed at which the flexibility mechanisms are implemented will depend on whether the Kyoto Protocol enters into force. However, the ability and willingness of countries to implement the flexible mechanisms in the absence of its entry into force suggest that the Protocol has already advanced domestic regulation of greenhouse gas emissions through the development of structures that anticipate Kyoto.

Notes

1. See Yamin, Part I in this volume, for a discussion of the role of DNAs in the CDM project cycle.
2. See Lefevere, Part II in this volume, for a discussion of the EU ETS and the Linking Directive Proposal.
3. See the Conclusion by Erik Haites in Part IV of this volume for a more detailed discussion of US initiatives.
4. Participating states include Connecticut, Vermont, New Hampshire, Delaware, Maine, New Jersey, Pennsylvania, Massachusetts and Rhode Island. Maryland and Pennsylvania are participating as observers. Cited in State of New York press release: 'Governor Announces Cooperation on Clean Air Initiative', 24 July 2003. See also the Conclusion by Erik Haites in this volume.

5. 'Regional Greenhouse Gas Initiative: Goals, Proposed Tasks, and Short-term Action Items', Adopted at Meeting of Commissioners, 29 September 2003.
6. Senate Bill 372.
7. House Bill 284-FN.
8. DEP Regulation 310 CMR 7.29.
9. H.R. 2569, H.R. 2980, and S. 1369.
10. See the Conclusion by Erik Haites in Part IV of this volume for details.
11. See Yamin, Part I in this volume, for a more detailed discussion of participation and eligibility requirements.
12. Yamin, Part I in this volume, explains the underlying policy reasons for scepticism in more detail.
13. See Yamin, Part I in this volume, on the evolution of the CDM, and the contribution by Kenber, Part III, Chapter 6 in this volume.

Chapter 6

The Clean Development Mechanism: a tool for promoting long-term climate protection and sustainable development?

Mark Kenber

6.1 Introduction

Despite the long negotiations and the compromises that many felt seriously weakened the final agreement, the months following the adoption of the Kyoto Protocol in late 1997 were times of heady enthusiasm among those who championed the cause of climate protection and sustainable development. In particular, those who saw the use of market mechanisms as an essential part of the environmental policy toolkit were excited by the inclusion of International Emissions Trading, Joint Implementation and the Clean Development Mechanism among the Protocol's provisions.

Although the details would not be finalized for another four years, perhaps the greatest enthusiasm surrounded the creation of the Clean Development Mechanism. Seen by many as a way of achieving two benefits for the price of one – cost-effective GHG emission reductions in the North and increased sustainable development investment in the South – estimates of the size of the market and corresponding investment flows reached as much as $17 billion a year by 2010.[1] Many also saw the promise of the CDM catalysing markets for sustainable energy and setting developing countries on the path to a low-carbon future. Likewise CDM would reconcile the goals of economic efficiency and climate protection, mirroring the growing interest in the pricing of environmental services as a tool for conservation.

Although 2004 has seen a rise in CDM interest and a growing number of projects, much of the initial enthusiasm has now waned. This has been the result of a number of factors, all of which have served to dampen expectations of both the likely volume of investment and the benefits that this will bring. The US Administration's announcement in March 2001 that it would not ratify the Kyoto Protocol has perhaps been the most important of these; this raised the spectre that the Protocol might not enter into force[2] and effectively removed the largest potential buyer from the market.[3] The uncertainty over if and when Russia will ratify the Protocol has led to greater caution over the eventual size of the market for CERs.

Technical issues surrounding the CDM itself have also resulted in greater realism. The complexity of the procedures for receiving approval by the CDM Executive Board – not least regarding additionality and appropriate baselines – and the resulting high transaction costs involved with CDM projects have also caused some reductions in aspirations regarding the *speed* at which the CDM market would be established on a sizeable scale.

Nevertheless, with European governments required to define their National Allocation Plans under the EU emissions trading system[4] and concomitant decisions about policy for other sectors, it is becoming clear that many will be looking to buy CERs and ERUs. In addition, there is a small but growing market for credits from CDM and CDM-like[5] projects for purposes other than compliance with Kyoto or national emission reduction targets. This non-Kyoto market may take on even greater importance if the Kyoto Protocol does not enter into force.

Therefore, despite the fact that the scale of the CDM market is considerably smaller than originally hoped and expected, with perhaps less than €1bn invested annually, until 2012 at least it seems likely the CDM will have an important role in international climate policy and emissions reductions. Whether it has greater longevity – for example, as part of a long-term global policy framework – will of course depend on the ability of nations to agree to the deeper emission cuts that the scientific evidence suggests is necessary, in particular the terms on which the United States seeks to re-engage with the climate regime.[6] However, it will also depend on the CDM's ability to meet its twin goals: to assist Parties not included in Annex I in achieving sustainable development and in contributing to the ultimate objective of the Convention,[7] and to assist Parties included in Annex I in achieving compliance with their quantified emission limitation and reduction commitments.[8] This chapter looks at the contribution the CDM can make to sustainable development, focusing on methodological tools to assess the sustainable development component of projects that are being deployed by CDM project proponents, national authorities and those active in the emerging carbon markets. The chapter closes by looking at what improvements might be made to the CDM, concluding that the biggest problem – low prices – result from effective lack of demand caused by lack of ratification of the Protocol by the United States. Other problems caused by lack of guidance on how the sustainable development components should be assessed and procedural complications leading to increased transaction costs will also need to be considered when the modalities of the CDM are next considered.

6.2 Assessing the CDM's contribution to sustainable development

The CDM's potential contribution to sustainable development in host countries has generally centred on the first of the two objectives mentioned above and in particular the specific mention of sustainable development. How this is defined and can be assessed is dealt with in the following subsection on direct contribution.

However, given the reference to the long-term objective of the Climate Convention and the role that climate change itself plays in sustainable development, any assessment of the CDM's contribution must recognize the wider role that projects and the mechanism itself can have. This may in fact be more significant than the direct sustainable development benefits and is discussed in the subsection on the contribution to sustainable development on p. 268. Finally, in the subsection on p. 269 we consider the role of additionality in the CDM's sustainable development contribution.

Direct contribution

How the CDM's contribution to sustainable development should be interpreted and implemented was the subject of considerable debate among negotiators and observers alike during the negotiation of the Kyoto Protocol and Marrakesh Accords. Host countries were concerned about their sovereignty and were unwilling to have externally determined sustainable development priorities imposed on them, while buyer Parties were happy to avoid trying to operationalize the concept.

As a result, in spite of demands by environmental groups that the sustainable development contribution should be codified and measured against a series of indicators,[9] the decisions contained in the Marrakesh Accords were both general and aimed at avoiding unwanted impacts rather than the active promotion of sustainable development. The main components are:

- affirmation by the host country government that the project activity meets its own sustainable development criteria;

- a requirement that an environmental impact assessment (EIA) be carried out if deemed necessary by project participants;
- a mandatory process of public consultation;
- a requirement that any project funding not result in a diversion of overseas development assistance (ODA).

Concern that CDM projects might be concentrated in a few larger countries with more developed economies also led Parties to charge the CDM Executive Board with the responsibility for ensuring that there is an equitable geographical spread of projects. This is seen to be of particular interest to Africa due to its weak energy infrastructure and difficult investment environments[10] but how this is to be implemented has not yet been elaborated.

The details of these four aspects of CDM project activity approval have already been covered in some depth by Yamin in Part I, so here we limit ourselves to a few comments on their relationship to the likely sustainable development outcomes of projects.

Host country approval

As noted above, the negotiations over how to implement the sustainable development contribution were fraught with difficulties and the final agreement specified only 'that it is the host Party's prerogative to confirm whether a clean development mechanism project activity assists it in achieving sustainable development'.

Consequently, non-Annex I countries can define the sustainability requirements for CDM projects in their country according to their own wishes which, in theory, should guarantee that projects at least do not undermine sustainable development goals. In practice this is unlikely to lead to many projects being rejected or modified nor to particularly stringent approval criteria as governments, especially countries short of foreign investment, will be reluctant to risk losing inflow of funds and the opportunity to build a portfolio of projects.

Environmental impact assessments

As with the debate over how to define a CDM project's contribution to sustainable development, the proposal by NGOs and a number of governments that there should be a basic set of environmental and social impacts that project participants should assess and report on was the subject of heated discussion. Once again, the imposition of a predetermined set of indicators was viewed by a number of Parties as an infringement on their sovereignty as was the idea of a mandatory (EIA) for all (large) projects.

As a result the final language is weak, but it was to be hoped that the requirement for EIAs will at least ensure that serious adverse impacts and major threats to sustainable development will be avoided. However, experience suggests that without considerable scrutiny and assurance that prevention and mitigation measures are put into place, such impact assessments may have only limited effect.

Public consultation

As set out in detail in Part I, the Marrakesh Accords provide specific opportunities for the public to comment on CDM projects at various stages of the project cycle. The Rio Declaration, Agenda 21, the Aarhus Convention and the World Summit on Sustainable Development (WSSD) agreement all recognize that important aspects of environmental responsibility and sustainable development include: public access to information, meaningful public participation in decision-making, and access to justice, including redress and remedy.

Active public participation can bring a number of benefits to the design of sustainable CDM projects:

- local communities can influence project design, bringing their knowledge of prevailing conditions and ensuring their priorities are met;
- project developers are able to recognize community needs and gain public support early in the project cycle, avoiding subsequent delays and financial risks associated with costly political opposition, legal action or local unrest;
- incorporating the expertise of citizens and NGOs on baseline methodology and additionality data increases the likelihood that projects provide real, measurable emissions reductions.

Despite these benefits, the CDM rules are relatively quiet on how the direct engagement of local stakeholders should be undertaken, beyond the requirement that documents be made available for comment for 30 days on the Internet. Again this resulted from the unwillingness of countries with diverse approaches to public participation to agree harmonized (and stricter) standards. Concerns elicited from the local population – likely to centre on a project's contribution to their sustainable development – need only be reported on in the project documentation, but how this is to be done and the influence it has is not defined.

Overseas development assistance

The CDM's role as promoter of sustainable development in host countries aligns it with many ongoing North–South development activities and the language in the Marrakesh Accords seeks to ensure that any ODA used for CDM projects complements and does not replace existing funding streams.

The reasoning for the agreement not to divert ODA to CDM was clear:

- Current ODA falls short of commitments made and the needs it is meeting will continue to exist; adding climate change as an issue should therefore bring additional financing.
- CER purchases are a commercial exchange and should be seen as such.
- The CDM may not be the most effective way to channel ODA resources, especially in the absence of strong sustainability standards.
- 'Millennium Commitments' to increase ODA, made in Monterrey in 2000, will be undermined if donors seek new ways of meeting those commitments, as the agreement was based on ongoing need, not on the additional generation of emissions reduction credits.

Contribution to sustainable development

The aspects of the Marrakesh Accords that deal directly with sustainable development do more to avoid social and environmental damage than they do in terms of active sustainable development promotion. Indeed, the primary focus of almost all the projects proposed to date has been on maximizing the generation of CERs rather than being sustainable development projects that, at the same time, reduce emissions.

This constitutes a rather limited view and it can be argued that the contribution to sustainable development should encompass a wider variety of emissions and non-emissions-related issues. Indeed, Article 12.5 of the Kyoto Protocol states that emission reductions should bring 'long-term benefits related to the mitigation of climate change', beyond any more immediate impacts. In particular the following have been identified as potential benefits of well-designed CDM projects:

- direct financial incentives for proving the competitiveness of paradigm-shifting technologies;
- development of supporting policy initiatives;
- increased understanding and acceptance of the importance and application of sustainable energy technologies;
- dissemination of best-practice techniques;
- strengthening of local institutional capacity, including: credit provision, extension services, technology development and training;
- increased foreign investment;
- increased access to sustainable energy services.

These, however, are by no means guaranteed in the vast majority of CDM projects. The efforts to operationalize sustainable development in the CDM described in section 6.3 are all attempts to give these benefits equal if not greater weight than the emissions reductions generated by a project. However, none of the benefits can be attributed to the CDM if additionality is not rigorously applied.

Additionality

The additionality requirement is key to the immediate climate impact of CDM projects. In terms of impact on global emissions reductions, the best the CDM can do is not to allow global emissions to rise: any reductions generated by projects merely offset an equal rise in the country that acquires the CERs.

If a project is not additional, no new emissions reduction takes place and Annex I emissions will rise without the corresponding reduction elsewhere. The extent of this increase will depend on the baseline and size of the project. Since GHG emissions contribute to climate change, itself inimical to sustainable development, that CDM projects be additional is clearly of central importance in a project's sustainable development contribution.

This importance is underlined when one considers that if a project itself is not additional, the other benefits that are generated – including most of those mentioned above – will also have occurred anyway and therefore cannot be attributed to the CDM. Since the long-term climate change benefits depend to a large extent on the multiplier and catalyst effects of CDM projects, without additionality these will be lost.

As stated in Bernow et al (2000):

> *With inadequate additionality safeguards, credits awarded to business-as-usual projects could be as much as 600 Mt C or a quarter of the OECD reduction requirement under the Kyoto Protocol. Arguably, a small flow of free-rider credits might be acceptable, if ... the CDM catalyzed development and adoption of technologies that could underpin a global transition away from carbon-intensive fuels and contribute to sustainable development. But, in the cases investigated here, it is not evident that the magnitude of potential free-rider credits is justified by the obtained benefits, such as the transfer of some renewable energy technologies to the host countries.*

6.3 Tools to assess CDM project eligibility and sustainability

Concerns over the ability of the CDM to at least balance emissions increases in industrialized countries and make a meaningful contribution to sustainable development have led to a number of independent initiatives that aim to give greater definition to this contribution. Likewise, host country governments, through their Designated National Authorities (DNAs) have worked to develop national project approval criteria. In one way or another, all attempt to identify a set of procedures, criteria and indicators that allow the sustainability of projects to be evaluated.

Host country approval criteria

In order to be registered with the CDM Executive Board, a project activity must receive the official approval of both host and investor country governments. As a result, many potential hosts have already developed a set of approval criteria or are in the process of doing so, often with the financial and/or technical assistance of international donors. In most cases, however, as a result of the competition for the reduced levels of investment likely to be directed at the CDM, governments have been unwilling to install too complex or strict conditions for approval.

At the time of writing 64 countries had established Designated National Authorities, the entities responsible for approving CDM projects. Most have done little in terms of establishing national CDM criteria beyond the rules established under the Marrakesh Accords. The most notable exception is South Africa which has a list of questions concerning project impacts that must be answered by project participants and an assessment framework with over twenty parameters and a numeric scoring system.

A number of countries have received capacity-building support that has included the development of national criteria. UNDP, for example, has worked with the Nicaraguan and Moroccan governments to develop CDM project approval criteria.[11] In Nicaragua, three basic requirements have been identified, each of which is based on an assessment matrix:

- contribution to the improvement of the social, economic and environmental conditions of Nicaragua;
- contribution to the implementation of environmental plans and strategies – biodiversity, climate change, desertification, among others;
- contribution to the implementation of national development plans and strategies.

Similarly, in Morocco, three basic conditions for approval have been defined:

- The project must be integrated into a developing country's main objectives and be a part of the defined priorities in the National Sustainable Development Strategy.
- The project must conform to current country laws and in particular those laws related to the environment and its preservation. It is particularly essential that an environmental impact study be realized in conformity with the national regulations on environmental impact studies.
- The project must use clean and confirmed technologies and avoid any outdated technologies.

A set of core indicators have also been established:

- contribution to the mitigation of global climate change;
- contribution to the sustainability of the local environment;
- contribution to the creation of employment;
- contribution to the durability of balance of payments;
- positive contribution to the macro-economic plan;
- effects on costs;
- contribution to technology autonomy;
- contribution to the sustainable use of natural resources.

However, in both cases, as in many other countries, neither the decision criteria nor how tradeoffs are to be assessed are made clear, leaving decisions to be made on an ad hoc basis.

Helio International and SouthSouthNorth Network

The first comprehensive independent proposal for operationalizing the CDM's twin goals was Helio International's Criteria and Indicators for Appraising Clean Development Mechanism (CDM) Projects, published in 1999.[12] These were designed to fulfil two functions: as a tool for influencing the then ongoing negotiations on the CDM's rules and procedures and as a framework to guide project developers.

The document first sets out 26 criteria, covering project selection (including eligible project types, baselines, supplementarity, regional equity and sustainable development), project participation (capacity building, alignment with national   priorities and sovereignty), project verification (auditing and reporting), project crediting (insurance and leakage) and project financial issues (additionality, ODA, share of benefits and cost-effectiveness). These criteria are designed to act as operational guidelines for implementing the provisions of the Kyoto Protocol's Article 12.

In the area of sustainable development, a set of eight indicators – reflecting the three-pillared approach to sustainability plus technology development – is also developed (see Table 6.1).

Each of these indicators is assessed and scored –1, 0 or 1, with –1 indicating a significant negative impact and 1 a significant positive impact. These indicators have subsequently been further elaborated and road-tested by SouthSouthNorth Network on projects in Brazil, South Africa, Indonesia and Bangladesh and criteria for accepting projects have been developed. They are currently being used to guide further project development.

Table 6.1 *Eight indicators of sustainable development*

Sustainability component	Indicator	Measurement
Environmental	1. Contribution to the mitigation of global climate change	Percentage reduction in greenhouse gas emissions below a baseline
	2. Contribution to local environmental sustainability	Percentage change in the emissions of the most significant local pollutant, including air pollution, solid or liquid waste
Social	3. Contribution to net employment generation	No of additional jobs created by the CDM project
Economic	4. Contribution to the sustainability of the balance of payments	Net foreign currency savings resulting from reduction of imports
	5. Contribution to macroeconomic sustainability	Reduction of direct government (national, provincial and local) investments (including budgets of state enterprises) made possible by the foreign private investment in the CDM project
	6. Cost effectiveness	Cost reductions implied by the CDM project
Technological	7. Contribution to technological self-reliance	Reduction of foreign expenditure on royalty payments, licence fees and imported technical assistance via a greater contribution of domestically produced equipment
	8. Contribution to the sustainable use of natural resources	Reduction in the depletion of non-renewable natural resources through the adoption of technologies with higher energy efficiency or through an increased deployment of renewable resources

The Gold Standard

The Gold Standard – an off-the-shelf project methodology for Joint Implementation (JI) and CDM projects – builds on the work of Helio and SouthSouthNorth, providing a set of criteria for assessing the environmental effectiveness of projects and basic procedures that need to be followed by project developers and the CDM's operational entities.

The Gold Standard aims to provide assurance that CDM projects will deliver real emissions reductions and a clear contribution to sustainable development, focusing on renewable energy and energy-efficiency projects that meet these criteria. Based on wide public consultation and building on the guidance given by the CDM Executive Board in its project design document (PDD), the Gold Standard reflects three design criteria:

- balance between environmental rigour, ease of application by project developers and operational entities and additional transactions costs;
- direct compatibility with the CDM and JI project cycles;
- global standards, readily applicable in a variety of local and national contexts and across different sectors.

In addition to the requirements of the CDM, the Gold Standard comprises three screens: project type; additionality and baselines; and sustainable development.

Project type

The project activities eligible for the Gold Standard were chosen on the basis of their adherence to the following characteristics:

- paradigm-shifting energy technologies;
- inherent additionality and sustainability attributes;
- widespread support from environmental NGOs.

As a result, a Gold Standard project activity must employ exclusively one or more of the technologies shown in Table 6.2.

Table 6.2 *Gold Standard technologies*

Renewable energy:

- Photovoltaic
- Solar thermal
- Ecologically sound biomass, biogas and liquid biofuels for heat, transport and electricity generation
- Wind
- Geothermal
- Small low-impact hydro, with a size limit of 15 MW, complying with World Commission on Dams guidelines

End-use energy-efficiency improvement:

- Industrial energy efficiency
- Domestic energy efficiency
- Energy efficiency in the transport sector
- Energy efficiency in the public sector
- Energy efficiency in the agricultural sector
- Energy efficiency in the commercial sector

Additionality and baselines

The Gold Standard takes an approach to additionality that recognizes the importance of both environmental and project additionality. The screens are designed to ensure that CERs are awarded neither to business-as-usual activities nor for reductions of emissions that have not occurred or result from a CDM project activity. Project developers must respond to two basic questions:

1. Would the project activity have occurred in the absence of the CDM?
2. Will the project activity result in lower greenhouse gas emissions than would have occurred in the absence of the project?

The answer to question 1 must be 'no' while the answer to question 2 must be 'yes' for a project to meet the Gold Standard.

In order to answer question 1, a project activity must not previously have been announced as going ahead without the CDM, prior to any payment being made for the implementation of the project, except in cases where the project activity was subsequently cancelled. A project developer must also show that the use of the CDM and/or receipt of carbon credits enables the project activity to overcome at least one barrier that would otherwise prohibit its implementation. These barriers are divided into five categories: financial, political, institutional, technological and economic.

Question 2 concerns the methodology and technical assumptions used to construct a baseline for the project activity. In order to reduce the risk of artificially inflating the number of CERs received by a project activity, baselines must be constructed in a conservative manner. In practical terms, where there is uncertainty over one or more numerical data sets (such as generator efficiencies, fuel types and resulting emissions factors, etc.) or more than one credible methodology, the more conservative number or approach should be used (i.e. that which produces the lowest baseline emissions). This includes the technical assumptions used and should take into account likely technological and policy development.

With regard to ODA, the Gold Standard simply states that such funding must only cover activities prior to registration of the project activity, while project operational costs and purchases of CERs are specifically excluded.

Sustainable development

The Gold Standard seeks to ensure that the sustainable development aspects of CDM project activities are maximized with the involvement of local stakeholders. The sustainable development standards comprise three elements: the use of a sustainability matrix, environmental impact assessment (EIA) procedures and stakeholder consultation.

The sustainability matrix provides a simple and participatory means of assessing a project's contribution to sustainable development. Potential impacts in three areas – environmental, social and economic – are assessed on the basis of existing data, stakeholder consultation and, where necessary, on-site measurement. The key variables are shown in Table 6.3. Projects must demonstrate that the overall contribution is positive, be non-negative in all components and have no major adverse impact on any variable.

Table 6.3 *Key variables in the components of the sustainability matrix*

Component
Local/regional/global environment
• Water quality and quantity
• Air quality (emissions other than GHGs)
• Other pollutants (e.g. toxicity, radioactivity, POPs, stratospheric ozone layer depleting gases)
• Soil condition (quality and quantity)
• Biodiversity (species and habitat conservation)
Social sustainability and development
• Employment (including job quality, fulfilment of labour standards)
• Livelihood of the poor (including poverty alleviation, distributional equity and access to essential services)
• Access to energy services
• Human and institutional capacity (including empowerment, education, involvement, gender)
Economic and technological development
• Employment (numbers)
• Balance of payments (sustainability)
• Technological self-reliance (including replicability, hard currency liability, skills development, institutional capacity, technology transfer)

Source: Gold Standard project design document, available at:
http://www.cdmgoldstandard.org.

The EIA component is designed to give consistency in the application of EIAs across projects and create an explicit link between the assessment, stakeholder consultation processes and evaluation of the project's contribution to sustainable development. In addition to when host country regulations demand one, or when the project participants deem one necessary, an EIA is required under the Gold Standard when initial stakeholder consultations or an environmental pre-screen show that there are likely to be significant impacts.

In addition to the requirements contained in the CDM PDD, the Gold Standard stakeholder consultation process requires at least two consultations – including one open meeting – before and after an EIA and any modifications to the project activity are carried out. Local policy-makers, local people directly impacted by the project and, if applicable, local NGOs must be involved and local and national NGOs that have endorsed the Gold Standard must be invited to participate in the initial consultation and in the main consultation process. The Gold Standard PDD contains a checklist to be used to guide the consultations and requires that non-technical project summaries and impact statements be made available to stakeholders in their own language. For a project activity to meet the Gold Standard there must be general support for and no significant opposition to the project once required changes have been incorporated.

CDM and JI project developers wishing to have their projects validated and verified to the Gold Standard should follow the same procedures as any other CDM or JI project and instruct the operational entity they employ to base their work on the Gold Standard PDD and technical appendices instead of the basic CDM PDD. The certificate of the operational entity that the Gold Standard has been met is sufficient to demonstrate compliance.

Sutter's Sustainability Check-Up for CDM Projects

Christophe Sutter (2003) assesses these and other approaches on the basis of four requirements – adjustability with regard to preferences, the relative nature of measurements, the validity of results and the comprehensiveness of approach – and finds all wanting on one or more aspects. Instead he proposes a multi-attributive assessment of CDM, based on multi-attribute utility theory (MAUT). This enables the development, weighting and assessment of a range of indicators that reflects the priorities of the project participants and stakeholders. Rather than define indicators a priori these are developed as part of the process of evaluating a given project.

This process has five steps:

1. Identification of sustainability criteria.
2. Defining indicators:
   (a) specifications of indicators;
   (b) utility functions of indicators.
3. Weighting the criteria.
4. Assessment of CDM project.
5. Project aggregation and interpretation of results.

As with the other approaches, the criteria and indicators reflect some fairly standard sustainable development issues. In the road-testing of the methodology in India and South Africa the criteria and indicators listed in Table 6.4 emerged.

Table 6.4 *Criteria and indicators in multi-attributive assessments of CDM projects*

Criteria	Indicators	Measurement
Social	Stakeholder participation	Qualitative, using five-step scale
	Improved service availability	Change in availability of services
	Capacity development	Qualitative, using five-step scale
	Equal distribution of project return	Quantitative share of turnover benefiting people below poverty line
Environmental	Fossil energy resources	Quantitative MWh coal saved/GHG reduction
	Air quality	Change relative to baseline (quantitative compilation and qualitative judgements)
	Water quality	Change relative to baseline (quantitative compilation and qualitative judgements)
	Land resources	Change relative to baseline (quantitative compilation and qualitative judgements)
Economic	Microeconomic efficiency	Internal rate of return (IRR)
	Technology transfer	Qualitative, using five-step scale
	Regional economy	Economic performance of project location
	Employment generation	Additional man-month per GHG reduction

Source: Sutter (2003).

In order to be able to sum and compare the value of the indicators, a utility function is constructed for each, with a resulting value on a continuous scale between −1 and 1. The indicator and its utility are then weighed by each participant/stakeholder so as to best reflect individual, local and national sustainable development priorities. At this point the methodology can be applied to specific projects with scores being ascribed by the evaluators using quantitative measurements and/or participatory assessment techniques as part of a stakeholder consultation. Finally, the scores are added up[13] and an overall total calculated.

A project is only considered acceptable if:

- its overall utility meets the minimum requirement (usually greater then zero);
- any criterion considered critical performs above its defined threshold.

Buyer criteria

In addition to the host country and independent standards and assessment frameworks already described, the existing funds and tender programmes have their own screening processes for accepting or rejecting projects. The two largest to date – the World Bank's Prototype Carbon Fund (PCF) and the Dutch government's CERUPT programme – both include such screens but do not go beyond the requirements of the Marrakesh Accords or the CDM Executive Board in terms of additionality testing or sustainable development criteria.

The Bank's Community Development Carbon Fund (CDCF) – a smaller and more recent counterpart to the PCF – represents an attempt to give the CDM a more direct sustainable development focus by investing in small-scale projects with local development benefits. This to be achieved by applying the following broad project selection criteria:

- consistency with the UNFCCC and/or the Kyoto Protocol;
- consistency with relevant national criteria;
- consistency with IBRD's Country Assistance Strategy;
- achievement of national and local environmental benefits;
- improvement in the quality of life of the poor – projects must provide measurable and certifiable benefits on local livelihoods;
- generation of independently documented benefits for poorer communities in developing countries;
- the CDCF management will work to achieve the goal of placing at least 25 per cent of the first tranche of funds in projects located in LDCs and other poor developing countries.

Despite these laudable aims, it is questionable, given the low prices offered to developers – a maximum of around €5 per tonne of CO_2e – whether significant non-carbon benefits will be achieved.

6.4 Evolution of the CDM market

Although the CDM is still at an early stage – the first projects are expected to be registered in 2004 – there has been enough pre-registration activity to enable a preliminary assessment of how the projects proposed match up to the attempts to operationalize and strengthen the sustainable development aspects of CDM projects outlined above. We start by giving an overview of the current state of the CDM market.[14]

The CDM market in 2004

In 2003 the international carbon market began to take shape, with Point Carbon reporting transactions totalling nearly 80 million tonnes of CO_2 equivalent, including EU and UK allowances, assigned amount units, certified emissions reductions and emission reduction units. Within this total, which is expected to double during 2004, forward CER trades, despite being lower than forecast earlier in the year, accounted for over 60 per cent. Two-thirds of these were credits to be generated by Prototype Carbon Fund and CERUPT projects.[15]

CDMWatch reports that, up to March 2004, 75 CDM projects had been proposed which expect to generate around 131 million CERs until 2012. Six of these had been made available for the 30-day public comment period required for validation.[16] CDM Watch's summary of these projects states that:

> *Overall, the CDM is dominated by a small number of large countries and a small number of high-volume projects. Non-CO_2 gas capture and destruction projects dominate in terms of credits being generated while renewables are the most common project type.*[17]

In terms of project type, 39 per cent involve renewable energy but, in terms of expected CERs, these are dwarfed by five projects that reduce emissions of non-CO_2 gases and account for nearly half the total. This compares with the 13 million CERs to be claimed by all the renewable energy projects together. A complete breakdown is shown in Table 6.5.

Table 6.5 *Breakdown of CDM project types*

Project type	No and share of total projects		Share of expected CERs (Mt CO_2e)	
Renewables	34	(39%)	13	(10%)
Non-CO_2 gas capture and destruction projects	20	(23%)	71	(54%)
Energy efficiency	13	(15%)	14	(11%)
Large hydroelectric	10	(11%)	9.5	(7%)
Fossil fuel-switching	6	(7%)	15	(11%)
Forest sequestration	2	(2%)	6	(5%)
Waste incineration	1	(1%)	0.5	(<1%)
Total*	87	(100%)	131	(100%)

* The 87 projects included in this table include different components of one project as separate projects where different types of activity take place as part of the same project.

Source: CDMWatch (2004).

While the 75 projects are spread over 26 countries, two countries – Brazil and India – play host to 27 projects representing 73 million of the total 131 million CERs. As expected, a number of big developing countries are beginning to dominate the market. If the three projects in Indonesia and South Korea are added to those in Brazil and India, then these four countries account for over 92 million or around 70 per cent of the 131 million CERs.

Sustainable development contribution

The size of the CDM market is still considerably below the levels expected when the CDM first emerged. As a result, inward investment – a potential driver for sustainable development – in host countries has also been lower than was hoped. As noted above, this situation is expected to change over the coming years as Annex I countries finalize their emission reduction policies; most will need a significant supply of CERs and ERUs to meet their obligations.

Cursory examination of the qualitative contribution that these early projects are likely to make – and any precedent this will set for subsequent projects – reveals a number of weaknesses, with particular concerns over additionality, the types of project and public participation.

Stricter additionality testing would certainly lead to a decline in the number of CERs being claimed by many of the current projects including the renewable energy activities. This was clear in the initial rejection by the CDM Executive Board of all the methodologies proposed by the first round of 15 projects. Moreover, CDMWatch argues that all the nine remaining renewables projects being developed under the Dutch CERUPT programme, such as the Suzlon wind farm – seeking validation at the time of writing – are business as usual. Yet these nine account for about 25 per cent of all renewables projects and are responsible for over 30 per cent of the CERs that renewables projects are claiming in total. As a result the climate change mitigation impact of these projects is likely to be limited.

The CDM is also being used to finance projects that appear inherently unsustainable, such as monoculture plantations in ecologically sensitive areas and large hydropower projects. While these project types do not compare with the non-CO_2 projects (see below) in terms of project numbers or volumes of credits, they are a significant contributor to some portfolios, including the two largest – the PCF and CERUPT. Of the 11 million CERs from the remaining 17 projects in the CERUPT portfolio, 5 million come from large hydro projects, while PCF's renewables projects generate less CERs than the single sinks component of its Plantar project in Brazil: 2.3 million CERs from the eight renewables projects compared to 4.3 million from Plantar's plantations.

Both types of project have aroused significant controversy; the Bujagali hydroelectric plant in Uganda was abandoned as a CDM project by the Dutch government while the Esti dam in Panama has been widely criticized over its lack of additionality. Likewise, a major PCF project, Plantar, in Brazil, has been the subject of a campaign of opposition by many Brazilian NGOs, workers and residents in the affected and neighbouring areas. Only after a letter with 53 signatory organizations and attempts to discredit these critics had failed,[18] did the PCF agree to the possibility of reviewing the project.

Low CER prices

The emergence of non-CO_2 projects, mainly landfill gas capture and destruction of industrial fluorinated gases (F-gases), has demonstrated the ability of the market to discover low-cost options, suggesting that efficiency gains are being harvested. Due both to the size of the projects and the high global warming potential of methane and the F-gases, the volumes of credits that will be generated are large while the costs are low.

The two largest projects provide almost a third of the total projected reductions. These manufacture the ozone-depleting gas HCFC-22. As part of the manufacturing process, a greenhouse gas covered by the Kyoto protocol, HFC-23, is emitted, but is easily captured and destroyed. The first approved CDM methodology was one to abate HFC-23 emissions from an HCFC factory in Ulsan, South Korea, that will yield 12.6 million CERs in the first commitment period. This same methodology was then used by an even bigger HFC-23 project in India which will generate 27 million CERs by 2012. By way of comparison, the nearly 30 renewable energy projects will yield only 18 million credits over the same period.

These projects are likely to be additional and lead to net emissions reductions. Although CERs can be generated for under €0.50 a tonne using technology that is widespread in Europe, the initial set-up costs of the thermal oxidation plants that carry out the process and the lack of regulation requiring them mean that there are no other such plants in any non-Annex I country. The likelihood that these projects will be approved by the CDM Executive Board, the fact that once a baseline methodology has been approved for one project others will be able to use this at no additional expense, their low cost and concomitant high returns are likely to result in proliferation of similar projects.

While destroying HFC-23 is clearly beneficial to the climate, the extra benefits brought by the CDM – possibly as high as selling the HCFC-22 itself – may have the perverse effect of delaying the planned 2040 phase-out of HCFC-22, itself a potent greenhouse gas. Both the lure of potential earnings and the danger of losing the ability to compete may motivate HCFC-22 producers to seek CDM status for their investments even where this is not planned. Countries such as China with large capacity had not planned to seek CDM credit for proposed facilities but may now be forced to do so. The result is a distortion of HCFC policy and large volumes of cheap credits pumped into the CDM. Market projections show that the volume of credit from HFC-23 projects could top 100 Mt CO_2e within a decade.

The fact that large volumes of CERs can be created at extremely low prices also has a significant knock-on effect on the market as a whole and on the development of projects with potentially greater sustainable development benefits. The sustainable energy and efficiency projects that policy-makers talked about when they created the CDM are now minimally represented in terms of the overall amounts of credits from projects now being proposed. The 33 renewables projects submitted to date amount to less than half of the credits claimed for the two HFC projects. With the relatively high transaction costs associated with the CDM registration process, there is a clear advantage for low-cost high-volume projects.

This is borne out by the experience to date that suggests that Annex I investors are primarily interested in using renewables projects to 'green' portfolios that consist mainly of low-cost CERs from supply-side efficiency investments and waste management projects that target non-CO_2 gases. Even with these sustainable energy projects, investors and buyers clearly prioritize low costs over sustainability and, as the first Kyoto commitment period draws nearer and investors' targets become ever more remote, cheap credits will continue to be the primary consideration. Current prices for CERs (as of year end 2003), compared with abatement costs and projected prices in the European emissions trading market, are shown in Figure 6.1.

It is evident that purchasers of CERs are receiving a significant discount to their abatement costs, even when the use of domestic flexible mechanisms, such as the EU ETS, are taken into account. This becomes even greater when one considers that the emission reductions from destroying HFCs may cost less than €1/t CO_2e.

The low prices being offered for CERs force developers to cut corners on additionality and sustainable development aspects of projects. The value of carbon credits provides the necessary financial resources for developers to invest in projects that achieve better outcomes in terms of climate protection and sustainable development compared to investments without carbon finance. Project quality is therefore linked to the amount of additional investment available. A low carbon price results in:

- reduced chances of a project being 'additional' because the carbon finance available to the project developer is insufficient to remove barriers to new and innovative projects;
- reduced chances of a project contributing towards sustainable development. Many project developers consider that sustainability testing will lead to increases in transaction costs that cannot be borne if they have to deliver a low carbon price.

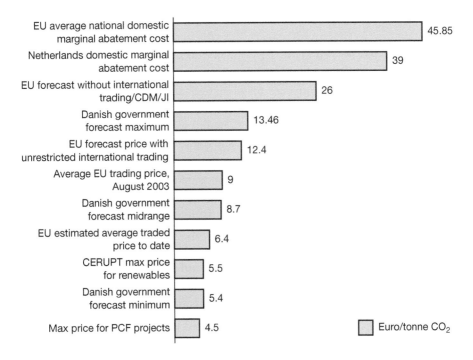

Figure 6.1 *Comparison of abatement costs, carbon credit prices and forecasts.*

Source: Pearson and Salter (2003).

The implications of this are clear: absent a higher CER price, the CDM will not be a significant driver for renewables or contributor to sustainable development in developing countries.[19] While high carbon prices will not guarantee project quality, low carbon prices certainly reduce and in some cases completely negate benefits to host countries. As a consequence, it is perhaps not surprising that projects so far seem to:

- generate few if any net emission reductions, increasing global emissions at a time when the need for deep cuts is becoming increasingly evident;
- result in the market being swamped by non-additional projects generating little new investment and maintaining low prices;
- promote the continued dependence on unsustainable energy sources and technologies and do little to enhance the market for sustainable energy technologies and other long-term climate solutions, despite the declarations in favour of renewable energy and energy efficiency made by many political and business leaders at the World Summit on Sustainable Development in Johannesburg;
- cause environmental and social damages to host country communities.

6.5 Future issues and options

This chapter has looked at the early evolution of the CDM and its potential contribution to sustainable development, considering both on-the-ground impacts and wider consequences for climate change mitigation. Although the first Kyoto commitment period is still more than three years off, early experience with CDM projects has revealed potential weaknesses in the system.

From an economic perspective there is no doubt that the market has been remarkably effective in seeking out low-cost emission abatement opportunities; the non-CO_2 projects put forward will generate large volumes of CERs at a cost well below the existing market price and below domestic abatement costs in Annex I countries. Together with the currently weak demand, the potential volume of these projects is such that they may dampen the price of carbon for a long period. This is clearly of benefit to the buyers of CERs and – inasmuch as it leads to greater reductions and the diffusion of sustainable energy technologies, and assuages concerns about the high costs of reducing emissions – may have positive impacts on emissions over the longer term.

However, low prices are also both the symptom and agent of a number of the more worrying aspects of the way in which the CDM has evolved. The availability of large volumes of cheap CERs removes the incentive for early domestic action, investments and technological innovation that will drive greater future reductions in industrialized countries. In conjunction with the withdrawal of the US and, to a lesser extent, Australia from the Kyoto Protocol, both potentially important sources of demand for CERs, and the relatively lax caps being imposed on domestic emitters by those countries that have remained committed, a key driver for structural changes in energy use and emissions levels is being undermined.

The concern over the additionality of many of the projects proposed to date also suggests that use of the CDM is likely to lead to a rise in global emissions.[20] Again, low carbon prices mean that the key incentive to invest in additional and potentially riskier low-carbon technologies is lost, resulting in at best the diffusion of existing good practice that arguably should already represent the norm. Indeed it appears that many investors and project developers do not even consider the value of CERs or ERPAs (emission reduction purchase agreements) in their project analysis.

The magnitude of the net rise in emissions is hard to estimate. The CDM Executive Board's rejection of many of the early baseline methodologies and widespread recognition of the problems with assessing additionality indicate that the potential is significant, even given the likely additionality of the non-CO_2 projects. Therefore, at a macro level, the CDM appears to be contributing little to current and future emission reductions; it can be argued that rather than facilitating Annex I party efforts to reduce emissions, the ability to receive

CERs from the use of already commercially viable and widely used technologies in fact makes these reductions less likely.

At the project level, few of the projects proposed to date would be likely to meet any of the Helio/SouthSouthNorth criteria, the Gold Standard criteria or pass Sutter's test. Stakeholder consultation rarely goes beyond the required 30-day posting of PDDs on the Internet and the rate of response, except in a few specific cases, is low. It is too early to judge if the projects implemented will lead to significant positive or negative sustainable development impacts, but the additionality problems and the low 'sustainability premium' implicit in carbon prices mean that it is debatable whether these would not have happened anyway. Lack of clear criteria – from either buyers or host countries – requiring concrete and measurable sustainable development benefits lead to these being treated as incidental gains rather than a central feature of projects.

Smaller community-based and designed projects – for example distributed energy systems or efficient boilers – can often make the biggest euro-for-euro contribution to local livelihoods and environmental protection. However, the high fixed transaction costs associated with the CDM project cycle and the minimal contribution of carbon finance mean that few are likely to be viable without concessional financing or as a supplier of specific niche markets.[21] The same is true for projects attempting to use innovative sustainable energy technologies that involve greater risks as a result of their novelty and the often weak financial stability of project developers.

From the discussion above it is clear that – in its present form at least – the CDM is unlikely to act as a significant new source of funds for renewable energy and sustainable development. The projects summarized in Table 6.5 represent a flow of carbon finance of approximately €500 million over their lifetimes, with renewables accounting for only €50–60 million; even if the associated project investments are fairly large, they pale into insignificance alongside total energy investments in developing countries.

The apparent trade-off between the twin objectives of the CDM – cost-efficient emission reductions and sustainable development contribution – highlights some of the more structural weaknesses inherent in the CDM and perhaps the Kyoto mechanisms in general. Foremost is the conflict between the investor and host country objectives. Annex I countries benefit from lax additionality requirements and the availability of low-cost emission reductions. On the other hand, globally applied strict sustainability criteria – such as those described in section 6.3 – serve non-Annex I countries in a twofold manner: they would obtain positive development outcomes and profit from carbon prices that are likely to rise considerably. While it is not the case that economic efficiency and sustainable development are incompatible, their complementarity is only likely if adequate market frameworks are in place. In this sense the CDM market is no different from any other: the rules and guidelines that are applied will determine the outcome.

Beyond this, it is also debatable whether a project-based approach is the most appropriate way to pursue sustainable development goals, except where projects are designed within wider sustainable development programmes or according to a strict set of preconditions that reflect more general priorities, such as renewables-based rural electrification. Again, whether CDM investments are driven by host-country sustainable development policies and measures or investor demand for low-cost emission reductions will depend on the rules that are established and the willingness of countries to agree to maintain high standards and prioritize sustainable development aspects.

If the CDM, or some modified version, is to play a useful – and perhaps more central – role in future climate policy, this agreement will be an essential precursor. Samaniego and Figueres,[22] for example, propose a sector-based CDM, in which the host government takes a policy and programme-led – and, hence, more structural – approach to cutting emissions, using additional domestic and/or external investments. While this approach would overcome some of the project-level problems and reduce transaction costs, as with the use of higher standards its viability will depend on guarantees of investment levels and/or higher carbon prices. Options for achieving this include:

- fixed spending on CDM proportional to GDP or emissions by Annex I countries;
- fixed emissions reductions from CDM investments proportional to emissions by Annex I countries;
- creation of sector-orientated carbon funds with support from international financial institutions;
- use of proceeds from increased emissions-related levies and/or other support to kick-start carbon markets.

If the United States ratified the Protocol, the CDM's potential to play an important role in bringing developing countries into a future global climate regime and catalysing investments in renewables and energy efficiency would increase enormously. This makes it even more important for all the options above to be taken into consideration to ensure this potential is achieved. Without ratification of the Kyoto Protocol of the United States and meaningful second commitment period targets, however, even the current low level of interest in the CDM is likely to dwindle. The continuing lack of entry into force of the Kyoto Protocol has tended to undermine confidence in the long-term future of the CDM at a time when developing countries were expected to gear up their attentions to its implementation. Yet until the Protocol enters into force, it is unlikely that developing countries will want to engage in discussions about the future of the climate regime. Securing the entry into force of the Protocol is thus a critical first step to advancing the next round of policy dialogue and negotiations on the future shape of the CDM.

Notes

1. Austin et al (1999).
2. The Protocol contains a double trigger for its entry into force; at least 55 countries must deposit their ratification – achieved in 2001 – and ratification by industrialized countries with emission reduction targets (Annex I countries) must include countries accounting for at least 55 per cent of total Annex I emissions in 1990 (Article 25). With the US accounting for 36.1 per cent and Russia 17.4 per cent – together 53.5 per cent – of these emissions non-ratification by both these countries effectively blocks entry into force.
3. The third US National Communication suggests that baseline, i.e. without policy, greenhouse gas emissions in 2010 might be as much as 39 per cent above their 1990 levels, compared to a Kyoto target reduction of –7 per cent. This is equal to an annual 'gap' of some 1200 million tonnes of CO_2 equivalent, or 6000 million tonnes over the 2008–2012, to be found from a combination of domestic measures, accounting for carbon sinks and acquisitions of allowances and project credits from overseas. Even if only 10 per cent of this were to come from the CDM, this would dwarf expected demand from the remaining Annex I countries.
4. The deadline for this was 31 March 2004.
5. Verified Emission Reductions (VERs) are credits from projects that are either awaiting CDM registration or designed to generate emission reductions in developing countries without following the full CDM procedures but, in most cases, meeting similar additionality and sustainable development requirements.
6. The link between mechanisms and the long-term evolution of the climate regime is discussed by Haites in the Conclusion to this volume.
7. The ultimate objective of the Convention, as stated in its Article 2, is 'the stabilization of greenhouse gas concentrations in the atmosphere at a level that would prevent dangerous anthropogenic interference with climate system. Such a level should be achieved within a timeframe sufficient to allow ecosystems to adapt naturally to climate change, to ensure that food production is not threatened and to enable economic development to proceed in a sustainable manner.'
8. Kyoto Protocol, Article 12, paragraph 2.
9. See, for example, Helio International, etc.
10. See, for example, Sokona and Nanasta (2000).
11. Kashyap (2003). The indicators presented here are only a sample of those developed by each country.
12. Thorne and Larovere (1999).
13. A multiplicative approach can also be used.
14. Much of the information in this section is drawn from the work of CDMWatch, available at: http://www.cdmwatch.org.
15. *Carbon Market Monitor, CDM Monitor*, available at: http://www.pointcarbon.com.
16. CDMWatch (2004) *Clean Development Mechanism Status Note*, March, available at: http://www.cdmwatch.org.
17. Ibid, p1.
18. 'Re letter from FASE-ES,' Plantar S/A, available at: http://www.cdmwatch.org with the full exchange of letters on this issue.
19. Michaelowa et al (2003).
20. Relative to the level of emissions that would have arisen had industrialized countries had to make all their emission reductions at home.

21. For example, for offsetting the missions associated with high-profile events and for other marketing purposes.
22. Samaniego and Figueres (2002).

References

Austin, D. and Faeth, P. (2001) *Financing Sustainable Development with the Clean Development Mechanism*, Washington, DC, World Resources Institute.

Austin, D., Faeth, P., Seroa Da Motta, R. et al (1999) *How Much Sustainable Development Can We Expect from the Clean Development Mechanism?*, Washington, DC, World Resources Institute.

Bernow, S., Kartha, S., Lazarus, M. and Page, T. (2000) *Cleaner Generation, Free Riders and Environmental Integrity: Clean Development Mechanism and the Power Sector*, Boston, Tellus Institute.

CDMWatch (2004) *Clean Development Mechanism Status Note*, March, available at: http://www.cdmwatch.org.

Karp, L. and Xuemei, L. (2000) *The Clean Development Mechanism and Its Controversies*, UCB Working Paper No 93, Berkeley, CA, University of California.

Kashyap, A. (2003) *Sustainability Assessment of CDM Projects: Methodologies and First-Hand Experiences*, presentation at UNFCCC COP-9, Milan.

Michaelowa, A. et al (2003) *CDM and JI: New Instruments for Financing Renewable Energy Technologies*, background paper prepared for the Bonn Renewables Conference, December, available at: http://www.renewables2004.de/.

Pearson, B. and Salter, L. (2003) *Fair Trade? Who is Benefiting from the CDM?*, CDMWatch/WWF.

Pearson, B. and Shao Loong, Y. (2003) *The CDM: Reducing Greenhouse Gas Emissions or Relabelling Business as Usual?*, available at: http://www.cdmwatch.org.

Point Carbon, Carbon Market Monitor and CDM Monitor (various issues), available at: www.pointcarbon.com.

Samaniego, J. and Figueres, C. (2002) 'Evolving towards a sector-based Clean Development Mechanism', in K. A. Baumert et al (eds), *Building on the Kyoto Protocol: Options for Protecting the Environment*, World Resources Institute.

Sokona, Y. and Nanasta, D. (2000) 'The Clean Development Mechanism: An African delusion?', *Change*, no 54, pp8–11.

Sutter, C. (2003) *Sustainability Check-Up for CDM Projects – How to Assess the Sustainability of International Projects under the Kyoto Protocol*, Berlin, WVB.

Thorne, S. and Larovere, E. (1999) *Criteria and Indicators for Appraising Clean Development Mechanism (CDM) Projects*, Paris, Helio International.

Chapter 7

Determination of baselines and additionality for the CDM: a crucial element of credibility of the climate regime

Axel Michaelowa[1]

7.1 Introduction

The Clean Development Mechanism (CDM) allows countries without emission targets to invest in greenhouse gas reduction projects and thus create Certified Emission Reductions (CERs). CERs are calculated by comparing emissions of the CDM project with emissions of a hypothetical 'baseline scenario' (see Figure 7.1). The baseline shall reflect the business-as-usual scenario (Michaelowa and Fages, 1999). There is a wealth of baseline literature (see reference list in Probase, 2002), particularly several discussion papers of the OECD. For a good summary of these, see OECD (2001). This chapter looks, at the way baseline and additionality issues which will define the environmental credibility of the CDM therefore the amount of investment, and public support, it is likely to attract. It focuses, in particular, on the experience gained from the early experience with methodologies submitted to the Executive Board of the CDM.

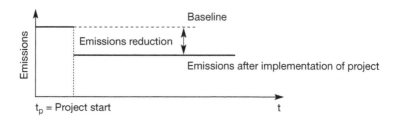

Figure 7.1 *Principle of the baseline.*

Contrary to many expectations, the international community has taken the fears about the integrity seriously and developed a complex international process to determine baselines and have them independently checked. These rules and processes are explained in Part I in detail with this chapter focusing on the practical way in which baselines and additionality issues will be approached. This is important because since 2003, the CDM Executive Board and its Methodology Panel have come a long way in defining the rules for baseline setting and additionality determination but it still remains to be seen how the EB's general guidance will be applied in practice when projects are validated.

Another question is now *which* projects can generate CERs. Shall any project be able to produce CERs if its emissions are below the baseline scenario defined for the project? Or shall it be tested whether the project would have happened anyway and thus is 'additional' to a business-as-usual development? Host countries do not have quantified emission limits, and therefore no assigned amount, from which the CERs would be deducted. Therefore both the host and the investor have an incentive to overstate the amount of emission reduction achieved by the CDM project as they can then enhance revenues (Michaelowa, 1998). If CERs are created that represent emission reductions that would have happened anyway, these 'fake' reductions will undermine the integrity of the Kyoto Protocol. In the international climate negotiations, this debate has been raging over several years.

7.2 Baseline determination

Baselines can be project-specific or standardised. In the latter case, the same baseline is used for an entire class of projects (for a thorough discussion see Probase, 2002). A mix of the two applies if partly standardised data, e.g. emissions factors, are chosen.

The Marrakesh Accords (UNFCCC, 2001) have set three baseline approaches but did not specify how to choose between them:

- existing actual or historical emissions;
- emissions of an 'economically attractive course of action, taking into account barriers to investment';
- the 'average emissions of similar projects undertaken in the previous five years, in similar ... circumstances, and whose performance is among the top twenty percent of their category'.

They also specify that baselines should be project-specific.

Small-scale project rules

For small-scale projects under the thresholds defined in the Marrakesh Accords, baseline rules have been fixed by the Executive Board in January 2003 (UNFCCC, 2003a).[2] These rules give an indication how large-scale project rules could look. Energy project types were differentiated strongly and for each type a methodology was defined. In some cases, project proponents can choose between several methodologies. I will describe the methodologies and data needs for the most important project types.

For an electricity generation system where all fossil fuel fired generating units use fuel oil or diesel fuel, the baseline is the annual kWh generated by the renewable electricity unit times an emission factor (kg CO_2/kWh) defined by the Executive Board for three load factor ranges and five size classes. They range from 0.8 to 2.4. In this case thus no data have to be collected.

For grids with different fuels, the emission factor calculation is relatively complex and needs up-to-date data. There are several options:

(a) the average of the 'approximate operating margin' and the 'build margin', where:
   (i) the 'approximate operating margin' is the weighted average emissions of all generating sources serving the system, excluding hydro, geothermal, wind, low-cost biomass, nuclear and solar generation;
   (ii) the 'build margin' is the weighted average emissions of recent capacity additions to the system. Either most recent 20 per cent of plants built or the five most recent plants are chosen, depending on which achieved a higher electricity generation in the last year where data are available;
(b) the weighted average emissions (in kg CO_2/kWh) of the current generation mix.

These data are often not publicly available and their collection has been difficult for many project developers. Sometimes they have resorted to some rough estimates which, however, so far have not been accepted by the CDM Executive Board.

BOX 7.1 HOW TO CALCULATE GRID EMISSION FACTORS

You install a hydro plant of 10 MW that generates 70 GWh p.a. The grid it serves has the following characteristics:

- 5000 MW hydro generating 35 TWh p.a.
- 10,000 MW coal generating 70 TWh p.a. with an emissions factor of 1.1 kg CO_2/kWh
- 3000 MW gas generating 15 TWh p.a. with an emissions factor of 0.5 kg CO_2/kWh
- 2000 MW oil generating 6 TWh p.a. with an emissions factor of 0.8 kg CO_2/kWh

The last 4000 MW built have the following characteristics:

- 1000 MW hydro generating 7 TWh p.a.
- 2000 MW coal generating 14 TWh p.a. with an emissions factor of 0.9 kg CO_2/kWh
- 1000 MW gas generating 6 TWh p.a. with an emissions factor of 0.4 kg CO_2/kWh

Option (a) is calculated as follows:

The approximate operating margin is $\dfrac{70 \times 1.1 + 15 \times 0.5 + 6 \times 0.8}{91} \times 0.981$ kg CO_2/kWh

The build margin is kg CO_2/kWh $\dfrac{7 \times 0 + 14 \times 0.9 + 6 \times 0.4}{27} \times 0.556$ kg CO_2/kWh

The average of the two is 0.769 kg CO_2/kWh.

Option (b) gives:

$\dfrac{35 \times 0 + 70 \times 1.1 + 15 \times 0.5 + 6 \times 0.8}{126} \times$ kg CO_2/kWh

To maximize CER volume, option (a) is chosen. Baseline emissions are 70 GWh × 769 t CO_2/GWh = 53,830 t CO_2

International procedures to derive baseline rules

In contrast to earlier expectations, no generic rules for baseline setting for large projects were defined by the CDM Executive Board. The Board decided, instead, to set up a 'Methodology Panel' whose task is to evaluate proposed baseline methodologies on a case-by-case basis. Project developers have to submit a new methodology together with their project design document unless an approved methodology already exists for that project type (UNFCCC, 2003c). Each methodology submission is evaluated by two independent experts. Then the Methodology Panel makes a recommendation to the Executive Board which takes the final decision.

In March 2003, the Executive Board agreed on some, fairly general, principles for baseline determination (UNFCCC, 2003b). Data sources and assumptions have to be specified in detail. Baselines are to be defined in relation to production, i.e. emissions per unit of production. This avoids the generation of CERs for reduction of production. Retrofit projects can only use historical emissions of the old plant if the production and lifetime of the plant are not extended. If one chooses the third approach under the Marrakesh Accords, the lower of two emissions factors has to be used:

1. production-weighted emissions of the best 20 per cent of similar projects done under similar conditions in the last five years;
2. production-weighted emissions of similar projects done under similar conditions in the last five years that belong to the best 20 per cent of all currently operating projects.

Until October 2004, 68 methodologies were submitted of which 16 have been accepted. While initially the success rate of methodology submissions was rather low, it has improved in the recent rounds (see Figure 7.2). The sectoral distribution is shown in Figures 7.3 and 7.4.

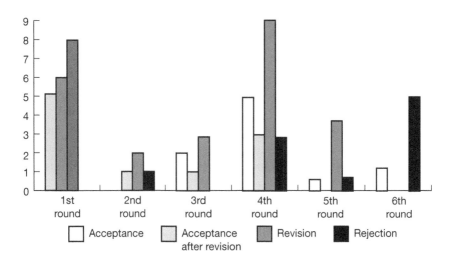

Figure 7.2 *Status of methodology submissions.*

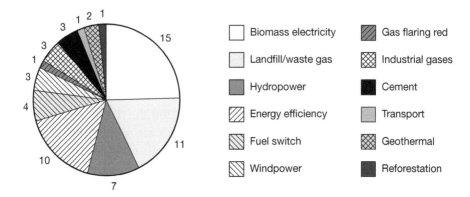

Figure 7.3 *Baseline methodology submissions according to sector.*

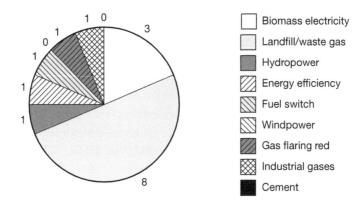

Figure 7.4 *Accepted baseline methodologies according to sector.*

The results show that different methodologies are approved for the same project types due to the fact that they are heavily circumscribed. However, the Executive Board recently started a consolidation of methodologies to avoid 'methodology proliferation' which could give rise to gaming due to 'methodology shopping'. The first two consolidated methodologies for renewable electricity generation (except biomass-based) and landfill gas collection were agreed in September 2004.

What would methodologies for large electricity projects look like under different approaches? While the approaches differ, they often involve the operating/build margins that have been defined for the small-scale projects. Ellis (2003) provides a good overview of the different methodologies submitted. Figure 7.5 sketches the approaches used so far.

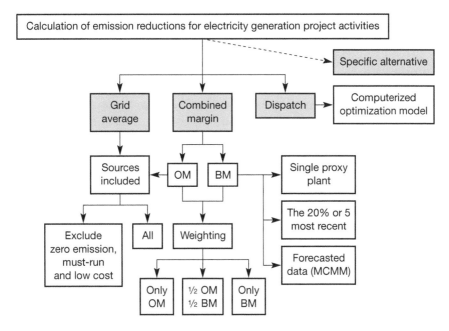

Figure 7.5 *Approaches used for electricity baselines.*

Source: Müller-Pelzer (2004).

Dispatch modelling, also known as the production cost model, is the most sophisticated approach. It can be applied ex post for verification or ex ante if a model is used to simulate the complex operations of the interconnected grid system responding to a volatile demand. Taking into account both short-term marginal costs and long-term marginal costs (also called marginal capacity costs), this approach covers adjustments of the current system as well as impacts on capacity addition. However, its implementation is costly and data requirements are high. Dispatch modelling was a favourite in the initial submissions to the EB but suffered serious setbacks as two methodologies using it in a 'black box' fashion were rejected outright.

The AT Biopower project methodology in Thailand was the first renewable electricity methodology to be approved. It uses a simple grid average (option (b) from above) and forecasts the emission factor for the entire crediting period. If, however, the actual emission factor proves to be lower, the latter has to be used. It could be argued that, in this respect, the developer has an unfavourable deal as the developer will still have to collect the actual data but without a chance to enhance CER generation. A more favourable outcome would have been achieved by using the ex-post approach from the outset. The methodology can only be used for plants where the biomass supply is at least twice the demand from the project.

The second electricity methodology, developed for the El Gallo hydro project in Mexico, uses a combined operating and grid margin (option (a) above) on the basis of ex post activity and grid data. It can only be used for projects below 60 MW.

The Vale de Rosario bagasse co-generation methodology is special inasmuch as it uses a combined operating and grid margin (option (a)) for the first crediting period and switches to a pure build margin for the remaining two seven-year crediting periods. Hydro is included in the operating margin as long as the load is fully covered by hydro. The methodology is restricted to cases where more than 80 per cent of installed capacity is hydro.

In the consolidated methodology for renewable electricity generation, project participants can choose the weight of the operating and build margin. They can also choose between the most recent statistics at time of document submission and an annual ex post up date.

Data needs for baselines – an underestimated problem

To calculate an electricity baseline, up-to-date and detailed grid data have to be available. This is only the case in a few countries like India and Chile, where the generation data for each power plant are posted on the Internet. In China, on the other hand, these data are a heavily guarded state secret and even a multi-man month effort could not report reliable grid emission factors. In other countries like Thailand, current grid data quality and availability is good but likely to deteriorate with liberalization, as independent power producers do not report their generation data. A necessary condition to become an attractive CDM host is thus that the Designated National Authority publishes and regularly updates the operating and build margin of the relevant grids. As this will entail costs, data collection and publication should be integrated into the ongoing multilateral and bilateral CDM capacity-building programmes (see Chapter 7 and Haites in the Conclusion to this volume).

7.3 Why baseline and additionality determination are not the same

Given the wide range of efficiencies, technologies and industrial practices found in most sectors, comparatively GHG-friendly performance will rarely be a sufficient measure of additionality for the CDM. Every sector and every industry has its more GHG-friendly facilities – the world's hydroelectric and nuclear facilities being notable examples in the electric sector. But the purpose of the CDM is not merely to identify low-emitting activities and reward them with a monetary payment. Its purpose is to stimulate additional low-emitting activities that would not have happened otherwise, thereby expanding the range of mitigation options available to Parties striving to meet their Kyoto targets. Any CDM project that purports to create GHG reductions when those reductions do not really reflect any extra achievement in exchange for credits has not legitimately earned those credits. If CDM decisions lose sight of this crucial point, then the CDM will be generating meaningless, unearned, tradable credits that displace real reductions in Annex I countries. To be effective, CDM investment must be clearly directed at new and additional effort (Kartha et al, 2003).

Many observers (e.g. Rentz, 1998) argue that once a baseline is set and the project emissions are below the baseline, the project is also additional. This is a fallacious argument as Figure 7.6 shows.

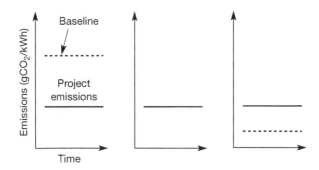

Figure 7.6 *Different baselines for the same CDM project.*

A new gas-fired power station is proposed as a CDM project. Let us assume that the baseline now is defined by the average grid emissions intensity. The high baseline on the left is due to a high share of the grid electricity based on old coal-fired power plants. In a grid with a high share of renewables, the baseline could be lower than the project emissions and the project would not earn any CERs. The baseline can also be equal to the project's emissions intensity (see middle of Figure 7.6). So the same project with the same economic parameters can earn highly different amounts of CERs depending on the grid it feeds.

The baseline can only determine additionality if it is defined by economic parameters. This is the case in the second baseline option of the Marrakesh Accord, where the baseline is the economically most attractive alternative. Using the same figure the left baseline would be chosen if a coal-fired plant would be the most profitable option. The baseline would equal project emissions if a gas-fired power plant is most attractive and it would be lower if a gas-fired co-generation plant is most attractive.

Need for additionality determination

Environmental credibility

As the CDM is the only path to increase the overall assigned amount of Annex B, the existence of non-additional CERs blows up the carbon constraints established by Kyoto through the addition of large amounts of 'tropical hot air'. Once non-additional emission rights have entered the system, they will remain in it due to the banking provision.[3]

No crowding out of real projects

Additionality determination avoids a downward trend on the market price due to the lower supply of CERs (see Figure 7.7). Thus real emission reductions in the Annex B countries and through the CDM are fostered.

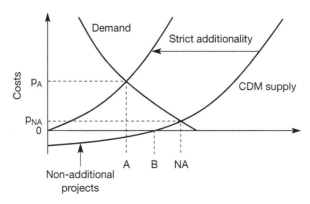

Figure 7.7 *CER price increase through additionality checks.*

An additionality requirement shifts the CDM supply curve to the left and leads to an increase of equilibrium price from p_{NA} to p_A. The amount of additional CDM projects increases from the distance B–NA to 0–A. Obviously, due to the competition between the Kyoto mechanisms and domestic action, the overall amount of CDM will fall from 0–NA to 0–A. However, CDM revenue rises if the demand curve is steep enough.

Real financial and technological flows to host countries

Non-additional projects by definition do not lead to additional financial flows to host countries. A positive revenue results only if the host country retains a part of the CERs. As current emission reduction purchase agreements show, generally CERs are given to the investor. In such a case, a non-additional project is a burden to the host country due to the approval transaction cost.

Incentives for host countries to take up policies and measures and targets

A strict additionality rule leaves profitable projects to the host countries and thus gives them an opportunity for trading if they take up an emission target. Otherwise a target and a continued CDM regime are equally attractive.

Problems related to additionality determination

Higher transaction costs

Additionality determination entails another check and thus increases the transaction costs of CDM projects. However, project developers will anyway calculate financial parameters of their projects to get external finance. If additionality tests are standardized (e.g. by applying thresholds), costs can be kept low. (For suggestions for standardization of financial parameter calculation and country risk factors see Greiner and Michaelowa (2003).)

Gaming of numbers or exaggeration of barriers

Project proponents can deliberately understate the economic attractiveness of their project or blow up barriers. Here the vigilance of validators becomes paramount which should be bolstered by the liability rules of the Marrakesh Accords. The experience with financial auditing shows that an overwhelming majority of audited companies did not 'cook the books'.

'Grubb's paradox' vanishes with rising CER prices

Michael Grubb (Grubb et al, 1999) succinctly stated the paradox that both overly cheap and overly expensive projects cannot make it into the CDM. This is illustrated by Figure 7.8.

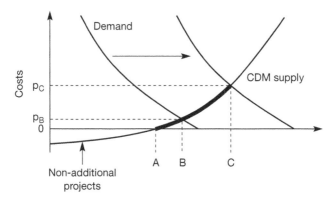

Figure 7.8 *Grubb's paradox.*

If non-additional projects are excluded, CDM is limited to A–B. However, an increase in demand increases CDM volume to A–C. As shown in Figure 7.2, an exclusion of non-additional projects leads to a CER price increase and thus increased CDM volume. Only in the case of an oversupply of hot air does the paradox bind as an increase in CER prices is impossible (see Figure 7.9).

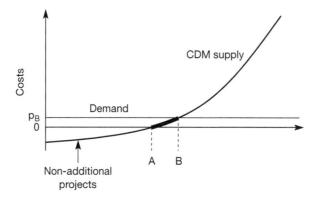

Figure 7.9 *Hot air supports Grubb's paradox.*

Additionality testing in the international CDM rules

Whereas since the Marrakesh Accords many observers thought that a separate additionality test would not be required, the CDM Executive Board decided in January 2003 that additionality should be tested for small-scale projects in the form of a barrier assessment (UNFCCC, 2003a, p19). The barriers tested are:

- investment barrier – a financially more attractive alternative would have led to higher emissions;
- technology barrier – a technologically less advanced alternative is less risky as the new technology has a lower market share and less uncertainty regarding its performance;
- common practice or political requirements would have led to higher emissions;
- other institutional, organizational or informational barrier.

For large projects, the Executive Board specified in March 2003 that an additionality test should be undertaken (UNFCCC, 2003b). After the outcry of project developers due to the rejection of eight methodologies for lack of additionality testing (JIQ, 2003), the Board decided in July 2003 that an 'explanation shall be made of how, through the use of the methodology, it can be demonstrated that a project activity is additional and therefore not the baseline scenario' (UNFCCC, 2003e). Unfortunately, the Board was not able to agree on the clear wording proposed by the Methodology Panel (UNFCCC, 2003d).

As the options for an additionality test specified by the Boad were not comparable, a consolidated additionality test was developed and adopted by the Board in October 2004 (UNFCCC 2004a). This test defines the following mandatory steps:

1. Project proponents that started projects before registration have to provide evidence that the 'incentive from the CDM was seriously considered in the decision to proceed with the project activity'.
2. Alternatives to the project have to be identified, including the continuation of the current situation. All alternatives have to comply with legal requirements, unless noncompliance with those requirements is 'widespread' in the host country.
3. Once the alternatives have been defined, an investment analysis has to be performed to determine that the project is not the most economically or financially attractive alternative. For the investment analysis, developers can choose between the analysis of investment parameters and a government bond benchmark. For projects involving only one potential project developer a company-internal benchmark can be used if the developer 'demonstrates that this benchmark has been consistently used in the past'. Alternatively, the developers can prove that the prohibitive barriers exist that prevent the project. The barrier test has not been specified clearly which means that it may become the open flank of the additionality test. Examples of barriers listed are:

   - unavailability of debt financing,
   - lack of access to international capital markets due to real or perceived risks
   - the project is the first of its kind in the host country.

4. Once the investment or barrier test is completed, an analysis of 'any other activities implemented previously or currently underway that are similar' to the CDM project is to be performed ('common practice test'). Projects are considered similar if they are in the same country/region and/or rely on a broadly similar technology, are of a similar scale, and take place in a comparable environment with respect to regulatory framework, investment climate, access to technology, access to financing, etc. Other CDM projects are not to be included in this analysis. I hope this will not require a thorough assessment of all CDM-relevant project types implemented in the larger host countries, which would be economically prohibitive for most projects. Databases collected by the DNAs and supported by CDM buyers could considerably reduce the documentation costs for project developers.

5. The last step is to prove how CDM registration helped making the project
   attractive or overcoming barriers.

The approved methodologies have a very different treatment of additionality.
Some calculate internal rates of return, others just say in one sentence that they
face a barrier. So now all depends on the interpretation of the additionality tests
by the validators who have to check the project design documents. A positive
note comes from the market leader in validation, Det Norske Veritas, which
recently stated that it rejects about 50 per cent of the projects submitted for validation. Another positive development is the wave of comments submitted on a
15 MW Indian wind power plant submitted under the small-scale rules. The
comments agreed that the additionality test was insufficient. It is likely that the
project will be withdrawn. Thus critical commenting by NGOs could become a
major weapon in the fight to preserve environmental integrity of the CDM.

7.4 Conclusions

Despite, or perhaps because of, fears that the CDM would undermine the environmental integrity of the CDM and related equity concerns about Annex I
Parties' excessive use of the 'low-hanging fruit' from developing countries, the
international rules for baseline determination and additionality testing of CDM
project proposals are surprisingly stringent. Additionality determination will
both help the climate and host countries. Implementing them is no more difficult than financial auditing. The fear of squeezing the CDM expressed by
Grubb's paradox is unfounded as long as there is demand for CERs which ultimately depends on the political will of Annex I Parties to take on reduction
commitments. In a situation of a huge oversupply of hot air non-additional
CDM projects can underbid the low price of hot air sales. In that situation they
would thus create revenues for host countries that would be otherwise foregone. Such a situation is only likely in the initial stage of the climate regime.

Rejection of several proposed methodologies by the EB of the CDM due to
lack of additionality testing as well as the decision on a consolidated additionality
test was a clear but simple message that business-as-usual projects would not be
accepted in the CDM. The long-term implications of this decision, however, on
the day-to-day practice of those interested in the CDM remains unclear and will
emerge only once projects begin to be registered. It also remains to be seen
whether baseline standardisation can be achieved to facilitate this. Certainly, the
signals coming from the Executive Board are encouraging in this regard. A hitherto neglected issue is data procurement to calculate baselines which will need
concerted efforts by Designated National Authorities and international donors.
Thus the CDM will continue to require more international cooperation between

developed and developing countries if it is to play a significant role in the future climate regime.

Notes

1. Programme International Climate Policy, Hamburg Institute of International Economics, Neuer Jungfernstieg 21, 20347 Hamburg, Germany, Tel +49 40 42834 309, Fax +49 40 42834 451, e-mail a-michaelowa@hwwa.de. A part of this paper has been presented as a background paper at the skillshare workshop on CDM in ASEAN, Jakarta, 18–19 March 2004 and I thank the EC-ASEAN Energy Programme for the related funding.
2. See Yamin, Part I, for details of the project cycle for small-scale projects.
3. Although the Marrakesh Accords contain restrictions on the amount of CERs that can be used for compliance by each Annex I Party, and also contain restrictions on banking of certain kinds of Kyoto units, as discussed in Yamin, Part I, these provisions still allow Annex I Parties to effectively increase their carbon budgets (assigned amount) in an unlimited fashion through use of the CDM.

References

Ellis, J. (2003) *Evaluating Experience with Electricity-generating GHG Mitigation Projects*, Paris, OECD.

Greiner, S. and Michaelowa, A. (2003) 'Defining investment additionality for CDM projects – practical approaches', *Energy Policy*, vol 31, pp1007–15.

Grubb, M., Vrolijk, C. and Brack, D. (1999) *The Kyoto Protocol: A Guide and Assessment*, London, Earthscan.

JIQ (2003) *The MethPanel Evaluation*, Groningen.

Kartha, S., Lazarus, M., Michaelowa, A. and Winkler, H. (2003) *Tradable Credits for Mozart? ... and for Milli Vanilli, too?*, Mimeo, Boston, Tellus Institute.

Michaelowa, A. (1998) 'Joint Implementation: the baseline issue', *Global Environmental Change*, vol 8, no 1, pp81–92.

Michaelowa, A. and Fages, E. (1999) 'Options for baselines of the Clean Development Mechanism', *Mitigation and Adaptation Strategies for Global Change*, vol 4, no 2, pp167–85.

Müller-Pelzer, F. (2004) *The Clean Development Mechanism: a comparative analysis of chosen methodologies for methane recovery and electricity generation*, Masters thesis, University of Cologne, Cologne.

OECD (2001) *Emissions Baselines – Estimating the Unknown*, Paris, OECD.

Probase (2002) *Procedures for Accounting and Baselines for JI and CDM Projects*, Final Report, Groningen.

Rentz, H. (1998) 'Joint Implementation and the question of additionality – a proposal for a pragmatic approach to identify possible Joint Implementation projects', *Energy Policy*, vol 4, pp275–9.

UNFCCC (2001) *Report of the Conference of the Parties on its 7th session*, Addendum, Part 2: Action taken by the Conference of the Parties, Vol II, FCCC/CP/2001/13/Add.2.

UNFCCC (2003a) Indicative simplified baseline and monitoring methodologies for selected small-scale CDM project activity categories, Appendix B1 of the simplified modalities and procedures for small-scale CDM project activities, Annex 6, *Report of the 7th meeting of the Executive Board*, Bonn.

UNFCCC (2003b) Clarifications on issues relating to baselines and monitoring methodologies, Annex 1, *Report of the 8th meeting of the Executive Board*, Bonn.

UNFCCC (2003c) Procedures for submission and consideration of a proposed new methodology, Annex 2, *Report of the 8th meeting of the Executive Board*, Bonn.

UNFCCC (2003d) *Report of the 5th meeting of the methodologies panel*, Bonn.

UNFCCC (2003e) *Report of the 9th meeting of the Executive Board*, Bonn.

UNFCCC (2004a) Tool for the demonstration and assessment of additionality, Annex 1, *Report of the 16th meeting of the Executive Board*, Bonn.

Chapter 8

Creating the foundations for host country participation in the CDM: experiences and challenges in CDM capacity building

Axel Michaelowa[1]

8.1 Introduction

When the CDM was agreed at the Kyoto conference, it was called a great surprise.[2] For the first time, a market mechanism for a global public good had been created. The CDM provides a mechanism to bridge the gap between the 39 Annex I Parties that have agreed to quantified emissions limits under the Protocol and the remaining approximately 140 countries who have not. Participation in the CDM is voluntary and thus cannot be forced on anyone. However, the CDM is not just about trading greenhouse gas emissions credits. Its equally important goal is the promotion of sustainable development of the host countries. Therefore each CDM project has to be approved by the host country. Ideally, each host country would have defined rules to assess the contribution of projects to its development. As sustainability can be understood in many different nuances, a wide range of rules is conceivable.

As the generation of Certified Emission Reductions (CERs) through CDM projects is happening in countries that are not subject to a quantified emissions limit, there is a strong incentive to artificially inflate the amount of CERs generated (see Chapter 7). Project developer, host country and investor/CER buyer have an interest to generate as much CERs as possible. This perverse incentive was one of the main reasons many NGOs opposed the CDM, believing it might sanction much business as usual resulting in no net environmental gains. The EU took up this criticism and pushed for checks and balances to avoid such 'tropical hot air'.[3] While the more market-oriented OECD countries did not wish an overly cumbersome procedure that in their view would have stifled the CDM, at COP-7 in Marrakesh it was possible to agree to a body of rules that goes a long way in protecting the environmental credibility of the CDM. The Marrakesh Accords specify a project cycle with a series of steps that require professional knowledge of all participants.[4]

As the CDM competes with domestic mitigation activities in Annex B countries, with project-based Joint Implementation (JI) and International Emissions Trading (IET), CDM host countries have to find their niche in the global greenhouse gas market. They need a clear picture of their strengths and weaknesses. To preserve environmental integrity, the CDM project cycle inevitably causes transaction costs that are higher than those of JI and IET. Mastering each step requires specialization. A host country that wants to be a competitive provider of CERs should build competence to master all steps to avoid costly outsourcing to service providers from industrialized countries. Obviously, this will be a challenge for many, if not the overwhelming majority of Non-Annex I countries. This chapter considers the contribution capacity-building initiatives can make to the successful implementation of the CDM in developing countries that want to take part in the global carbon markets. The success of such initiatives will be vital not only in the long term in engaging developing countries in the CDM but also in enhancing their role in designing the future climate regime to the extent that this will incorporate reliance on trading mechanisms that are Kyoto consistent.[5]

8.2 Capacity requirements to successfully implement the CDM

The Marrakesh Accords allocate tasks within the CDM project cycle to different players (see Figure 8.1). These tasks require specific knowledge, financial resources and official decisions.

Governments

Host country governments can decide which projects can be forwarded to the CDM Executive Board for registration. The approval decision is undertaken by the 'Designated National Authority' (DNA) that has to be notified to the UNFCCC Secretariat. The seemingly simple task of defining a DNA has been

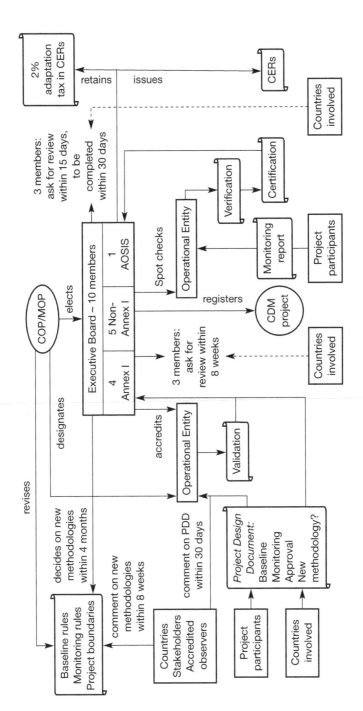

Figure 8.1 *Tasks and actors within the CDM process.*

complex and time-consuming in many host countries; in the three years since the Marrakesh Accords only a third of Non-Annex B countries have notified the UNFCCC Secretariat of their DNA. As shown in the regional breakdown of DNAs in Figure 8.2, Latin America was leading DNA setup but many of those DNAs are not fully operational (Michaelowa, 2003). For a good overview of possible DNA activities and their history in Latin America see Figures (2002).

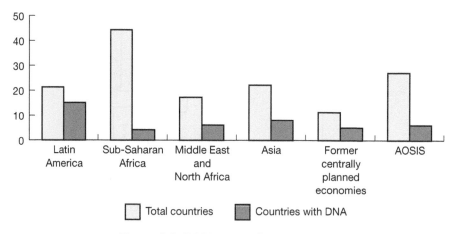

Figure 8.2 *DNAs according to region.*

For developing countries, another necessary condition for CDM participation is the ratification of the Kyoto Protocol. As the case of Indonesia shows they ratified the Protocol only in October 2004, this is not self-evident for developing countries given that there may be interest groups that fear ratification would mean the first step on the path to taking up an emissions target. Both the Indonesian coal lobby as well as forces linked to the oil-producing Middle East were blocking ratification in Indonesia for the last two years for these reasons.

Project developers

Project developers have to understand the technical options for greenhouse gas reduction and their costs. Moreover, they should be aware of the issues related to additionality determination (see Chapter 7) to avoid pursuing project ideas that will eventually be rejected because they are business as usual. Once a project is registered and starts operating, they should be able to apply a monitoring methodology.

Consultants

Consultants play a decisive role in writing the documentation necessary for registration of a CDM project. In this context, they have to know the fine print of CDM rules and be able to collect data for determination of the baseline. In the early stage of CDM development, they also have to submit new baseline methodologies. This is an art that requires in-depth understanding of the politics surrounding the CDM, especially with regard to the opinions of the CDM Methodology Panel.[6]

Validators and certifiers

Validators have to be officially accredited by the Executive Board and eventually confirmed by the COP. The Marrakesh Accords define the competences necessary to become a validator. This list is long and essentially limits validation to companies that have accumulated experience in other types of certification. So far, these requirements have been an effective deterrent for companies from developing countries.[7]

Financial sector

Banks and insurance companies are important players to enable project developers to undertake a CDM project. Although a number are becoming active in the CDM, the majority currently lack detailed knowledge about the CDM which means that they will have difficulties in evaluating requests for loans or project finance.

Lawyers

Contracts with an unusually long duration are characteristic for the sale of CERs. The structure of an Emission Reduction Purchase Agreement (ERPA) depends on the financial structure of the CDM project, the type of host country and the greenhouse gas exposure of the buyer. For example, for a project with a crediting period of 21 years that takes place in a host country likely to take up an emissions target within the next ten years, the contract has to define what happens when the emission target kicks in. Many buyers do not want to take all the CERs produced by a project but still retain an option for the acquisition of the remainder. Even without such clauses, the contract template developed by the International Emissions Trading Association (IETA) encompasses 30 pages (IETA 2004). Lawyers that draft ERPAs have thus to be aware of many issues surrounding international climate policy. So far, only one or two law firms, none of which are from a developing country, have this knowledge.

NGOs

NGOs play an important role in safeguarding the environmental and social integrity of the CDM process. At several levels of the CDM project cycle, they can submit comments or raise concerns. If domestic consultations have been inadequate or there are international ramifications allowing stakeholders outside a country an interest in the project, it is possible that a host country DNA might hear NGOs concerns when project proposals are submitted to the EB of the CDM for approval. Internationally, comments can also be made during a month when a new baseline methodology has been submitted to the Executive Board. Further comments can be submitted to the validator once a project developer has asked for validation.[8] NGOs can access the UNFCCC website to submit comments. However, in the first year of methodology development, NGO comments have been astonishingly rare; most comments have come from researchers or consultants. This is most likely due to the low availability of NGO personnel with the necessary technical knowledge. Moreover, if such staff are available, they are more likely to work on more 'glamorous' policy aspects of the CDM than technical assessments of CDM projects. An exception is the NGO CDMWatch which has been instrumental in coordinating NGO responses to doubtful projects.

8.3 Donor activities

Many Annex I Parties with Kyoto targets (called Annex B countries for short) have recognized that the CDM projects will not fall from the sky and that the CDM cannot commence operations in developing countries without concerted efforts to that effect. Thus many have started programmes to enhance capacity in host countries. These programmes have different target groups and designs which are set out below.

Spreading the message

Before the development of CDM projects can be contemplated, different target groups have to get a feeling for what the CDM is and what incentive it offers for them. All capacity-building activities of the late 1990s and many of the currently ongoing programmes have concentrated on this aspect.

Funded by Switzerland and increasingly other countries (Finland, Canada, Germany, Australia, Austria and Italy), the World Bank started its programme of National Strategy Studies (NSS) in 1998 which ran until the end of 2003. This programme was aimed at both CDM and JI host countries; the latter were the focus of the initial years while after 2000, almost all studies were on CDM. It consisted of writing a study estimating the CDM potential of the specific country with the following elements:

- description of the CDM;
- estimate of demand and supply on the international greenhouse gas market;
- estimate of costs and scope of greenhouse gas abatement options in the host country;

- institutional requirements for CDM;
- description of a project pipeline.

Each NSS would be financed by one Annex B country; the World Bank played a relatively limited role in financing, but asserted an important one when it came to content. After initial experiences with uncoordinated writing of the chapter on the international greenhouse gas market, they hired Swiss consultants Gruetter to develop an easy-to-operate general equilibrium model of the global market. This 'CERT model' was useful in giving some idea about market prices and overall demand and prevented the wasting of too many precious human resources on guessing about international markets. However, in the NSS programme the World Bank also promoted its own agenda, particularly concerning the development of a project pipeline for its Prototype Carbon Fund (PCF). This role of the World Bank sometimes led to conflict with the financing country and to long delays in publication of the studies. Therefore Germany made its financial contribution with the condition that it would retain control over the content and could decide unilaterally when a report would be fit for publication.

A team of host-country consultants would always be in charge of writing the report while consultants from the Annex B country financing the NSS would support them. Given the high degree of expert knowledge, however, the Annex B consultants often played a major role which, of course, limited the degree of capacity building that occurred. Annex B consultants also often took the major share of the available funding so that essentially the NSS programme built the capacity of Annex B consultants at least as strongly as the capacity of host-country consultants.

Usually, a NSS took 18 months to complete but in some cases it dragged on for three years. The main reasons for delay were insufficient ownership by the host country, conflicts about the allocation of financial resources, lack of competence and slow allocation of experts by the Annex B consultancy. An instructive case is Indonesia that had decided to separate its NSS into an energy part and a forestry part. The former was financed by Germany, the latter by Australia. While initially both parts were to be started in early 2000 and published jointly, it became clear quickly that the Australian part would have difficulties. Eventually, the German NSS was published in September 2001 and the Australian one more than two years later.

NSS were carried out in 16 CDM host countries (see Table 8.1). Interestingly, smaller countries were quicker in negotiating NSS terms with the World Bank. Larger host countries were sometimes sceptical and the negotiations took a long time. A prime example is India that only started its NSS in 2003.

Table 8.1 *NSS completion*

Year	Country
1998	Argentina
1999	Uzbekistan
2000	Colombia, Kazakhstan
2001	Bolivia, Indonesia I – Energy, South Africa, Zimbabwe
2002	Egypt, Thailand
2003	Chile, Indonesia II – LULUCF, Peru, Uruguay
2004	China, India, Vietnam

The NSS programme was an important catalyst for CDM institution building in host countries but it became obvious that it would not be sufficient. Moreover, many UN organizations were in dire straits because of funding cutbacks and looking for new opportunities to justify their budgets to preserve staff. So a donor competition started which involved UNCTAD, UNDP, UNEP and UNIDO. Moreover, several Annex B governments started their own efforts (see Table 8.2). Even today, these efforts are continuing despite having been superseded by targeted institution building and project development support.

Table 8.2 *Selected CDM awareness-building programmes*

Agency	Name	Duration	Host countries	Comments
UNCTAD	Carbon Market E-Learning Center	2001–ongoing	Not specified	Online, fee-based training course on CDM. Dubious quality and semi-commercial character
UNDP	RAB	1999–2002	Algeria, Morocco, Tunisia	Workshops, development of project pipeline
UNIDO	Concept for Developing National Capacity to Implement Industrial Clean Development Mechanism Project in Africa	1998–2001	Ghana, Kenya, Nigeria, Senegal, Tanzania, Zambia, Zimbabwe	Evaluation of CDM potential in industrial sector
UNIDO sector	Capacity Mobilization to enable Industrial Projects under the Clean Development Mechanism	2001–2002	Indonesia, Malaysia, Philippines, Thailand, Vietnam	Evaluation of CDM potential in industrial

Agency	Name	Duration	Host countries	Comments
EC ASEAN Energy Facility	CDM-ASEAN	2003–2005	ASEAN	Two regional workshops, background papers
EU Commission	CAPSSA	2001–2002	South Africa, Senegal, Zambia	Workshops, small strategy studies, development of sustainability criteria
EU Synergy		2003–2004	China	Workshops
EU Synergy	IRIS	2003–2004	India, Morocco	Workshops
EU TACIS		2004–2006	Caucasus, Central Asia	Workshops, strategy studies
IGES (Japan)	ICS	2003–2006	Cambodia, India, Indonesia, Philippines	Well-funded programme aiming to expand to other countries. Workshops outside the capital and organized by different organizations. Focus on waste management, renewable energy, and small-scale projects
UK		2003–2004	India	Workshops
USAID		1998–2000	India	Workshops and US study tour with industry leaders

A common feature of all these programmes was the writing of studies and the holding of general CDM workshops. The most effective of these programmes was the US effort in India which took place before the US withdrew its support for the Protocol and the CDM. The US/India initiative involved secondment of two US experts to the renowned Indian research institute TERI. They focused on the Indian industry associations with which they did several CDM workshops. The apex of that programme was the sending of a 50-strong Indian company CEO delegation to Washington which was received by Vice President Gore in the White House! Afterwards, CDM awareness of Indian business was extremely high and it was instrumental in changing the negative attitude of the Indian bureaucracy towards CDM. The only flaw was that the US programme had spread the impression that CDM revenues would amount to billions of dollars that could be reaped easily. Because most models show that US withdrawal from the Protocol leads to a collapse of the carbon price, for CDM capacity-building donors arriving in India after March 2001 these early impressions of the CDM's potential gains were hard to explain and did not facilitate their work.[9]

At the other end of the effectiveness scale were the UNIDO efforts. A lot of money was spent on studies on the CDM potential in the industrial sector in African and ASEAN countries without any follow-up. In Indonesia, the study was written by a notoriously unreliable consultant of the local coal industry!

Supporting institutions

After the Marrakesh Accords had been agreed and it became clear that many countries had problems in setting up their DNAs, the focus of donor activities shifted. From 2002 onwards, several activities have focused on DNA building (see Table 8.3).

Table 8.3 *CDM institution-building programmes*

Agency	Name	Duration	Host countries	Comments
UNDP/UNEP	RAB	2003–2004	Morocco	Follow-up of earlier programme, now centred on making approved DNA fully operational
UNEP	CD4CDM	2002–2005	North Africa and Middle East: Egypt, Jordan, Morocco Sub-Saharan Africa: Côte d'Ivoire, Mozambique, Uganda Asia: Cambodia, Philippines, Vietnam Latin America: Bolivia, Ecuador, Guatemala	Largest CDM institution-building programme to date. Intense preparation of stakeholders on national and regional level
World Bank	CF Assist	2003–	Not specified	Training modules for establishing DNAs; book on legal issues
DANIDA (Denmark)		2003	Malaysia	Support for the Energy Secretariat of the DNA
GTZ (Germany)	CAPP	2002–2006	Indonesia, Mongolia, Tunisia	Local consultant works with ministries and other stakeholders until DNA is agreed

The institution-building programmes show that host country ministries often feel the CDM could bring them new possibilities to collect rents from the private sector. Therefore inter-ministerial conflicts arise about who will be in charge of the DNA and whose desk(s) project proposals will have to pass. It is key to have a local consultant who understands this political game and manages to minimize the possibilities for rent seeking.

The German programme in Indonesia benefited from such a well-connected consultant; even so, it took almost two years to get the formal decision about the setup of the DNA.

Many host countries do not understand the long-term institutional commitment necessary to develop a consistent CDM strategy. They hope that donor funds will pay for the operational costs of the DNA. Only rarely do they try to assess the level of costs involved. In Indonesia, the German programme stressed the need for a calculation of the DNA budget which came to $180,000 per year and suggested a fee of 0.5 per cent of expected CERs be levied from project proponents.

Supporting project proposals

For those host countries that have set up their DNAs, the challenge is to get project proposals developed. Particularly those funds or Annex B countries that want to procure CERs have been less than happy with the slow pace of project development and the low share of attractive proposals among the project idea notes (PINs) submitted. For capacity-building programmes launched in response to this problem, see Table 8.4. The first entity to discover this was the World Bank's PCF that experienced difficulties in finding sufficient interesting proposals. Only about 10 per cent of the PINs were followed up.

The Dutch were the first to launch a tender for procurement of CERs. Their CERUPT programme subsidized the cost of PDD development. However, the Netherlands discontinued CERUPT after one round because they felt it was too inflexible. CERUPT essentially built the capacity of Annex B consultancy companies and did not help host-country companies. Last year, DANIDA started a similar programme in Thailand. Canada subsidized some small-scale PDDs in India but the subsidy was not sufficient to finalize them. Other procurement programmes (Austria, Finland, Sweden) do not subsidize PDDs at all. The Japanese NEDO has consistently financed CDM feasibility studies over the years but the transformation of those studies into CDM proposals has been very difficult. Most studies were done by Japanese consultants so that the capacity-building component is rather limited.

Future directions

After the establishment of a DNA and support of the first PDDs within each promising project category, CDM capacity building has other targets.

Sustainability criteria

Astonishingly, host countries so far have put relatively low importance on defining criteria that CDM project proposals should fulfil regarding their contribution to sustainable development. This may be due to the fact that stakeholders interested in sustainable development often have little clout in government decision-making in developing countries. Donors interested in an

Table 8.4 *CDM project development programmes*

Agency	Name	Duration	Host countries	Comments
EU Synergy	Business opportunities for CDM project development in the Mediterranean	2004–2005	Israel, Jordan, Lebanon, Malta, Morocco, Syria	Development of 20 project idea notes (no PDDs!), regional workshops
Canada	–	2003	India	Subsidization of six small-scale PDDs
DANIDA (Denmark)	–	2003–2004	Thailand	Tender for project proposals. Selected proposals receive subsidy for PDD development and to negotiate ERPA
GTZ (Germany)	CDM India	2003–2006	India	Tenders for project proposals. Selected proposals receive subsidy for PDD development
NEDO (Japan)		1999–2004	Not specifically targeted	Support of feasibility studies; more than 150 studies have been supported
SENTER (Netherlands)	CERUPT	2002	Not specifically targeted	Selected proposals receive subsidy for PDD development and to negotiate ERPA
Netherlands	SouthSouth North	2001-2004	Bangladesh, Brazil, Indonesia, South Africa	Unique efforts of horizontal capacity building through exchange of experience between four host countries. Project proposals suffer from being small and partly very complicated.

environmentally and socially credible CDM thus should try to strengthen those stakeholders and support the elaboration of criteria. The German programme in Indonesia has put a strong emphasis on criteria definition. In this context, the methodology developed by Sutter (2003) can be usefully applied.

Specific training courses for project developers and consultants

As the methodologies become clearer, project developers and consultants have to be trained in depth. Detailed, multi-day training courses for the most attractive project types should be developed and implemented. This can only be done once a country has defined the priority sectors. SouthSouthNorth has developed a modular programme for training courses including very effective role-play exercises where participants learn to understand the interests of the different stakeholders.

Baseline data collection

With a growing number of accepted baseline methodologies, data needs to calculate a baseline scenario correctly are becoming clearer. It is obvious that some data could be centrally provided by the DNA, for example electricity grid emission factors for the operating and build margin (for a discussion of the underlying concepts see Chapter 7). Collection of these data by the project developers would cause high transaction costs and involve unnecessary repetition. While some countries (India) already have high-quality raw data available on websites, other countries have not managed to collect them despite long efforts (China). Future capacity building could thus focus on assistance in data collection and continuous updating.

Training of domestic Operational Entities

So far, of 24 submitted applications for the status of Operational Entity (OE), only three come from Non-Annex B countries. This means that industrialized-country fee levels, which reflect their higher costs, have to be expected for validation and verification even if these OEs set up branches in some host countries. To keep some of the value added within the host country, capacity building could support the setup of OEs, especially in countries that already have a flourishing certification industry.

8.4 Challenges

Five years of experience with CDM capacity building shows a lot of challenges have to be overcome to make the programmes really effective. This section outlines some important lessons.

Workshop tourism

The number of CDM workshops held globally since Kyoto must now have reached more than a hundred, not including the side events at the UNFCCC negotiations. This has given rise to several undesirable consequences. Many developing country and also Annex B CDM experts spend their time reworking presentations for these workshops instead of working on new challenges: the development of project proposals or institutions. In some cases, participation in such workshops can result in attendance by ill-prepared officials whose work does not relate directly to CDM. The proliferation of workshops exaggerates fragmentation of capacity building efforts and can be addressed by the following actions:

- Workshops should no longer be held by each donor separately, but coordinated in advance among themselves, preferably as part of a broader pattern of capacity-building coordination. This coordination could be achieved, for example, through donor meetings at each session of the UNFCCC negotiations.
- To ensure that not all the same familiar faces attend capacity-building workshops, activities should be thoroughly focused on hitherto not involved target groups and workshops should be held in locations where no workshop has been held so far.
- Regional workshops in developing countries should be prioritized wherever possible as these can reduce travel needs as well as provide developing country participants with a chance to meet others in their field thus building opportunities for South–South engagement.

Ideally, one or two events worldwide could be used to train DNA officials and Operational Entities. Point Carbon's and IETA's Carbon Market Fairs would be ideal as most current DNAs will be represented there anyway to market their countries but other certification-related events may be preferable to engage a broader constituency from developing countries.

Printed CDM manuals

Several capacity-building programmes have developed a printed CDM manual. The widest distribution is achieved by the CD4CDM guide (UNEP, 2003) which has been translated into Vietnamese and Khmer. UNDP (2003) and UNCTAD (2002) are other attempts. The problem with such material is that it becomes obsolete very quickly and users are not aware of that. Web-based manuals that are constantly updated might be preferable but some hard copies are useful as not everyone has access to web-based sources in developing countries – where simple facts such as power cuts can be highly disruptive. Again to avoid proliferation of resources, some of these could also be shared by different capacity-building initiatives.

Donor competition

As shown above, many donor programmes are active in the same countries and with similar targets. Resources could be spent much more efficiently if donors either pooled their resources or clearly separated tasks. While it is clear that those countries looking for CERs will not coordinate their procurement, at least the underlying capacity building can be coordinated. Otherwise, we may see the perverse outcome that the same project proposal is subsidized several times by different donors.

Sourcing cheap CERs

The Marrakesh Accords contain provisions specifying that public funding for CDM projects is not to result in the diversion of ODA and is to be separated from and not count towards the financial commitments of Annex I Parties.[10] Nevertheless there is a perception that some donors are doing capacity building only because they see it as a necessary condition to procure cheap CERs. Of course, host countries are very sensitive about CDM initiatives that tend to go against the spirit of the Marrakesh Accords and thus become wary about the donor's intentions about a CDM capacity-building initiative at its outset. The Danish programme in Thailand has stalled over this issue and Japanese activities are often viewed with some degree of uneasiness in this regard. A clear commitment separating CDM capacity-building initiatives from the eventual purchase of CERs, such as that made by Germany, is needed to avoid these misunderstandings and to build longer-term partnerships based on trust.

Weaning host countries from foreign aid

In the long term, there is a distinct danger that host-country institutions could become dependent on capacity-building funds from abroad. This can already be seen in the case of Cambodia, which had been involved in three capacity-building programmes and whose DNA staff is aware of the fact that they would have to close down immediately if these funds stopped. So any capacity-building programme must have a component for developing a coherent exit strategy for the donor(s) which is linked to the development of a stable financial environment for the subsequent functioning of the DNA – whether it is from the country's own public resources or otherwise (e.g. CDM project).

8.5 Conclusions

Capacity building is a necessary, but not sufficient, condition for a flourishing CDM. Over the last seven years about 30 million euros have been spent to support awareness generation of CDM, to set up designated national authorities, to develop sustainability criteria and to subsidize project design documents. Several large programmes have just started and so a further expansion of capacity building can be expected, focusing on in-depth training of project developers, consultants and financial sector representatives. The collection of data to determine baselines and help frame discussions about sustainability criteria are other relevant topics. While initially multilateral programmes prevailed, bilateral activities have superseded them. This is due to the fact that several Annex B countries see capacity building as an important step towards the procurement of CERs. Expressed as a share of the current CDM market volume of about 600 million euros, capacity building-activities cost between 5 and 10 per cent. In the early years, development of the CDM is likely to need substantial public resources which will require greater collaboration and partnerships among and between public and private actors as well as donors and recipient countries.

Notes

1. Programme International Climate Policy, Hamburg Institute of International Economics, Neuer Jungfernstieg 21, 20347 Hamburg, Germany, Tel +49 40 42834 309, Fax +49 40 42834 451, e-mail: a-michaelowa@hwwa.de. I thank the German Technical Cooperation GTZ for challenging opportunities to do CDM capacity building in several Asian and North African countries.
2. See Kenber, Part III, Chapter 6 in this volume for more details of the negotiating history.
3. See Anderson and Bradley, Part III, Chapter 4 in this volume for discussion of the temperate (Russian and other CEE countries) 'hot air' problem.
4. See Yamin, Part I of this volume for an explanation of the CDM rules and the CDM project cycle.
5. See the Conclusion by Haites in Part IV of this volume on the role of mechanisms, linkages with mechanisms in non-Kyoto Parties and the evolution of the climate regime.
6. See Yamin, Part I for details about the structure, work programme and various panels established by the Executive Board of the CDM.
7. See Yamin, Part I for details of COP and EB action to encourage further applicants from developing countries.
8. Yamin, Part I describes the various reviews which the EB of the CDM can undertake in the CDM project cycle in response to such concerns.
9. Part III, Chapter 2 by Blanchard illustrates the volume of the carbon market under Kyoto and with the impact of US exit from the Protocol.
10. See Yamin, Part I for further details.

References

Figueres, C. (2002) *Establishing National Authorities for the CDM: A Guide for Developing Countries*, Washington, DC, CSDA.
IETA (2004) CDM Emission Reduction Purchase Agreement V.2.0, Geneva.
Michaelowa, A. (2003) 'CDM host country institution building', *Mitigation and Adaptation Strategies for Global Change*, vol 8, pp201–20.
Sutter, C. (2003) *Sustainability Check-Up for CDM Projects: How to Assess the Sustainability of International Projects under the Kyoto Protocol*, Berlin, Wissenschaftlicher Verlag.
UNCTAD (2002) *A Layperson's Guide to the CDM*, Geneva.
UNDP (2003) *CDM Manual*, New York.
UNEP (2003) *CDM Information and Guidebook*, Roskilde.

Part IV

Conclusion: Mechanisms, linkages and the direction of the future climate regime

Erik Haites

This concluding chapter of the book covers three broad topics. First it reviews proposals for domestic greenhouse gas emissions trading programmes outside the European Union and the United States. Then ways of linking such trading programmes are analysed. Finally, the future direction of the international climate change regime is discussed. The European Union emissions allowance trading scheme is discussed in Part II and Part III, Chapter 3. Greenhouse gas emissions trading initiatives in the United States and in developing countries are reviewed in Part III, Chapter 5.

IV.1 Domestic GHG trading programmes

The European Union has adopted a Directive establishing a greenhouse gas emission allowance trading scheme within the Community (European Commission, 2003b) which is discussed in detail in Part II of this book. The Directive requires each Member State to implement an emission allowance trading scheme by 1 January 2005. The Directive specifies many of the features of the national trading programmes, but leaves some choices, primarily the allocation of allowances, to each Member State.[1] Initially, the programme will cover the CO_2 emissions of specified industrial sources.[2] The rest of this section looks at proposals for emissions trading in Canada, Japan, Norway and Switzerland.

Canada[3]

The Climate Change Plan for Canada (Canada, 2002) sets out a three-step approach for achieving Canada's Kyoto Protocol commitment. One of the initiatives outlined in the Climate Change Plan is a domestic emissions trading (CDET) programme for large final emitters (LFEs). The CDET is expected to cover all firms in sectors that have average annual greenhouse gas emissions per facility of 8 kt CO_2e or more and average annual emissions of 20 kg CO_2e or more per $1000 of output. These sectors include thermal electricity, oil and gas, mining and manufacturing. The estimated 500 to 700 firms in this group are forecast to account for about half of Canada's total greenhouse gas emissions in 2010. Consideration is being given to excluding firms whose annual emissions are less than a minimum threshold.

A 'backstop and covenant' system for the LFE companies will establish the CDET. The 'backstop' will take the form of legislation and regulations that will define emission intensity targets, emission measurement methodologies, emission reporting requirements, provisions for emissions trading (including the use of domestic offset credits and Kyoto Protocol units), penalties for non-compliance and conditions for the negotiation of a voluntary covenant. A firm that can demonstrate it would be disadvantaged by the backstop requirements due to early emission reduction actions or foreign competition will be eligible to negotiate a 'covenant' (agreement) with the government that modifies its backstop emission intensity target or other backstop requirements.

The overall target of the 'backstop and covenant' system is to reduce the projected 'business as usual' emissions of large final emitters by 55 Mt CO_2e per year.[4] A key feature of the proposed programme is that targets will be expressed as emission intensities and compliance will be assessed relative to the target emission intensities. Reducing the projected 2010 'business as usual' emission intensities by 15 per cent is estimated to be sufficient to achieve the 55 Mt CO_2e per year reduction target. An emission intensity target covering all greenhouse gases will be set for each process (product) produced by each of the industries covered. The government has proposed that the emission intensity targets decline by 3 per cent per year over the 2008–2012 period.[5] Conceptually, the schedule would have the structure shown in Table 9.1.

A covenant might modify the emission intensity targets for some (or all) processes in a particular industry or for an individual firm. An industry able to demonstrate that compliance with the proposed targets would impose costs that would significantly weaken its competitive position relative to foreign firms could negotiate less stringent targets as part of a covenant. An individual firm that implemented early action to reduce its greenhouse gas emissions could negotiate a covenant with less stringent targets if it was able to demonstrate that its early action placed it at a disadvantage under the backstop targets.

Table 9.1 *Hypothetical backstop intensity targets*

Industry/process	2008	2009	2010	2011	2012
Thermal Electricity Generation					
Process A (t CO_2e/MWh)	0.9116	0.8858	0.8600	0.8342	0.8084
Process B (t CO_2e/MWh)	0.318	0.309	0.300	0.291	0.282
Manufacturing Industry 1					
Process 1A (t CO_2e/unit)	21.2	20.6	20	19.4	18.8
Process 1B (t CO_2e/unit)	0.0106	0.0103	0.0100	0.0097	0.0094
Process 1C (t CO_2e/unit)	1.484	1.442	1.400	1.358	1.316
Manufacturing Industry 2 (etc.)					

At the end of each year every participant will receive a free allocation of allowances equal to its actual output for each process (product) multiplied by the corresponding emission intensity target as established by the backstop regulation or covenant. The allocation will cover all establishments operated by the firm that are subject to the backstop and covenant system. Thus the allocation to a firm would involve a calculation such as that in Table 9.2.

Table 9.2 *Allocation of allowances to Firm Q for 2010*

	Actual output during 2010 by Firm Q at its three production establishments			Intensity target 2010	Allowance allocation for 2010
	Site 1 (units)	Site 2 (units)	Site 3 (units)	(t CO_2e/unit)	(000 t CO_2e)
Process 1A	1,139		5,378	20	130
Process 1B	250,666	867,234		0.0100	11
Process 1C	5,295	29,748		1.400	49
Total					190

Since actual output cannot be known until after the end of the year, the final allocation cannot be made until that time. To provide liquidity for the trading programme, a procedure for early distribution of some of the allowances is being studied. For example, allowances might be distributed based on 90 per cent of the previous year's output. After the final allocation has been made each firm will be required to surrender eligible emission commodities equal to its actual emissions for the year. Eligible emission commodities include: allowances issued to LFE participants, domestic offset credits and Kyoto Protocol units.

To illustrate the compliance process assume that Firm R produces only one product, that the emission intensity target for this product in 2010 is 0.71 t CO_2e/tonne of product, and that the projected output for 2010 was 100,000 tonnes of product. The firm is responsible for meeting the emission intensity target, but the government bears the risk that output by LFE firms will be higher than projected, leading to a reduction from 'business as usual' emissions of less than 55 Mt CO_2e per year. The cases shown in Table 9.3 illustrate the compliance possibilities.

Table 9.3 *Compliance by Firm R in 2010*

	Actual emission intensity (t CO_2e/t)	Actual output, 2010 (tonnes)	Allowance allocation (kt CO_2e)	Actual emissions (kt CO_2e)	
1	0.71	100,000	71	71	Precise compliance by the firm, actual emissions as projected.
2	0.71	125,575	89	89	Precise compliance, the government is responsible for the extra 18 kt CO_2e of actual emissions.
3	0.71	81,396	58	58	Precise compliance, the government benefits because actual emissions are 13 kt CO_2e lower than projected.
4	0.80	100,000	71	80	Intensity higher than target, firm must buy 9 kt CO_2e allowances to comply, no impact on government.
5	0.80	81,396	58	65	Intensity higher than target, firm must buy 7 kt CO_2e of allowances to comply, actual emissions 13 kt below projection.
6	0.65	100,000	71	65	Intensity lower than target, firm has 6 kt CO_2e of surplus allowances, no impact on government.
7	0.65	125,575	89	82	Intensity lower than target, firm has 7 kt CO_2e of surplus allowances, government responsible for 18 kt CO_2e of extra emissions.

Note: Emission intensity target is 0.71 t CO_2e/tonne and projected output is 100,000 tonnes for projected total emissions of 71 kt CO_2e in 2010.

If the actual emission intensity is higher than the target, the firm must purchase allowances to achieve compliance, regardless of its actual output. Conversely, if the actual emission intensity is lower than the target, the firm will have surplus allowances regardless of its actual output. Regardless of the firm's actual emission intensity, the government will face an emission reduction obligation if the firm's actual output is higher than the projected output. Lower than projected output by the firm reduces the emission reduction burden of the government, regardless of the firm's actual emission intensity.

The emission intensity targets are being established through negotiations between the government and the associations for the affected industries. These negotiations are confidential. It is believed that the negotiations define the major processes (products) and emission intensity targets for an industry. Firms that have unique processes (products) are then addressed in a manner consistent with the industry agreement.

The government has promised a price cap of CDN$15/t CO_2e (estimated to be about US$10/t CO_2e) for allowances during the first commitment period. It proposes to implement the price cap by entering into annual agreements with LFE companies for the sale of price assurance mechanism (PAM) units at C$15/t CO_2e.[6] The PAM units can only be used for compliance during the year they are issued by the firm to which they are issued; they cannot be banked or traded.

Japan[7]

Japan is examining a domestic greenhouse gas emissions trading (JDET) programme to help meet its emissions limitation commitment under the Kyoto Protocol. The JDET will be implemented in steps: the first step (2002–2004) is to have voluntary pilot programmes. Based on a policy review in 2004, a full-scale JDET could be implemented at the second step (2005–2007) or the third step (2008–2012). The Ministry of Environment (MoE) and Mie Prefecture have launched a pilot project to assess the potential of a JDET market.

The MoE recognizes that an accurate calculation of the greenhouse gas emissions of each participant is critical to a JDET. To address this need, the MoE has issued guidelines for calculating greenhouse gas emissions by the participants of a JDET programme. The reporting unit is a corporation. Subsidiaries may be included if the corporation holds more than 50 per cent control. Emissions of all of the gases covered by the Kyoto Protocol – CO_2, CH_4, N_2O, HFC, PFC and SF_6 – are to be calculated. The calculation covers both direct emissions from fossil fuel combustion, production processes and vehicles and the indirect emissions associated with purchased electricity and heat.

When reporting its emissions, a corporation must provide its greenhouse gas emissions, the reporting period, the organization(s) and activity(ies) covered by the calculation, relevant management indexes (number of employees, sales, output, number of stores, etc.) emissions intensity (per unit of output or input, per monetary amount) and other relevant information. Public reports are subject to verification by an external inspector. The inspector prepares a verification report that identifies errors, analyses risks that affect the calculation and recommends revisions.

Norway[8]

A Royal Commission report in December 1999 recommended that Norway implement a domestic emissions trading system for greenhouse gases.[9] Following the Commission's recommendation, the government published a White Paper in June 2001, proposing rules for the system.

The proposed system would cover all greenhouse gases, as many sectors as possible and carbon sinks. To achieve coverage of a large percentage of Norway's total emissions, the system would include large industrial sources as well as importers or wholesalers of fuels to other sectors. The only sources excluded from the system are:

- N_2O and CH_4 from combustion and from agriculture;
- CO_2 from the use of lime in agriculture and from solvents;
- HFCs and PFCs as substitutes for CFCs; and
- Halons and SF_6 except from production of magnesium.

The system would be mandatory for between 100 and 200 entities and cover about 80 per cent of Norway's total greenhouse gas emissions.

The Commission was not able to agree on a method for distributing allowances. A majority recommended that the allowances be auctioned. A minority recommended that industries exposed to international competition and currently exempt from the CO_2 tax receive allowances equal to 95 per cent of their 1990 emissions free.[10] Another minority stated that the method of distribution is a political decision.[11]

The Commission concluded that mechanisms for trading allowances would develop on their own. The Central Norwegian Securities Depository might be well suited to perform the registry function, but a number of legal questions would need to be investigated first. The importance of compatibility of the national registry with international rules was recognized.

The system would allow international trade in allowances. Companies would be allowed to buy Kyoto Protocol units or to generate credits through investment in Norwegian sources not regulated by the trading programme. If the international rules, such as supplementarity and the commitment period reserve, restrict trade, the Commission suggested that the right to import or export Kyoto Protocol units be auctioned or sold domestically by the Norwegian government.

During 2005–2007, the government proposes to retain the Norwegian CO_2 tax at its current level and to introduce a limited emissions trading programme for sectors exempted from the tax which account for about 30 per cent of total emissions.[12] This includes the metal industry and other process industries, which have strongly objected to being included in a mandatory emissions trading programme before 2008. Participants in the emissions trading programme will be allocated allowances free of charge. Beginning in 2008, Norwegian authorities plan to replace the carbon tax system and limited emissions trading programme with a comprehensive emissions trading programme encompassing most sources and gases.

Norwegian representatives have participated actively in discussions on the development of the EU ETS. A Norwegian emissions trading programme might be linked to the EU ETS bilaterally or through the European Economic Area.

Switzerland[13]

The Swiss CO_2 law aims to reduce emissions from fossil fuel combustion by fostering energy efficiency and the use of renewable energy.[14] The objective of the law is to reduce CO_2 emissions by 10 per cent from 1990 levels over the period 2008 through 2012.[15] The emission reductions are to be achieved through voluntary actions by industry and consumers. If the emission reduction goal cannot be achieved by voluntary action, the government can impose a tax on fossil fuels beginning in 2004.

The tax rate will depend on the gap between target and actual emissions. The tax is limited to a maximum of 210 Swiss francs per metric tonne of CO_2 (approximately US$130/t CO_2). Tax revenues from the business sector will be recycled through a reduction of employer social security contributions while tax revenues from households will be recycled on a per capita basis.

Article 9 of the CO_2 law allows large companies, groups of companies or companies in energy-intensive sectors to be exempted from the tax if they commit themselves to reduce their emissions. Companies with such commitments could participate in emissions trading. A directive released in July 2001 clarifies the rules of the emissions trading system.[16]

Each (group of) company(ies) negotiates its own absolute CO_2 emissions commitment for 2010.[17] The average reduction should be 15 per cent from 1990 emissions, but commitments may differ from this based on the remaining reduction potential, the profitability of the measures to reduce CO_2 emissions and expected output growth. Allowance allocations are adjusted *ex post* for differences between projected and actual output growth. The allowance allocation will be the 2010 commitment multiplied by five to cover the years 2008 through 2012. Allowances will be distributed free.

During 2008–2012 participants must retire allowances equal to their actual emissions annually. But compliance is assessed for the 2008–2012 period as a whole. Participants that do not meet their targets must pay the carbon tax with interest for all of their emissions since the introduction of the carbon tax. Non-compliant participants also will not benefit from any redistribution of the carbon tax revenue. If a group of companies does not meet its target, the group is held liable. How the penalties are to be borne by the members of the group must be agreed by the group itself.

Only allowances that are 'unlikely to be used for compliance' are tradable. The directive does not specify how this provision is to be implemented. Trading within a group of firms (intra-group trading) is unrestricted. Trading between different groups (inter-group trading) is subject to regulation by the groups involved. Inter-group transactions need to be reported to a registry to be set up and maintained by the group.[18]

The directive states that the use of the Kyoto mechanisms should be supplemental to domestic measures, but it is not clear how this will be implemented.[19] Emission reductions achieved through energy efficiency and renewable energy CDM and JI projects are eligible. The directive is ambiguous with respect to the possible use of international emissions trading (IET) for compliance.[20] The eligibility of CDM or JI projects that reduce non-CO_2 emissions or that enhance sequestration by sinks is not clear.

IV.2 Links among domestic GHG emissions trading programmes

This section reviews the rationale for linking domestic greenhouse gas emissions trading programmes, describes different ways in which they can be linked and then discusses issues raised by linking trading programmes.

Rationale for linking domestic greenhouse gas emissions trading programmes

Emissions trading allows a specified emission limit for a set of sources to be achieved at a lower cost than policies that mandate reductions at each source to achieve the same aggregate limit. Emissions trading does this by providing an economic incentive to sources with low-cost emission reduction opportunities to implement larger reductions and sources facing high-cost emission reductions to buy surplus allowances or credits from low-cost sources.

A source that has relatively low-cost emission reduction opportunities can profit by implementing measures whose cost per tonne of emissions reduced is less than the market price of an allowance or credit. A source faced with relatively high-cost emission reduction opportunities can reduce its compliance cost by purchasing credits or allowances rather than implementing higher-cost measures to reduce its own emissions.

The cost savings that can be achieved by an emissions trading programme depend upon the differences in the compliance cost per tonne of the participating sources. Typically, the larger the number of participants, the greater the diversity in the compliance costs and the larger the potential cost savings.

Greenhouse gas emission reductions have virtually the same climate change benefit regardless of where the emission reduction occurs. This is because the climate impacts of a greenhouse gas emission persist for decades to millennia, while the greenhouse gas molecules could be anywhere in the atmosphere within a year or two regardless of where they are emitted.

The reason for linking domestic greenhouse gas emissions trading programmes is that it offers the potential for larger cost savings, an objective of each of the domestic programmes. When programmes are linked, the cost savings depend upon the differences in the compliance costs of all participants rather than those in each programme separately.[21] Since the environmental benefits do not depend on the location of the reductions, linking greenhouse gas emissions trading programmes is environmentally sound. Other potential benefits of linking include increased market liquidity, a more competitive allowance market for programmes with few participants and more efficient technology development.

Ways domestic greenhouse gas emissions trading programmes can be linked

Two emissions trading programmes are linked if a participant in either programme can use, directly or indirectly, an allowance or credit from either programme for compliance purposes. An indirect link involves an exchange of an allowance or credit from the seller's trading programme for an internationally recognized unit that is sold to a buyer in another programme and which, in turn, is exchanged for a credit or allowance in the buyer's trading programme.

A unilateral link can be created by a trading programme if the administrator accepts credits or allowances from another programme for compliance purposes. A credit or allowance used in this way usually needs to be cancelled in its own registry to ensure that it is not used twice. When two trading programmes are linked there is usually a net flow into one of the programmes.[22] If the programme creating the unilateral link is the recipient of a net inflow of credits or allowances, a unilateral link could yield all of the cost savings of a full link.[23]

Links through international agreements

If two (or more) countries are parties to an international emissions trading agreement, their domestic emissions trading programmes may be linked by that agreement.[24] Two international emissions trading agreements are expected to enter into force:

- the European Union emission allowance trading scheme (EU ETS) that begins in January 2005 and will apply to all Member States; and
- international emissions trading (IET) under Article 17 of the Kyoto Protocol beginning in January 2008.[25]

One or both of these international agreements will apply to almost all domestic greenhouse gas emissions trading programmes currently being implemented or contemplated. The 25 Member States of the EU as of May 2004 include 23 Annex B Parties to the Kyoto Protocol.[26] Two additional Annex B Parties may join the EU in 2007 and other Annex B Parties may link their emissions trading programmes to the EU ETS bilaterally or through the European Economic Area.[27] If the Kyoto Protocol enters into force, all Annex B Parties to the Protocol could use IET to link their domestic emissions trading programmes.[28] This would cover at least 33 countries.[29]

The EU ETS is described in detail in Part II and its linking provisions are discussed at length in Chapter 3, Part III. A few key points warrant repetition here. Article 12 of the Directive establishing the EU ETS specifies that allowances can be used throughout the European Community thus linking the trading programmes of all Member States.[30] Article 25 allows links to be established between the EU ETS and the emissions trading programmes of Annex B Parties that are not members of the European Union.[31]

All Member States except Cyprus and Malta are also Annex B Parties, so the emissions trading programmes will be linked both through the ETS and IET under the Kyoto Protocol, if the Protocol enters into force. Special provisions are proposed in the draft regulations governing the registries under the EU ETS to deal with the complications raised by EU Member States that are non-Annex B Parties.[32]

International emissions trading (IET) under the Kyoto Protocol allows one Annex B Party to transfer some of its allowable emissions, assigned amount units (AAUs) or acquired emission reduction units (ERUs), certified emission reductions (CERs) and removal units (RMUs), to another Annex B Party.[33] This increases the allowable emissions in the recipient country and reduces those of the seller country. An Annex B Party may allow legal entities to participate in IET. Each Annex B Party also may establish rules for participation of its legal entities in IET.

Domestic action is supposed to constitute a significant element of the effort made by each Annex B Party to meet its emission limitation commitment. To help meet this undertaking Annex B Parties could, if they so wish, limit net purchases of Kyoto units (AAUs, CERs, ERUs and RMUs).[34] Each Annex B Party must maintain a minimum holding of Kyoto units in its national registry – a commitment period reserve – and may need to limit sales of such units to comply with this requirement.[35]

In principle, an Annex B Party that implements a domestic emissions trading programme could use national AAUs as the allowances for the domestic trading programme and allow them to be traded freely on the domestic and international markets. The commitment period reserve requirement would then be enforced by the independent transaction log.[36] Alternatively, an Annex B Party with a domestic emissions trading programme could use national allowances. Kyoto units would be exchanged for national allowances, and vice versa, under specified conditions. The supplementarity and commitment period reserve requirements could then be enforced through the exchange process.

The EU ETS will use EU allowances and will allow the exchange of CERs and ERUs, but not AAUs and RMUs, for EU allowances beginning in 2005 for CERs and 2008 for ERUs (European Commission 2003a). From 2008, the use of CERs and ERUs will be constrained to a percentage, to be specified by each Member State in its National Allocation Plan, of the allowance allocation to each installation.

Possible unilateral links

American legislators are aware that they can decide unilaterally whether to allow the use of Kyoto units for compliance with domestic emission limitation obligations even if the United States does not ratify the Kyoto Protocol. The Climate Stewardship Act would have allowed the use of emission permits from countries that met specified conditions (United States Senate, 2003).[37] Annex B Parties eligible to participate in IET probably would have been the only countries to meet those conditions. Thus, if legislation with provisions similar to those of the Climate Stewardship Act is passed in the United States, it could be a buyer of Kyoto units even if it does not ratify the Protocol.[38]

More broadly, any national or subnational greenhouse gas emissions trading programme in the United States, Australia or any other non-Annex B Party could decide unilaterally to allow the use of Kyoto units for compliance purposes.

Issues raised by linking trading programmes[39]

The designs of the existing and proposed domestic greenhouse gas emissions trading programmes vary significantly. Baron and Bygrave (2002) and Haites and Mullins (2001) review the designs of emissions trading programmes to determine which features preclude links between programmes. The design features that pose the greatest difficulty when linking programmes are the allocation method, the point of imposition, the non-compliance penalty, the banking of pre-2008 allowances into the commitment period and trading restrictions.

Allocation method

Allowances can be distributed by auction; be free, based on output; or be free, based on historic activity. The method used affects the total cost and/or aggregate emissions as well as the distribution of costs among participants. Adoption of different distribution methods can create competitive distortions when participants can sell their products in other countries. This is why the EU ETS specifies how allowances are to be distributed, establishes criteria for the distribution of allowances, and allows the Commission to reject allocation plans judged to be inconsistent with those criteria.[40] Linking can accentuate or attenuate competitiveness impacts due to differences in the distribution methods adopted by the programmes being linked.

Point of imposition

Energy related CO_2 emissions can be regulated at the point of release to the atmosphere (downstream) or at any point in the distribution chain for fossil fuels (upstream) by regulating the carbon content of the fuels. Emissions due to fossil fuel use by electricity generators can be regulated directly in a downstream design by requiring them to participate in the trading programme. Emissions due to electricity generation can be regulated indirectly by holding customers responsible for the emissions associated with the electricity they use. Linking trading programmes with direct coverage poses no problems. If one or more of the programmes being linked has indirect coverage, linking may require cumbersome arrangements to ensure that all emissions are properly accounted for in the combined regime.

Non-compliance penalty

Emissions trading programmes require effective enforcement of compliance including penalties sufficient to deter non-compliance. A non-compliance penalty defined strictly in financial terms might be lower than the market price if the programme is linked with other programmes. Linking should not create a situation where non-compliance may be rewarded. This requires not only appropriate non-compliance penalties, but also confidence in the monitoring, verification and enforcement regime, the integrity of the registry and the level of the 'safety valve' price.

Banking into the commitment period

A government may authorize participants in its emissions trading programme to bank pre-2008 allowances into the Kyoto commitment period to encourage early reductions. Linking a programme that does not allow banking into the commitment period with one that allows such banking may lead to an inflow into the programme that allows banking into the commitment period. Due to the increased obligations during the commitment period, countries are likely to limit banking into the commitment period, if it is allowed at all. The UK rule for direct entry participants that limits such banking to the extent of over-compliance by the participant effectively addresses this issue.

Trading restrictions

The UK programme has a 'gateway' to limit flows from the rate-based (allocations tied to output) sector to the absolute (reductions in total emissions) sector. The existence of such restrictions could complicate attempts to link different programmes. The impact on trading restrictions of linkages with other programmes must be examined on a case-by-case basis.

Summary

Differences in design present few technical barriers to linking emissions trading programmes. But linking programmes clearly is easier if the designs are similar. Almost all domestic greenhouse gas emissions trading programmes currently being implemented or contemplated will be linked by the EU ETS, international emissions trading under the Kyoto Protocol, or both. These two international agreements adopt quite different strategies to linking domestic emissions trading programmes.

The EU ETS specifies many of the design features of the Member State emissions trading programmes, including the participating sources, the emissions covered, the point of imposition and the non-compliance penalty. Member States have limited discretion in the distribution of allowances and may choose to allow banking into the 2008–2012 period. The programme designs become even more homogenous in 2008. The opt-out provision is no longer available, differences related to banking into the 2008–2012 period no longer apply and the Commission is required to specify a harmonized distribution of allowances for the 2008–2012 period.

In contrast, Annex B Parties have complete freedom in the design of their domestic emissions trading programmes, if any, under the Kyoto Protocol. An Annex B Party that wants to link its domestic emissions trading programme to the IET provisions must evaluate the potential consequences of such a link given the design of its domestic programme. A link may attenuate or accentuate competitive distortions arising from the distribution of allowances. A link could lead to double coverage or exemptions of some emissions due to the choice of the point of imposition. And a link could create problems due to the safety valve price, if one is adopted, and the non-compliance penalty. Since each Annex B Party is responsible for meeting its national emissions limitation commitment, it has an incentive to ensure that linking its domestic emissions trading programme to the IET provisions does not make meeting this commitment more difficult.

The programmes of the EU Member States, and possibly a few other countries such as Norway, will be linked through the EU ETS from 2005. IET allows each Annex B Party to link its domestic emissions trading programme to those of other Annex B Parties beginning in 2008 if the Kyoto Protocol enters into force. The proposed trading programmes in Canada, Japan and Switzerland will not begin until 2008, so they can be linked with the programmes of other Annex B Parties from the outset.

Would linking the programmes in Canada, Japan and Switzerland with the EU ETS be desirable? If the Kyoto Protocol does not enter into force but those countries maintained their domestic trading programmes, this would be the easiest way to link the programmes. Even if the Protocol enters into force, links to the EU ETS could improve liquidity in the other programmes because the allowances would trade in a much larger market.[41] However, the trading programme designs proposed in Canada and Switzerland are quite different from those of the EU ETS and a link might require greater harmonization with the EU ETS design at a time when the programmes are just being launched.[42]

Linking emissions trading programmes that are already operating creates winners and losers in each programme. The buyers in the programme with the high price prior to linking and the sellers in the low-price programme benefit from the link. Conversely, the sellers in the high-price programme and the buyers in the low-price programme lose as a result of the link. The creation of winners and losers may weaken support for linking trading programmes. If programmes are linked before the distribution of allowances has been decided, it is very difficult to identify the winners and losers.

IV.3 Direction of the future climate change regime

The United Nations Framework Convention on Climate Change (UNFCCC), which has been ratified by 187 countries (plus the European Union), has as its ultimate objective 'stabilization of greenhouse gas concentrations in the atmosphere at a level that would prevent dangerous anthropogenic interference with the climate system'. The Intergovernmental Panel on Climate Change (IPCC) notes that what constitutes 'dangerous anthropogenic interference with the climate system' is a value judgement determined through socio-political processes, taking into account considerations such as development, equity and sustainability, as well as uncertainties and risk.[43]

Carbon dioxide (CO_2) is the most abundant of the greenhouse gases and the largest contributor to climate change. The IPCC notes that stabilization of the CO_2 concentration of the atmosphere *at any level* requires ultimate reduction of global net emissions to a small fraction of current emissions.[44] A stable concentration requires that emissions to the atmosphere be balanced by removals from the atmosphere. Natural processes remove CO_2 from the atmosphere at a much slower rate than anthropogenic emissions to the atmosphere.[45]

Thus to limit climate change global emissions of greenhouse gases must be reduced significantly from the current level. The Kyoto Protocol, if it enters into force, will reduce the rate of growth of global emissions, but global emissions will continue to increase rather than fall. The future climate change regime then must achieve larger total emission reductions from more countries – more stringent commitments and broader participation.

Both the UNFCCC and the Kyoto Protocol contain provisions to initiate negotiations on future emission limitation commitments. The Conference of the Parties to the UNFCCC (COP) is required to periodically examine the obligations of the Parties in light of the objective of the Convention.[46] The Kyoto Protocol requires that consideration of commitments for subsequent periods be initiated by 2005.[47] Other provisions of both the UNFCCC and the Kyoto Protocol could also be used to initiate negotiations on future emissions limitation commitments.[48]

More stringent commitments

Emission limitation commitments must be accepted voluntarily by the affected countries. This means that each country's commitment must be considered fair and reasonable relative to those of other countries.[49] The emission limitation commitments in Annex B of the Kyoto Protocol range from 92 per cent to 110 per cent of the country's base year emissions. Other provisions provide additional differentiation. Some countries have flexibility in the choice of the base year. Current members of the European Union have redistributed their joint commitment.[50] And countries differ in terms of their potential to use sinks.

Given that existing commitments by developed countries are differentiated, future commitments by a larger and more diverse group of countries are likely to be differentiated further. Differentiation is likely to be achieved by varying the nature of the commitment, the stringency of the target and the timing of the commitment.

A variety of possible commitments have been proposed, especially for developing countries.[51] Here proposals for future commitments are grouped into three categories: absolute emission limits, emissions intensity limits and different types of commitments. In the first two categories, differentiation is achieved through variations in the stringency of the commitment. In the third category, differentiation is reflected in the nature of the commitments adopted by different countries. Some proposals provide further differentiation through the date at which a country adopts the proposed target.

Absolute emission limits

Proposals for absolute emission limits typically establish a target for global emissions and then distribute this target among countries using a formula that reflects one or more equity principles. The focus is on the distribution of the global emissions target rather than on how the global target is set.[52] Principles that have been proposed as the basis for establishing national commitments include: historical responsibility;[53] current contributions – emissions per capita;[54] ability to afford reductions – per capita GDP; and multi-sector emissions convergence.[55] Evaluations of such rules find that some seemingly reasonable criteria have rather extreme outcomes (Rose et al, 1998), that they are problematic for at least some countries (Winkler et al, 2002), and that they need to be constantly revised because of the economic needs of participants (Edmonds et al, 1995).

One of the main concerns with absolute emission limits is cost uncertainty associated with meeting a commitment established five to ten years earlier.[56] This economic risk is particularly large for developing countries with high, and highly variable, rates of economic growth.[57] International emissions trading, including the project-based mechanisms, reduces the economic risk by minimizing the cost of compliance across all sources. The economic risk can be reduced further by negotiating smaller, more frequent changes to commitments and by a 'safety valve' on the price of units.[58]

Emission intensity limits

Several analysts (Hargrave, 1998; Baumert et al, 1999; Frankel, 1999; Kim and Baumert, 2002) have proposed emission intensity targets.[59] In each case a country's emissions limit would be a function of an emission intensity target and its GDP. How the emissions intensity target is determined varies across the proposals. The proposals also differ in whether the limit is set *ex ante* based on forecast GDP or *ex post* based on actual GDP. Differentiation requires that the emission intensity targets vary across countries (Lisowski, 2002). To reduce global emissions, the emission intensity must become negative – emissions decline as GDP rises.

Historical data suggest that emission intensity is less variable than total emissions, so emission intensity targets involve less economic uncertainty than absolute emission limits.[60] Emission intensity varies widely across countries, reflecting factors such as the energy sources available, the structure of the economy and climate. This means that an emission intensity limit must be negotiated for each country. Since historical data reveal both increases and decreases in emission intensity with both decreases and increases in GDP, negotiating the future emission intensity for a country could be difficult (Müller et al, 2001).

Different types of commitments

Other types of commitments have been proposed, including the following:

- *Optional emissions budget.* An emissions budget is negotiated with no penalties for non-compliance, but the ability to sell allowances equal to the difference between actual emissions and the budget (Philibert, 2000; Cavard et al, 2001). Trading would occur *ex post*, but could occur *ex ante* if the country agreed to make the budget binding (incur penalties for non-compliance).
- *Sectoral targets.* Commitments could be established for specific sectors rather than the country as a whole (Cavard et al, 2001; Philibert and Pershing, 2001; Samaniego and Figueres, 2002). Suitable sectors should account for a substantial share of national emissions, be capable of implementing mitigation actions and be able to measure their emissions accurately.
- *Policies and measures.* Countries agree to implement specific policies and measures, rather than to achieve emissions targets (Philibert and Pershing, 2001). Developing countries could pledge to implement sustainable development policies that reflect national priorities and circumstances. The incremental cost of the policies that also reduce greenhouse gas emissions could be eligible for funding under the UNFCCC (Winkler et al, 2002).

Some proposals provide further differentiation through the date at which countries should adopt a commitment or a more stringent commitment. Typically the obligation to adopt a commitment is triggered by reaching a specified level of per capita emissions, per capita GDP or historic contribution. Kinzig and Kammen (1998), for example, propose that a country adopt a commitment when its per capita emissions reach 0.73 tonnes of CO_2.

All of the proposals, including absolute emission limits and emission intensity limits, could be used simultaneously by different groups of countries to produce an equitable set of commitments.[61] Indeed, different types of commitments with variations in stringency and variable implementation dates may be the only set of future commitments acceptable to a large number of countries.

Broader participation

Countries must accept their commitments voluntarily. Actions to limit greenhouse gas emissions impose costs on a country in the short term. The climate change mitigation benefits of those actions occur globally over the next several centuries. This creates a strong incentive to be a 'free rider', to avoid incurring costs to limit greenhouse gas emissions while benefiting from the actions implemented by other countries.

To secure broader participation and reduce free riding, a future climate change regime could include incentives to participate and possible penalties for non-participation. Possible incentives for participants include commitments that provide 'hot air' and shared research results.[62] Trade sanctions are a possible penalty for non-participation.

Commitments that provide 'hot air'

The Kyoto Protocol commitments for a number of transition economy countries exceed current projections of 'business as usual' emissions during 2008–2012. This leaves them with surplus AAUs ('hot air') that can be sold to other Annex B Parties.[63] The intent was to set commitments for transition economy countries that they could meet at minimal cost. The weaker than anticipated economic performance of some countries, especially the Russian Federation and the Ukraine, has increased the estimated amount of 'hot air' significantly.

Regardless of the factors that contributed to its origin, the existence of 'hot air' may be a precedent for the future climate change regime if the Kyoto Protocol enters into force. Other countries may want 'hot air' as a condition of their participation. This would affect the nature of the commitments since 'hot air' requires a quantitative commitment – an absolute emission limit or an emission intensity limit – together with international emissions trading and/or banking of surplus units.

'Hot air' reduces the aggregate reduction from the 'business as usual' emissions of the participating countries. There must be a net reduction of aggregate emissions to yield a climate change benefit and for the 'hot air' to have any value. Thus there is a limit to the amount of 'hot air' available for any negotiation. And it represents a subsidy to the recipients from countries with commitments below their 'business as usual' emissions.[64] Thus 'hot air' has limited potential as an incentive to induce broader participation.

Shared research results

The Kyoto Protocol requires all Parties to cooperate in the promotion of effective modalities for the development, application and diffusion of, or access to, environmentally sound technologies, know-how, practices and processes pertinent to climate change, including the formulation of policies and programmes for the effective transfer of environmentally sound technologies.[65] Mechanisms for implementing this provision have not yet been agreed. Discussions have focused on technology transfer rather than research and development.

Modelling results suggest that cooperation on research and development, including technology transfer, can enhance cooperation on climate change if the two are linked.[66] Barrett (2001) proposes collectively funded research and development and coordinated adoption of national standards to drive adoption of lower emitting technologies. All Parties to the climate change agreement, including developing countries, would help fund the research and the research results would be available free to all Parties.[67]

Possible trade sanctions.[68]

Some multilateral environmental agreements (MEAs) including CITES, the Montreal Protocol and the Basle Convention, define 'specific' trade measures, usually against non-parties. The Montreal Protocol imposes trade restrictions on goods made with, but not containing, ozone-depleting substances. So far no trade measure taken pursuant to an MEA has been challenged in the World Trade Organization (WTO) by a non-party. The legal ambiguity surrounding the possibility of such a challenge raises uncertainty over the effectiveness of such measures.

WTO rules allow border tax adjustments for environmental taxes or charges on products (e.g. ozone depleting substances) or physically incorporated inputs (e.g. chemicals in plastic products), but not on production processes (e.g. CO_2 emissions during production) or non-physically incorporated inputs (energy used in production).[69] This has meant border tax adjustments for production processes or methods (PPMs) used in the exporting country were not allowed. However, the shrimp-turtle case seems to signal an evolution of the WTO towards dealing with PPMs.

Barrett (2001) states that trade restrictions are the most obvious enforcement mechanism for an international climate change agreement, but they are difficult to apply for climate change due to the very large number of goods affected, the difficulty of calculating the appropriate border tax for each product and likely inconsistency with the international trade agreements.

Buck and Verheyen (2001) conclude that countries willing to act on climate change can use measures allowed under WTO law to put economic pressure on climate change laggards. Brack and Gray (2003) note that compliance with the MEA can be interpreted as a condition for access to the domestic market of an MEA party and so be consistent with the shrimp-turtle decision of the WTO. Finally, Aldy et al (2001) notes that WTO rules do not forbid any trade measure, they simply allow retaliation against countries that impose measures inconsistent with WTO law, so a country that ignores climate change and is subjected to trade sanctions becomes more ostracized by retaliating against countries that are taking action.

In summary, several multilateral environmental agreements specify trade measures that may be taken against non-parties. At least one agreement imposes trade restrictions based on production processes or methods (PPMs). The WTO may be moving to greater acceptance of trade restrictions on non-parties to an MEA even if they are based on PPMs. However, considerable legal uncertainty remains in all of these areas. Nevertheless, it appears that the parties to the Kyoto Protocol could amend the Protocol to include specific trade measures to be taken against non-parties provided that the measures are related to climate change.

Modelling results indicate that implementing trade sanctions against non-parties does not guarantee cooperation. Their effectiveness depends upon the specific circumstances. For example, Kemfert et al (2004) find that broad trade sanctions against American products are more costly for the Kyoto Parties than for the United States while a boycott of American coal exports could be a credible trade threat.[70] Trade sanctions are more likely to be effective if the economic benefits of the trade agreement are larger than the cost of the climate change mitigation agreement.

Possible agreements with non-parties

The United States currently is the largest emitter of greenhouse gases, accounting for over 20 per cent of the world total. The American process for ratification of international agreements is relatively difficult with the result that it is not a Party to many agreements. Yet the United States is often prepared to act in a manner consistent with the provisions of agreements to which it is not a Party. The future climate regime could allow for agreements with non-parties to accommodate situations where a country does not ratify the agreement but is willing to implement emission reduction measures.

The Convention on the Conservation of Migratory Species of Wild Animals (CMS or Bonn Convention) includes such a mechanism.[71] The Bonn Convention aims to conserve terrestrial, marine and avian migratory species through Memoranda of Understanding (MoU) among all states the species migrates through, whether or not the states are Parties to the Convention. The United States, which is not a Party to the Bonn Convention, is a party to one of the four Memoranda of Understanding, the MoU on South East Asian Marine Turtles.

The ability to negotiate agreements with non-parties becomes more important if the future climate regime includes possible trade sanctions or other provisions aimed at non-parties. A negotiated agreement then allows a compromise where the non-party implements agreed measures and in return is not subject to the sanctions and is not required to ratify the Protocol.

International emissions trading

International emissions trading and the project-based mechanisms should continue to be part of the climate change regime. These mechanisms help reduce the cost of meeting the aggregate emission reduction commitment of the Parties. These mechanisms also reduce the economic uncertainty for individual Parties of meeting their commitments because they all have access to the global market for allowances and credits. Finally, the mechanisms reduce competitiveness concerns for industries subject to international competition because firms in participating countries are likely to face the same marginal cost for their emissions.

The mechanisms may also help in the negotiation of future commitments because they will provide information on which countries have been net buyers of allowances and credits and which countries have been net sellers. A country that has been a net seller could be pressed to justify why it should not accept a more stringent target for the next period.

International emissions trading requires quantitative commitments. These could be absolute or intensity emission limits for the country as a whole or for specified sectors. In principle, all countries with quantified national or sectoral emission limits could trade in a single market. In practice, a gateway that restricts the flow of allowances or credits into the absolute segment of the market, as in the United Kingdom, may be desirable. International emissions trading will encourage the adoption of domestic emissions trading programmes by countries with quantified commitments. That will help reduce compliance costs in those countries.

A 'safety valve' would be a desirable feature of emissions trading under the future climate regime. The 'safety valve' is simply a price at which a Party can sell additional allowances. The price is set somewhat above, perhaps double, the expected price and serves to cap the compliance cost.[72] Each Party would be permitted to sell additional allowances to participants in its domestic emissions trading programme at the 'safety valve' price.[73]

Summary

Equity and efficiency are inextricably linked in a climate change agreement (Carraro, 2000). An agreement in which the burden is equitably shared is more likely to be signed by a large number of countries. An increasing number of participating countries reduces the burden for each signatory and, with emissions trading, reduces total compliance costs.

Equity will require differentiated commitments – different types of commitments with variations in stringency and timing – among countries. In addition adoption of differentiated commitments is likely to require links with other policies or incentives and possible threats. International emissions trading and the project mechanisms can reduce the cost of achieving the commitments and so improve efficiency. Since equity and efficiency are linked, these mechanisms also improve equity.

One of the main concerns in negotiating commitments is the cost of compliance. If the commitments are negotiated five to ten years in advance, estimates of the future compliance cost are uncertain. The uncertainty is particularly large for developing countries with high, and highly variable, rates of economic growth. The uncertainty causes countries to be cautious about the commitments they will accept. Reducing the uncertainty may lead to negotiation of more stringent commitments. The uncertainty can be reduced by allowing international emissions trading, establishing a 'safety valve' and negotiating smaller changes to commitments more frequently.

Notes

1. The allocation plans must meet specified criteria and may be rejected by the Commission. Further harmonization of the national trading programme designs is specified after 2008.
2. The Commission may recommend that coverage be extended to other gases at the end of 2004 and any Member State may propose expansion of the programmes to additional gases and sources beginning in 2008.
3. For a more extensive discussion of Canada's proposed emissions trading programme see Haites (2003b).
4. Other measures in the Climate Change Plan are expected to reduce emissions by the large final emitters by a further 36 Mt CO_2e per year.
5. The 2008 target would be 106 per cent of the 2010 intensity and the 2012 target would be 94 per cent of the 2010 intensity.
6. Each January LFE companies will be able to enter into a forward contract for issuance of a maximum quantity of PAM units valid for compliance with obligations for that calendar year at \$15/t CO_2e. When compliance is determined, late the next year, the government will issue PAM units equal to the lesser of the firm's emissions gap and the contracted quantity. The firm's emissions gap is its actual emissions less any permits issued to the firm for that year and any valid banked permits held by the firm. The cost of the forward contract has not yet been determined.
7. Jung (2003) provides a more extensive description of emissions trading developments in Japan.
8. For a more extensive description of Norway's proposed programme see Haites and Mullins (2001).
9. See Schreiner (2000) and Norway (2000).
10. The Confederation of Norwegian Business and Industry has since proposed that firms in these sectors be allowed to choose either 1990 or 1998 as their base year and that they receive a free allocation equal to 84 per cent of their base year emissions.
11. A majority of the Commission recommended that if allowances are distributed free, the allocation be based on historic emissions and the allowances not be tradable. A minority recommended that some of the allowances distributed free be tradable.
12. Tjernshaugen (2002).
13. This section draws heavily from Haites and Mullins (2001) and is based on Janssen and Springer (2001).
14. Switzerland (1999).
15. The Swiss commitment is to reduce emissions of all greenhouse gases covered by the Kyoto Protocol by 8 per cent from 1990 levels during 2008–2012. The goal of the CO_2 law is to reduce CO_2 emissions from fossil fuel combustion by 10 per cent from 1990 levels during the same period. The law further specifies that emissions from the combustion of fossil fuels for heat and power production (industry, commerce and households) shall be reduced by 15 per cent and emissions from the combustion of automotive fuels (transport sector) must be cut by 8 per cent.
16. Switzerland (2001).

17. The directive specifies that companies or groups of companies must have annual emissions of at least 250,000 metric tonnes of CO_2 to take on such commitments.

18. The directive does not contain explicit provisions on banking and borrowing. However, since allowances are allocated for the five-year period and retired annually, banking and borrowing within the 2008–2012 period are allowed.

19. The directive was released before the resumed session of COP-6 in Bonn. Prior to that session quantitative limits on use of the mechanisms was proposed. However, Parties agreed that 'the use of the mechanisms shall be supplemental to domestic action and domestic action shall thus constitute a significant element of the effort made by each Party...' See Yamin, Part I.

20. Clause 79 states that emission reductions achieved abroad may be taken into account 'in an appropriate way' through use of the three Kyoto mechanisms. Clause 55 indicates that emission reductions achieved through energy efficiency and renewable energy projects will be recognized as part of a reduction commitment regardless of whether the Kyoto Protocol enters into force. Clause 55 also states that emission reductions achieved via JI or CDM projects will be recognized as part of a reduction commitment. IET is not mentioned in Clause 55.

21. Assume there are emissions trading programmes in countries A and B. Each will achieve cost savings based on the range of emission control costs of its participants. Given their aggregate targets, the mix of participants in each programme, and the range of control costs on the participants in each programme, the market price for a credit or allowance is likely to be different in the two programmes. Assume that the market price is higher in country A. Then linking the programmes enables additional cost savings if sources in country B implement more reductions and sell the extra credits/allowances to participants in country A. This should continue until the market price of a credit/allowance is the same in both countries.

22. Units will flow from the programme with the lower market price to the programme with the higher market price.

23. If the programme creating the unilateral link has a lower market price than the other programme, there will be no net inflow and no outflow since its units are not accepted by the programme with the higher price, so there will be no cost saving due to the unilateral link.

24. An international emissions trading agreement would allow units to be traded between any two parties to the agreement, so the domestic emissions trading programmes of those parties would be fully linked by such an agreement. A unilateral link generally would be established by a trading programme not covered by an international emissions trading agreement.

25. As of March 2004, entry into force of the Kyoto Protocol depends upon ratification of the Kyoto Protocol by the Russian Federation. International emissions trading under the Kyoto Protocol also would be governed by subsequent decisions such as the Marrakesh Accords.

26. Cyprus and Malta have ratified the Kyoto Protocol, but do not have emission limitation commitments and so are not Annex B Parties.

27. Bulgaria and Romania, both Annex B Parties, are negotiating accession to the European Union in 2007. The European Economic Area negotiates implementation of EU laws and regulations by Iceland, Liechtenstein and Norway. These countries could agree to implement the EU ETS. Alternatively, any of these countries as well as Canada, Japan and Switzerland could negotiate a bilateral arrangement with the EU ETS.

28. Annex B Parties are the 38 countries (plus the European Union) listed in Annex B to the Kyoto Protocol with emission limitation commitments that ratify the Protocol.

29. As of March 2004, 32 of the 38 countries listed in Annex B have ratified the Kyoto Protocol. The countries that have not ratified are Australia, Croatia, Liechtenstein, Monaco, the Russian Federation and the United States. Australia and the United States have indicated they do not plan to ratify the Protocol. To enter into force without their participation the Protocol must be ratified by the Russian Federation. If Russia ratifies the Protocol, it would cover at least 33 Annex B Parties.

30. European Commission (2003b), Article 12.

31. The Linking Directive requires that after the entry into force of the Kyoto Protocol, the Commission examine whether it could be possible to conclude agreements with Annex B countries that have not ratified the Protocol to provide for the recognition of allowances between the Community scheme and mandatory greenhouse gas emissions trading schemes capping absolute emissions established within those countries.

32. European Commission (2003c), Article 30.

33. Unless explicitly mentioned otherwise, CERs should be interpreted to include tCERs and lCERs issued for afforestation and reforestation projects under the Clean Development Mechanism.

34. Use of purchased Kyoto units is intended to be supplemental to domestic emission reduction and sink enhancement actions.

35. This requirement is designed to limit potential non-compliance due to overselling and so is more likely to affect Annex B parties that are net sellers than those that are net buyers.

36. The independent transaction log checks each proposed international transfer of Kyoto units to ensure that it is valid. Violation of the commitment period reserve is one of the items checked. If a proposed transaction would lead to a violation of the commitment period reserve requirement, the seller's national registry is notified. The proposed transaction can then be prohibited by either of the Annex B governments involved. If the transaction is consummated despite violating the reserve requirement, the units cannot be used for compliance purposes until the seller is again in compliance with its commitment period reserve requirement.

37. Section 312 of the Climate Stewardship Act would have allowed a participant to submit for compliance purposes the tradable allowances from another nation's market in greenhouse gas emissions if (a) the Secretary of Commerce certifies that the other nation's system for trading in greenhouse gas emissions is complete, accurate and transparent; (b) the other nation has adopted enforceable limits on its greenhouse gas emissions which the tradable allowances were issued to implement; and (c) the American participant using the allowance certifies that the tradable allowance has been retired unused in the other nation's market.

38. Of course, it would only make sense to buy Kyoto units if the price of Kyoto units was lower than the price of American allowances.

39. For a more extensive discussion of issues raised by linking emissions trading programmes see Haites (2003a).

40. The EU ETS specifies the maximum share of allowances that may be auctioned. Participants must receive an absolute quantity of free allowances rather than an allocation tied to output. The allocation plan must satisfy 11 criteria and is subject to review by other member states and the Commission. These are discussed in more detail by Mullins in Part III, Chapter 3 in this volume.

41. The export market for allowances from non-ETS programmes may be limited. For example, a participant with surplus allowances in the Canadian programme may

be able to exchange them for Canadian AAUs. The draft Linking Directive allows participants in ETS programmes to exchange CERs and ERUs, but not AAUs, for EU allowances. Thus the market for the Canadian AAUs is limited to participants in non-ETS programmes and Annex B governments, most of which are currently purchasing only CERs and ERUs. Canada is likely to be a net buyer, so the surplus allowances probably could be sold to other participants in the Canadian programme. And as a net buyer, the price of Canadian allowances is likely to be very similar to the price of Kyoto units. Thus the limited market for Canadian AAUs may not be a serious disadvantage in practice. But a link with the ETS would allow Canadian allowances to be traded among all participants in all ETS programmes, a much larger market.

42. The design of a Japanese emissions trading programme, if any, is not yet known.
43. IPCC (2001), p2.
44. IPCC (2001), p90.
45. The scope for increasing the rate of removals, by planting more trees for example, is very limited relative to the quantity of CO_2 in the atmosphere.
46. Article 7.2(a).
47 Article 3.9.
48. For example, Article 4.2(d) of the UNFCCC and Articles 9 and 13.4 of the Kyoto Protocol.
49. Ashton and Wang (2003).
50. The overall commitment is 92 per cent of the base year (1990) emissions for the 15 Member States in 2003. The commitments for individual Member States have been redistributed to range from 78 per cent (Luxembourg) to 127 per cent (Portugal) of 1990 emissions.
51. Bodansky (2003) assesses different types of commitments qualitatively and Blanchard (2002) assesses three types of commitments quantitatively.
52. Pershing and Tudela (2003) evaluate options for setting a long-term global target.
53. La Rovere et al (2002) review the Brazilian proposal for commitments based on relative responsibility for global warming.
54. Aslam (2002).
55. Jansen et al (2001).
56. The Kyoto Protocol commitments for 2008–2012 were agreed in 1997. Future commitments might be agreed five to ten years in advance.
57. McKibbin and Wilcoxen (2002) propose a structure that reduces the economic risk as well as potential disruptions due to new entrants and withdrawals.
58. Instead of negotiating revisions to commitments covering a five-year period several years in advance, smaller revisions could be negotiated more frequently. For example, commitments could be reviewed, and possibly be revised, every two or three years. Given the frequency of the reviews, the changes could be small. Commitments would remain in effect until revised, so a review need not result in a revision. A 'safety valve' is a price at which additional units are issued. This effectively caps the price and hence the compliance cost.
59. Bouille and Girardin (2002) reviews the development of the voluntary intensity target proposed by Argentina.
60. Ellerman and Wing (2003) demonstrate that a hybrid emission limit based on an absolute emission cap and an intensity target can shift the uncertainty between abatement costs and emissions.
61. Philibert and Pershing (2001), Baumert and Llosa (2002) and Bodansky (2003).

62. The climate change mitigation benefits cannot be restricted to participants. The incentives to participate must be capable of being limited to participants.
63. The surplus AAUs can also be banked to help meet future emission limitation commitments. Without banking and international emissions trading 'hot air' would not be a concern. In other words, if AAUs could be used only by the country to which they are issued for the period in which they are issued there would be no 'hot air'. If a country's actual emissions were less than its commitment, it would be in compliance but it could not sell or bank its surplus AAUs. The AAUs would not be valid for any other country or period.
64. Bohm and Carlén (2000) find that transfers of cash are more efficient than larger allocations of allowances ('hot air') in inducing developing country participation, assuming that developing countries are risk averse.
65. Article 10(c).
66. Kemfert et al (2004) summarizes key results from this literature.
67. Barrett proposes that each Party contribute in proportion to its UN assessment.
68. For a more extensive discussion of the possible use of trade sanctions see Kemfert et al (2004).
69. In other words, a country that imposes a tax on ozone-depleting substances could adjust tariffs or other border taxes to impose an equivalent tax on imported substances and products that contain such substances. But adjustments for energy taxes are not allowed.
70. Kemfert et al (2004) find that the effect of threats/incentives imposed by Kyoto Protocol Parties on the USA would be to reduce its target for 2010 from the unilateral commitment of a 3.6 per cent reduction from 'business as usual' emissions to a 13 to 16 per cent reduction from 'business as usual' emissions. Such a target would represent a small reduction from 2000 emissions, but would still be above its 1990 emissions and well above its proposed Kyoto Protocol target.
71. CITES, the Montreal Protocol and the Basle Convention also include provisions that allow trade with non-parties. The Bonn Convention is used as an example because the USA is not a party to that Convention but is a party to an MoU under the Convention.
72. Negotiating a specific price, such as $20/t CO_2e, caps the compliance cost. Negotiating a formula such as double the average price during the previous year caps the rate of price increase and the total cost.
73. Use of the safety valve might be constrained by limits on banking to prevent them from being acquired in anticipation of higher future prices. Individual purchasers might be allowed to purchase only the number of units needed to achieve compliance.

References

Aldy, J. E., Orszag, P. R. and Stiglitz, J. E. (2001) *Climate Change: An Agenda for Global Collective Action*, prepared for the conference on 'The Timing of Climate Change Policies', Pew Center on Global Climate Change, Arlington, Virginia, October.
Ashton, J. and Wang, X. (2003) 'Equity and climate in principle and practice', in *Beyond Kyoto: Advancing the International Effort against Climate Change*, Arlington, VA, Pew Center on Global Climate Change, chapter 4, pp61–84.
Aslam, M.A. (2002) 'Equal per capita entitlements: a key to global participation on climate change?', in K. Baumert, O. Blanchard, S. Llosa and J. F. Perkaus (eds),

Building on the Kyoto Protocol: Options for Protecting the Climate, Washington, DC, World Resources Institute, chapter 8.

Baron, R. and Bygrave, S. (2002) *Towards International Emissions Trading: Design Implications for Linkages*, Paris, Organization for Economic Cooperation and Development (OECD) and International Energy Agency (IEA), October.

Barrett, S. (2001) 'Towards a better climate treaty', *Policy Matters*, vol 29.

Baumert, K. A. and Llosa, S. (2002) 'Conclusion: building an effective and fair climate protection architecture', in K. Baumert, O. Blanchard, S. Llosa and J. F. Perkaus (eds), *Building on the Kyoto Protocol: Options for Protecting the Climate*, Washington, DC, World Resources Institute, chapter 10.

Baumert, K. A., Bhandari, R. and Kete, N. (1999) 'What might a developing country climate commitment look like?', in *Climate Notes*, Washington, DC, World Resources Institute.

Blanchard, O. (2002) 'Scenarios for differentiating commitments: a quantitative analysis', in K. Baumert, O. Blanchard, S. Llosa and J. F. Perkaus (eds), *Building on the Kyoto Protocol: Options for Protecting the Climate*, Washington, DC, World Resources Institute, chapter 9.

Bodansky, D. (2003) 'Climate commitments: assessing the options', in *Beyond Kyoto: Advancing the International Effort against Climate Change*, Arlington, VA, Pew Center on Global Climate Change, chapter 3, pp37–59.

Bohm, P. and Carlén, B. (2000) *Cost-Effective Approaches to Attracting Low-Income Countries to International Emissions Trading: Theory and Experiments*, Research Paper, Department of Economics, Stockholm University, Stockholm.

Bouille, D. and Girardin, O. (2002) 'Learning from the Argentine voluntary commitment', in K. Baumert, O. Blanchard, S. Llosa and J. F. Perkaus (eds), *Building on the Kyoto Protocol: Options for Protecting the Climate*, Washington, DC, World Resources Institute, chapter 6.

Brack, D. and Gray, K. (2003) *Multilateral Environmental Agreements and the WTO*, London, Royal Institute of International Affairs and International Institute for Sustainable Development, September.

Buck, M. and Verheyen, R. (2001) *International Trade Law and Climate Change – A Positive Way Forward*, Bonn, Stabsabteilung der Friedrich-Ebert-Stiftung, July.

Canada (2002) *Climate Change, Achieving Our Commitments Together: Climate Change Plan for Canada*, Ottawa, Government of Canada, November.

Carraro, C. (2000) *Costs, Structure and Equity of International Regimes for Climate Change Mitigation*, Nota Di Lavoro 61.2000, Milan, Fondazione Eni Enrico Mattei.

Cavard, D., Cornut, P. and Menanteau, P. (2001) 'How could developing countries participate in climate change prevention: the clean development mechanism and beyond', *Cahier de Recherche*, No 21bis, Grenoble, Institut d'économie et de politique de l'énergie, Grenoble, forthcoming in *Energy Studies Review*.

Edmonds, J., Wise, M. and Barnes, D. W. (1995) 'Carbon coalitions: the cost and effectiveness of energy agreements to alter trajectories of atmospheric carbon dioxide emissions', *Energy Policy*, vol 23, no 4/5, pp309–35.

Ellerman, A. D. and Wing, I. S. (2003) 'Absolute versus intensity-based emission caps', *Climate Policy*, vol 3, supplement 2, ppS7–S20.

European Commission (2003a) Proposal for a Directive of the European Parliament and of the Council amending the Directive establishing a scheme for greenhouse gas emission allowance trading within the Community, in respect of the Kyoto Protocol's project mechanisms, COM(2003)403 final, Brussels, 23 July 2003.

European Commission (2003b) Directive 2003/87/EC of the European Parliament and of the Council of 13 October 2003 establishing a scheme for greenhouse gas emission allowance trading within the Community and amending Council Directive 96/61/EC, *Official Journal of the European Union*, Brussels, I. 275, 25 October 2003, pp32–46.

EU Registry Regulation (Regulation for a standardized and secured system of registries pursuant to Article 19(3) of Directive 2003/87/EC and Article 6(1) of Decision 280/2004/EC (not yet been adopted).

Frankel, J. A. (1999) *Greenhouse Gas Emissions*, Policy Brief 52, Washington, DC, Brookings Institution.

Haites, E. (2003a) 'Harmonisation between national and international tradable permit schemes', in *Greenhouse Gas Emissions Trading and Project-based Mechanisms*, Proceedings of OECD Global Forum on Sustainable Development: Emissions Trading and CATEP Country Forum, 17–18 March 2003, Paris, OECD.

Haites, E. (2003b) 'Emissions trading in the United States and Canada', Y. G. Kim, E. F. Haites, S. Sorrell, T. Y. Jung, P. E. Morthorst, M. K. Lee and J. S. Lim, *Domestic Greenhouse Gas Emissions Trading Schemes*, Seoul, Korea Environment Institute, chapter 7.

Haites, E. and Mullins, F. (2001) *Linking Domestic and Industry Greenhouse Gas Emission Trading Systems*, Paris and Geneva, EPRI, International Energy Agency (IEA) and International Emissions Trading Association, October.

Hargrave, T. (1998) *Growth Baselines: Reducing Emissions and Increasing Investments in Developing Countries*, Washington, DC, Center for Clean Air Policy.

Intergovernmental Panel on Climate Change (IPCC) (2001) *Climate Change 2001: Synthesis Report*, Cambridge, Cambridge University Press.

Jansen, J. C., Battjes, J. J., Ormel, F. T., Sijm, J. P. M., Volkers, C. H. J., Ybema, R., Torvanger, A., Ringius, L. and Underdal, A. (2001) *Sharing the Burden of Greenhouse Gas Mitigation*, Final report of the joint CICERO-ECN project on the global differentiation of emission mitigation targets among countries, CICERO Center for International Climate and Environmental Research, Oslo, Norway.

Janssen, J. and Springer, U. (2001) The *Swiss CO_2 Emissions Trading System*, St Gallen, Switzerland, Institute for Economy and the Environment, University of St Gallen, September.

Jung, T. Y. (2003) 'Domestic greenhouse gas emissions trading program in Japan', in Y. G. Kim, E. F. Haites, S. Sorrell, T. Y. Jung, P. E. Morthorst, M. K. Lee and J. S. Lim, *Domestic Greenhouse Gas Emissions Trading Schemes*, Seoul, Korea Environment Institute, chapter 5.

Kemfert, C., Haites, E. and Missfeldt, F. (2004) 'Can Kyoto protocol parties induce the United States to adopt a more stringent greenhouse gas emissions target?', *Interdisciplinary Environmental Review Journal*, forthcoming.

Kim, Y. G. and Baumert, K. A. (2002) 'Reducing uncertainty through dual-intensity targets', in K. Baumert, O. Blanchard, S. Llosa and J. F. Perkaus (eds), *Building on the Kyoto Protocol: Options for Protecting the Climate*, Washington, DC, World Resources Institute, chapter 5.

Kinzig, A. P. and Kammen, D. M. (1998) 'National trajectories of carbon emissions: analysis of proposals to foster the transition to low-carbon economies', *Global Environmental Change*, vol 8, no 3, pp183–208.

La Rovere, E. L., Valente de Macedo, L. and Baumert, K. A. (2002) 'The Brazilian proposal on relative responsibility for global warming', in K. Baumert, O. Blanchard,

S. Llosa and J. F. Perkaus (eds), *Building on the Kyoto Protocol: Options for Protecting the Climate*, Washington, DC, World Resources Institute, chapter 7.

Lisowski, M. (2002) 'The emperor's new clothes: redressing the Kyoto Protocol', *Climate Policy*, vol 2, pp161–77.

McKibbin, W. and Wilcoxen, P. J. (2002) *Climate Change Policy after Kyoto*, Washington, DC, Brookings Institution Press.

Müller, B., Michaelowa, A. and Vrolijk, C. (2001) *Rejecting Kyoto: A Study of Proposed Alternatives to the Kyoto Protocol*, Climate Strategies, available at: http://www.climate-strategies.org.

Norway, Commission for a National Trading System for Greenhouse Gases (2000) *A Quota System for Greenhouse Gases: A Policy Instrument for Fulfilling Norway's Emission Reduction Commitments under the Kyoto Protocol*, Oslo. (An English summary is available at: http://odin.dep.no/md/engelsk/publ/rapporter/022021-020006/index-dok000-b-n-a.html.)

Pershing, J. and Tudela, F. (2003) 'A long-term target: framing the climate effort', in *Beyond Kyoto: Advancing the International Effort against Climate Change*, Arlington, VA, Pew Center on Global Climate Change, chapter 2, pp11–36.

Philibert, C. (2000) 'How could emissions trading benefit developing countries', *Energy Policy*, vol 28, no 13, pp947–56.

Philibert, C. and Pershing, J. (2001) 'Considering the options: climate targets for all countries', *Climate Policy*, vol 1, no 2, pp211–27.

Rose, A., Stevens, B., Edmonds, J. and Wise, M. (1998) 'International equity and differentiation in global warming policy: an application to tradeable emission permits', *Environmental and Resource Economics*, vol 12, pp25–51.

Samaniego, J. and Figueres, C. (2002) 'Evolving to a sector-based clean development mechanism', in K. Baumert, O. Blanchard, S. Llosa and J. F. Perkaus (eds), *Building on the Kyoto Protocol: Options for Protecting the Climate*, Washington, DC, World Resources Institute, chapter 4.

Schreiner, P. (2000) 'The Norwegian approach to greenhouse gas emissions trading', *RECIEL*, vol 9, no 3, pp239–51.

Switzerland (1999) Federal Law on the Reduction of CO_2 Emissions (CO_2 law), Federal Assembly of the Swiss Confederation, 8 October.

Switzerland (2001) *Guidelines on Voluntary Measures to Reduce Energy Consumption and CO_2 Emissions*, Bern, Bundesamt für Umwelt, Wald und Landschaft (BUWAL), July.

Tjernshaugen, A. (2002) 'Norway proposes combination of trading and taxes', *Cicerone*, 2/2002, CICERO, Oslo.

United States Senate (2003) Climate Stewardship Act of 2003, Senate Bill 139, 108th Congress, 1st Session, Introduced by Senators McCain and Lieberman, Washington, DC, 9 January.

Winkler, H., Spalding-Fecher, R., Mwakasonda, S. and Davidson, O. (2002) 'Sustainable development policies and measures: starting from development to tackle climate change', in K. Baumert, O. Blanchard, S. Llosa and J. F. Perkaus (eds), *Building on the Kyoto Protocol: Options for Protecting the Climate*, Washington, DC, World Resources Institute, chapter 3.

APPENDICES

1   Documents related to the EU emission allowance trading Scheme   353

2   EU Emission Allowance Trading Scheme Directive   355

3   EU Directive 2004/101/EC   371

4   EU Guidelines on Allocations of Allowances   383

Appendix 1

Documents related to the EU Emission Allowance Trading Scheme

- Directive 2003/87/EC of the European Parliament and of the Council of 13 October 2003 establishing a scheme for greenhouse gas emission allowance trading within the Community and amending Council Directive 96/61/EC
- Directive 2004/_/EC of the European Parliament and of the Council of 13 September 2004 amending Directive 2003/87/EC establishing a scheme for greenhouse gas emission allowance trading within the Community, in respect of the Kyoto Protocol's project mechanisms (the Linking Directive)
- European Commission Green Paper on greenhouse gas emissions trading within the European Union of 8 March 2000, COM(2000)87
- European Commission Proposal for a Directive of the European Parliament and of the Council establishing a scheme for greenhouse gas emission allowance trading within the Community and amending Council Directive 96/61/EC of 23 October 2001, COM(2001)581
- European Commission Proposal for a Directive of the European Parliament and of the Council amending the Directive establishing a scheme for greenhouse gas emission allowance trading within the Community, in respect of the Kyoto Protocol's project mechanisms of 23 July 2003, COM(2003)403
- DG Environment Non-Paper, The EU Emissions Trading Scheme: How to develop a National Allocation Plan, prepared for the 2nd meeting of Working 3, Monitoring Mechanism Committee, 1 April 2003
- Communication from the Commission on guidance to assist Member States in the implementation of the criteria listed in Annex III to Directive 2003/87/EC establishing a scheme for greenhouse gas emission allowance trading within the Community and amending Council Directive 96/61/EC, and on the circumstances under which *force majeure* is demonstrated of 7 January 2004, COM(2003)830
- Commission Decision 2004/156/EC of 29 January 2004 establishing guidelines for the monitoring and reporting of greenhouse gas emissions

pursuant to Directive 2003/87/EC of the European Parliament and of the Council
- Council Decision 2002/358/CE of 25 April 2002 concerning the approval, on behalf of the European Community, of the Kyoto Protocol to the United Nations Framework Convention on Climate Change and the joint fulfilment of commitments thereunder
- Decision No 280/2004/EC of the European Parliament and of the Council of 11 February 2004 concerning a mechanism for monitoring Community greenhouse gas emissions and for implementing the Kyoto Protocol

Appendix 2

EU Emission Allowance Trading Scheme Directive

DIRECTIVE 2003/87/EC OF THE EUROPEAN PARLIAMENT AND OF THE COUNCIL

of 13 October 2003

establishing a scheme for greenhouse gas emission allowance trading within the Community and
amending Council Directive 96/61/EC

(Text with EEA relevance)

THE EUROPEAN PARLIAMENT AND THE
COUNCIL OF THE EUROPEAN UNION,

Having regard to the Treaty establishing the European
Community, and in particular Article 175(1) thereof,

Having regard to the proposal from the Commission (^{1}),

Having regard to the opinion of the European Economic
and Social Committee (^{2}),

Having regard to the opinion of the Committee of the
Regions (^{3}),

Acting in accordance with the procedure laid down in
Article
251 of the Treaty (^{4}),

Whereas:

(1)   The Green Paper on greenhouse gas emissions trading within the European Union launched a debate
across Europe on the suitability and possible functioning of greenhouse gas emissions trading within
the European Union. The European Climate Change
Programme has considered Community policies and
measures through a multi-stakeholder process,
including a scheme for greenhouse gas emission
allowance trading within the Community (the
Community scheme) based on the Green Paper. In
its Conclusions of 8 March 2001, the Council
recognised the particular importance of the
European Climate Change Programme and of work
based on the Green Paper, and underlined the urgent
need for concrete action at Community level.

(2)   The Sixth Community Environment Action
Programme established by Decision No
1600/2002/EC of the European Parliament and of
the Council (^{5}) identifies climate change as a priority
for action and provides for the establishment of a
Community-wide emissions trading scheme by
2005. That Programme recognises that the
Community is committed to achieving an 8% reduction in emissions of greenhouse gases by 2008 to
2012 compared to 1990 levels, and that, in the
longer term, global emissions of greenhouse gases
will need to be reduced by approximately 70%
compared to 1990 levels.

(3)   The ultimate objective of the United Nations
Framework Convention on Climate Change, which
was approved by Council Decision 94/69/EC of 15
December 1993 concerning the conclusion of the
United Nations Framework Convention on Climate
Change (^{6}), is to achieve stabilisation of greenhouse
gas concentrations in the atmosphere at a level
which prevents dangerous anthropogenic interference with the climate system.

(4)   Once it enters into force, the Kyoto Protocol, which
was approved by Council Decision 2002/358/EC of
25 April 2002 concerning the approval, on behalf of
the European Community, of the Kyoto Protocol to
the United Nations Framework Convention on
Climate Change and the joint fulfilment of commitments thereunder (^{7}), will commit the Community
and its Member States to reducing their aggregate
anthropogenic emissions of greenhouse gases listed
in Annex A to the Protocol by 8% compared to
1990 levels in the period 2008 to 2012.

(5)   The Community and its Member States have agreed to
fulfil their commitments to reduce anthropogenic
greenhouse gas emissions under the Kyoto Protocol
jointly, in accordance with Decision 2002/358/EC.
This Directive aims to contribute to fulfilling the commitments of the European Community and its
Member States more effectively, through an efficient
European market in greenhouse gas emission
allowances, with the least possible diminution of economic development and employment.

(6)   Council Decision 93/389/EEC of 24 June 1993 for a
monitoring mechanism of Community CO_2 and
other greenhouse gas emissions (^{8}), established a
mechanism for monitoring greenhouse gas emissions
and evaluating progress towards meeting commitments in respect of these emissions. This mechanism
will assist Member States in determining the total
quantity of allowances to allocate.

(7)   Community provisions relating to allocation of
allowances by the Member States are necessary to
contribute to preserving the integrity of the internal
market and to avoid distortions of competition.

(^{1}) OJ C 75 E, 26.3.2002, p. 33.

(^{2}) OJ C 221, 17.9.2002, p. 27.

(^{3}) OJ C 192, 12.8.2002, p. 59.

(^{4}) Opinion of the European Parliament of 10 October 2002 (not yet
published in the Official Journal), Council Common Position of 18
March 2003 (OJ C 125 E, 27.5.2003, p. 72), Decision of the
European Parliament of 2 July 2003 (not yet published in the Official
Journal) and Council Decision of 22 July 2003.

(^{5}) OJ L 242, 10.9.2002, p. 1.

(^{6}) OJ L 33, 7.2.1994, p. 11.

(^{7}) OJ L 130, 15.5.2002, p. 1.

(^{8}) OJ L 167, 9.7.1993, p. 31. Decision as amended by Decision
1999/ 296/EC (OJ L 117, 5.5.1999, p. 35).

(8) Member States should have regard when allocating allowances to the potential for industrial process activities to reduce emissions.

(9) Member States may provide that they only issue allowances valid for a five-year period beginning in 2008 to persons in respect of allowances cancelled, corresponding to emission reductions made by those persons on their national territory during a three-year period beginning in 2005.

(10) Starting with the said five-year period, transfers of allowances to another Member State will involve corresponding adjustments of assigned amount units under the Kyoto Protocol.

(11) Member States should ensure that the operators of certain specified activities hold a greenhouse gas emissions permit and that they monitor and report their emissions of greenhouse gases specified in relation to those activities.

(12) Member States should lay down rules on penalties applicable to infringements of this Directive and ensure that they are implemented. Those penalties must be effective, proportionate and dissuasive.

(13) In order to ensure transparency, the public should have access to information relating to the allocation of allowances and to the results of monitoring of emissions, subject only to restrictions provided for in Directive 2003/4/EC of the European Parliament and of the Council of 28 January 2003 on public access to environmental information ([1]).

(14) Member States should submit a report on the implementation of this Directive drawn up on the basis of Council Directive 91/692/EEC of 23 December 1991 standardising and rationalising reports on the implementation of certain Directives relating to the environment ([2]).

(15) The inclusion of additional installations in the Community scheme should be in accordance with the provisions laid down in this Directive, and the coverage of the Community scheme may thereby be extended to emissions of greenhouse gases other than carbon dioxide, *inter alia* from aluminium and chemicals activities.

(16) This Directive should not prevent any Member State from maintaining or establishing national trading schemes regulating emissions of greenhouse gases from activities other than those listed in Annex I or included in the Community scheme, or from installations temporarily excluded from the Community scheme.

(17) Member States may participate in international emissions trading as Parties to the Kyoto Protocol with any other Party included in Annex B thereto.

(18) Linking the Community scheme to greenhouse gas emission trading schemes in third countries will increase the cost-effectiveness of achieving the Community emission reductions target as laid down in Decision 2002/358/EC on the joint fulfilment of commitments.

(19) Project-based mechanisms including Joint Implementation (JI) and the Clean Development Mechanism (CDM) are important to achieve the goals of both reducing global greenhouse gas emissions and increasing the cost-effective functioning of the Community scheme. In accordance with the relevant provisions of the Kyoto Protocol and Marrakesh Accords, the use of the mechanisms should be supplemental to domestic action and domestic action will thus constitute a significant element of the effort made.

(20) This Directive will encourage the use of more energy-efficient technologies, including combined heat and power technology, producing less emissions per unit of output, while the future directive of the European Parliament and of the Council on the promotion of cogeneration based on useful heat demand in the internal energy market will specifically promote combined heat and power technology.

(21) Council Directive 96/61/EC of 24 September 1996 concerning integrated pollution prevention and control ([3]) establishes a general framework for pollution prevention and control, through which greenhouse gas emissions permits may be issued. Directive 96/61/EC should be amended to ensure that emission limit values are not set for direct emissions of greenhouse gases from an installation subject to this Directive and that Member states may choose not to impose requirements relating to energy efficiency in respect of combustion units or other units emitting carbon dioxide on the site, without prejudice to any other requirements pursuant to Directive 96/61/EC.

(22) This Directive is compatible with the United Nations Framework Convention on Climate Change and the Kyoto Protocol. It should be reviewed in the light of developments in that context and to take into account experience in its implementation and progress achieved in monitoring of emissions of greenhouse gases.

([1]) OJ L 41, 14.2.2003, p. 26.
([2]) OJ L 377, 31.12.1991, p. 48.

([3]) OJ L 257, 10.10.1996, p. 26.

(23) Emission allowance trading should form part of a comprehensive and coherent package of policies and measures implemented at Member State and Community level. Without prejudice to the application of Articles 87 and 88 of the Treaty, where activities are covered by the Community scheme, Member States may consider the implications of regulatory, fiscal or other policies that pursue the same objectives. The review of the Directive should consider the extent to which these objectives have been attained.

(24) The instrument of taxation can be a national policy to limit emissions from installations temporarily excluded.

(25) Policies and measures should be implemented at Member State and Community level across all sectors of the European Union economy, and not only within the industry and energy sectors, in order to generate substantial emissions reductions. The Commission should, in particular, consider policies and measures at Community level in order that the transport sector makes a substantial contribution to the Community and its Member States meeting their climate change obligations under the Kyoto Protocol.

(26) Notwithstanding the multifaceted potential of market-based mechanisms, the European Union strategy for climate change mitigation should be built on a balance between the Community scheme and other types of Community, domestic and international action.

(27) This Directive respects the fundamental rights and observes the principles recognised in particular by the Charter of Fundamental Rights of the European Union.

(28) The measures necessary for the implementation of this Directive should be adopted in accordance with Council Decision 1999/468/EC of 28 June 1999 laying down the procedures for the exercise of implementing powers conferred on the Commission ([1]).

(29) As the criteria (1), (5) and (7) of Annex III cannot be amended through comitology, amendments in respect of periods after 2012 should only be made through co-decision.

(30) Since the objective of the proposed action, the establishment of a Community scheme, cannot be sufficiently achieved by the Member States acting individually, and can therefore by reason of the scale and effects of the proposed action be better achieved at Community level, the Community may adopt measures, in accordance with the principle of subsidiarity as set out in Article 5 of the Treaty. In accordance

with the principle of proportionality, as set out in that Article, this Directive does not go beyond what is necessary in order to achieve that objective,

HAVE ADOPTED THIS DIRECTIVE:

Article 1

Subject matter

This Directive establishes a scheme for greenhouse gas emission allowance trading within the Community (hereinafter referred to as the 'Community scheme') in order to promote reductions of greenhouse gas emissions in a cost-effective and economically efficient manner.

Article 2

Scope

1.   This Directive shall apply to emissions from the activities listed in Annex I and greenhouse gases listed in Annex II.

2.   This Directive shall apply without prejudice to any requirements pursuant to Directive 96/61/EC.

Article 3

Definitions

For the purposes of this Directive the following definitions shall apply:

(a) 'allowance' means an allowance to emit one tonne of carbon dioxide equivalent during a specified period, which shall be valid only for the purposes of meeting the requirements of this Directive and shall be transferable in accordance with the provisions of this Directive;

(b) 'emissions' means the release of greenhouse gases into the atmosphere from sources in an installation;

(c) 'greenhouse gases' means the gases listed in Annex II;

(d) 'greenhouse gas emissions permit' means the permit issued in accordance with Articles 5 and 6;

(e) 'installation' means a stationary technical unit where one or more activities listed in Annex I are carried out and any other directly associated activities which have a technical connection with the activities carried out on that site and which could have an effect on emissions and pollution;

(f) 'operator' means any person who operates or controls an installation or, where this is provided for in national legislation, to whom decisive economic power over the technical functioning of the installation has been delegated;

(g) 'person' means any natural or legal person;

([1]) OJ L 184, 17.7.1999, p. 23.

(h) 'new entrant' means any installation carrying out one or more of the activities indicated in Annex I, which has obtained a greenhouse gas emissions permit or an update of its greenhouse gas emissions permit because of a change in the nature or functioning or an extension of the installation, subsequent to the notification to the Commission of the national allocation plan;

(i) 'the public' means one or more persons and, in accordance with national legislation or practice, associations, organisations or groups of persons;

(j) 'tonne of carbon dioxide equivalent' means one metric tonne of carbon dioxide (CO_2) or an amount of any other greenhouse gas listed in Annex II with an equivalent global-warming potential.

Article 4

Greenhouse gas emissions permits

Member States shall ensure that, from 1 January 2005, no installation undertakes any activity listed in Annex I resulting in emissions specified in relation to that activity unless its operator holds a permit issued by a competent authority in accordance with Articles 5 and 6, or the installation is temporarily excluded from the Community scheme pursuant to Article 27.

Article 5

Applications for greenhouse gas emissions permits

An application to the competent authority for a greenhouse
gas emissions permit shall include a description of:

(a) the installation and its activities including the technology used;

(b) the raw and auxiliary materials, the use of which is likely to lead to emissions of gases listed in Annex I;

(c) the sources of emissions of gases listed in Annex I from the installation; and

(d) the measures planned to monitor and report emissions in accordance with the guidelines adopted pursuant to Article 14.

The application shall also include a non-technical summary of the details referred to in the first subparagraph.

Article 6

Conditions for and contents of the greenhouse gas emissions permit

1.   The competent authority shall issue a greenhouse gas emissions permit granting authorisation to emit greenhouse gases from all or part of an installation if it is satisfied that the operator is capable of monitoring and reporting emissions.

A greenhouse gas emissions permit may cover one or more installations on the same site operated by the same operator.

2.   Greenhouse gas emissions permits shall contain the following:

(a) the name and address of the operator;

(b) a description of the activities and emissions from the installation;

(c) monitoring requirements, specifying monitoring methodology and frequency;

(d) reporting requirements; and

(e) an obligation to surrender allowances equal to the total emissions of the installation in each calendar year, as verified in accordance with Article 15, within four months following the end of that year.

Article 7

Changes relating to installations

The operator shall inform the competent authority of any changes planned in the nature or functioning, or an extension, of the installation which may require updating of the greenhouse gas emissions permit. Where appropriate, the competent authority shall update the permit. Where there is a change in the identity of the installation's operator, the competent authority shall update the permit to include the name and address of the new operator.

Article 8

Coordination with Directive 96/61/EC

Member States shall take the necessary measures to ensure that, where installations carry out activities that are included in Annex I to Directive 96/61/EC, the conditions of, and procedure for, the issue of a greenhouse gas emissions permit are coordinated with those for the permit provided for in that Directive. The requirements of Articles 5, 6 and 7 of this Directive may be integrated into the procedures provided for in Directive 96/61/EC.

Article 9

National allocation plan

1.   For each period referred to in Article 11(1) and (2), each Member State shall develop a national plan stating the total quantity of allowances that it intends to allocate for that period and how it proposes to allocate them. The plan shall be based on objective and transparent criteria, including those listed in Annex III, taking due account of comments from the public. The Commission shall, without prejudice to the Treaty, by 31 December 2003 at the latest develop guidance on the implementation of the criteria listed in Annex III.

For the period referred to in Article 11(1), the plan shall be published and notified to the Commission and to the other Member States by 31 March 2004 at the latest. For subsequent periods, the plan shall be published and notified to the Commission and to the other Member States at least 18 months before the beginning of the relevant period.

2.   National allocation plans shall be considered within the committee referred to in Article 23(1).

3.   Within three months of notification of a national allocation plan by a Member State under paragraph 1, the Commission may reject that plan, or any aspect thereof, on the basis that it is incompatible with the criteria listed in Annex III or with Article 10. The Member State shall only take a decision under Article 11(1) or (2) if proposed amendments are accepted by the Commission. Reasons shall be given for any rejection decision by the Commission.

Article 10

Method of allocation

For the three-year period beginning 1 January 2005 Member States shall allocate at least 95% of the allowances free of charge. For the five-year period beginning 1 January 2008, Member States shall allocate at least 90% of the allowances free of charge.

Article 11

Allocation and issue of allowances

1.   For the three-year period beginning 1 January 2005, each Member State shall decide upon the total quantity of allowances it will allocate for that period and the allocation of those allowances to the operator of each installation. This decision shall be taken at least three months before the beginning of the period and be based on its national allocation plan developed pursuant to Article 9 and in accordance with Article 10, taking due account of comments from the public.

2.   For the five-year period beginning 1 January 2008, and for each subsequent five-year period, each Member State shall decide upon the total quantity of allowances it will allocate for that period and initiate the process for the allocation of those allowances to the operator of each installation. This decision shall be taken at least 12 months before the beginning of the relevant period and be based on the Member State's national allocation plan developed pursuant to Article 9 and in accordance with Article 10, taking due account of comments from the public.

3.   Decisions taken pursuant to paragraph 1 or 2 shall be in accordance with the requirements of the Treaty, in particular Articles 87 and 88 thereof. When deciding upon allocation, Member States shall take into account the need to provide access to allowances for new entrants.

4.   The competent authority shall issue a proportion of the total quantity of allowances each year of the period referred to in paragraph 1 or 2, by 28 February of that year.

Article 12

Transfer, surrender and cancellation of allowances

1.   Member States shall ensure that allowances can be transferred between:

(a) persons within the Community;

(b) persons within the Community and persons in third countries, where such allowances are recognised in accordance with the procedure referred to in Article 25 without restrictions other than those contained in, or adopted pursuant to, this Directive.

2.   Member States shall ensure that allowances issued by a competent authority of another Member State are recognised for the purpose of meeting an operator's obligations under paragraph 3.

3.   Member States shall ensure that, by 30 April each year at the latest, the operator of each installation surrenders a number of allowances equal to the total emissions from that installation during the preceding calendar year as verified in accordance with Article 15, and that these are subsequently cancelled.

4.   Member States shall take the necessary steps to ensure that allowances will be cancelled at any time at the request of the person holding them.

Article 13

Validity of allowances

1.   Allowances shall be valid for emissions during the period referred to in Article 11(1) or (2) for which they are issued.

2.   Four months after the beginning of the first five-year period referred to in Article 11(2), allowances which are no longer valid and have not been surrendered and cancelled in accordance with Article 12(3) shall be cancelled by the competent authority.

Member States may issue allowances to persons for the current period to replace any allowances held by them which are cancelled in accordance with the first subparagraph.

3.   Four months after the beginning of each subsequent five-year period referred to in Article 11(2), allowances which are no longer valid and have not been surrendered and cancelled in accordance with Article 12(3) shall be cancelled by the competent authority.

Appendix 2 361

Member States shall issue allowances to persons for the current period to replace any allowances held by them which are cancelled in accordance with the first subparagraph.

Article 14

Guidelines for monitoring and reporting of emissions

1. The Commission shall adopt guidelines for monitoring
and reporting of emissions resulting from the activities listed in Annex I of greenhouse gases specified in relation to those activities, in accordance with the procedure referred to in Article 23(2), by 30 September 2003. The guidelines shall be based on the principles for monitoring and reporting set out in Annex IV.

2. Member States shall ensure that emissions are monitored in accordance with the guidelines.

3. Member States shall ensure that each operator of an installation reports the emissions from that installation during each calendar year to the competent authority after the end of that year in accordance with the guidelines.

Article 15

Verification

Member States shall ensure that the reports submitted by operators pursuant to Article 14(3) are verified in accordance with the criteria set out in Annex V, and that the competent authority is informed thereof.

Member States shall ensure that an operator whose report has not been verified as satisfactory in accordance with the criteria set out in Annex V by 31 March each year for emissions during the preceding year cannot make further transfers of allowances until a report from that operator has been verified as satisfactory.

Article 16

Penalties

1. Member States shall lay down the rules on penalties applicable to infringements of the national provisions adopted pursuant to this Directive and shall take all measures necessary to ensure that such rules are implemented. The penalties provided for must be effective, proportionate and dissuasive. Member States shall notify these provisions to the Commission by 31 December 2003 at the latest, and shall notify it without delay of any subsequent amendment affecting them.

2. Member States shall ensure publication of the names of operators who are in breach of requirements to surrender sufficient allowances under Article 12(3).

3. Member States shall ensure that any operator who does not surrender sufficient allowances by 30 April of each year to cover its emissions during the preceding year

shall be held liable for the payment of an excess emissions penalty. The excess emissions penalty shall be EUR 100 for each tonne of carbon dioxide equivalent emitted by that installation for which the operator has not surrendered allowances. Payment of the excess emissions penalty shall not release the operator from the obligation to surrender an amount of allowances equal to those excess emissions when surrendering allowances in relation to the following calendar year.

4. During the three-year period beginning 1 January 2005, Member States shall apply a lower excess emissions penalty of EUR 40 for each tonne of carbon dioxide equivalent emitted by that installation for which the operator has not surrendered allowances. Payment of the excess emissions penalty shall not release the operator from the obligation to surrender an amount of allowances equal to those excess emissions when surrendering allowances in relation to the following calendar year.

Article 17

Access to information

Decisions relating to the allocation of allowances and the reports of emissions required under the greenhouse gas emissions permit and held by the competent authority shall be made available to the public by that authority subject to the restrictions laid down in Article 3(3) and Article 4 of Directive 2003/4/EC.

Article 18

Competent authority

Member States shall make the appropriate administrative arrangements, including the designation of the appropriate competent authority or authorities, for the implementation of the rules of this Directive. Where more than one competent authority is designated, the work of these authorities undertaken pursuant to this Directive must be coordinated.

Article 19

Registries

1. Member States shall provide for the establishment and maintenance of a registry in order to ensure the accurate accounting of the issue, holding, transfer and cancellation of allowances. Member States may maintain their registries in a consolidated system, together with one or more other Member States.

2. Any person may hold allowances. The registry shall be accessible to the public and shall contain separate accounts to record the allowances held by each person to whom and from whom allowances are issued or transferred.

3. In order to implement this Directive, the Commission shall adopt a Regulation in accordance with the procedure referred to in Article 23(2) for a standardised and secured system of registries in the form of standardised electronic databases containing common data elements to track the issue, holding, transfer and cancellation of allowances, to   provide for public access and confidentiality as appropriate and to ensure that there are no transfers incompatible with obligations resulting from the Kyoto Protocol.

Article 20

Central Administrator

1. The Commission shall designate a Central Administrator to maintain an independent transaction log recording the issue, transfer and cancellation of allowances.

2. The Central Administrator shall conduct an automated check on each transaction in registries through the independent transaction log to ensure there are no irregularities in the issue, transfer and cancellation of allowances.

3. If irregularities are identified through the automated check, the Central Administrator shall inform the Member State or Member States concerned who shall not register the transactions in question or any further transactions relating to the allowances concerned until the irregularities have been resolved.

Article 21

Reporting by Member States

1. Each year the Member States shall submit to the Commission a report on the application of this Directive. This report shall pay particular attention to the arrangements for the allocation of allowances, the operation of registries, the application of the monitoring and reporting guidelines, verification and issues relating to compliance with the Directive and on the fiscal treatment of allowances, if any. The first report shall be sent to the Commission by 30 June 2005. The report shall be drawn up on the basis of a   questionnaire or outline drafted by the Commission in   accordance with the procedure laid down in Article 6 of Directive 91/692/EEC. The questionnaire or outline shall be sent to Member States at least six months before the deadline for the submission of the first report.

2. On the basis of the reports referred to in paragraph 1, the Commission shall publish a report on the application of this Directive within three months of receiving the reports from the Member States.

3. The Commission shall organise an exchange of information between the competent authorities of the Member States concerning developments relating to issues of allocation, the operation of registries, monitoring, reporting, verification and compliance.

Article 22

Amendments to Annex III

The Commission may amend Annex III, with the exception of criteria (1), (5) and (7), for the period from 2008 to 2012 in the light of the reports provided for in Article 21 and of the experience of the application of this Directive, in accordance with the procedure referred to in Article 23(2).

Article 23

Committee

1. The Commission shall be assisted by the committee instituted by Article 8 of Decision 93/389/EEC.

2. Where reference is made to this paragraph, Articles 5 and 7 of Decision 1999/468/EC shall apply, having regard to the provisions of Article 8 thereof.

The period laid down in Article 5(6) of Decision 1999/468/EC shall be set at three months.

3. The Committee shall adopt its rules of procedure.

Article 24

Procedures for unilateral inclusion of additional activities and gases

1. From 2008, Member States may apply emission allowance trading in accordance with this Directive to activities, installations and greenhouse gases which are not listed in Annex I, provided that inclusion of such activities, installations and greenhouse gases is approved by the Commission in accordance with the procedure referred to in Article 23(2), taking into account all relevant criteria, in particular effects on the internal market, potential distortions of competition, the environmental integrity of the scheme and reliability of the planned monitoring and reporting system.

From 2005 Member States may under the same conditions apply emissions allowance trading to installations carrying out activities listed in Annex I below the capacity limits referred to in that Annex.

2. Allocations made to installations carrying out such activities shall be specified in the national allocation plan referred to in Article 9.

3.   The Commission may, on its own initiative, or shall, on request by a Member State, adopt monitoring and reporting guidelines for emissions from activities, installations and greenhouse gases which are not listed in Annex I in accordance with the procedure referred to in Article 23(2), if monitoring and reporting of these emissions can be carried out with sufficient accuracy.

4.   In the event that such measures are introduced, reviews carried out pursuant to Article 30 shall also consider whether Annex I should be amended to include emissions from these activities in a harmonised way throughout the Community.

Article 25

Links with other greenhouse gas emissions trading schemes

1.   Agreements should be concluded with third countries listed in Annex B to the Kyoto Protocol which have ratified the Protocol to provide for the mutual recognition of allowances between the Community scheme and other greenhouse gas emissions trading schemes in accordance with the rules set out in Article 300 of the Treaty.

2.   Where an agreement referred to in paragraph 1 has been concluded, the Commission shall draw up any necessary provisions relating to the mutual recognition of allowances under that agreement in accordance with the procedure referred to in Article 23(2).

Article 26

Amendment of Directive 96/61/EC

In Article 9(3) of Directive 96/61/EC the following subparagraphs shall be added:

'Where emissions of a greenhouse gas from an installation are specified in Annex I to Directive 2003/87/EC of the European Parliament and of the Council of 13 October 2003 establishing a scheme for greenhouse gas emission allowance trading within the Community and amending Council Directive 96/61/EC (*) in relation to an activity carried out in that installation, the permit shall not include an emission limit value for direct emissions of that gas unless it is necessary to ensure that no significant local pollution is caused.

For activities listed in Annex I to Directive 2003/87/EC, Member States may choose not to impose requirements relating to energy efficiency in respect of combustion units or other units emitting carbon dioxide on the site.

Where necessary, the competent authorities shall amend the permit as appropriate.

The three preceding subparagraphs shall not apply to installations temporarily excluded from the scheme for greenhouse gas emission allowance trading within the Community in accordance with Article 27 of Directive 2003/87/EC.

(*) OJ L 275, 25.10.2003, p. 32.'

Article 27

Temporary exclusion of certain installations

1.   Member States may apply to the Commission for installations to be temporarily excluded until 31 December 2007 at the latest from the Community scheme. Any such application shall list each such installation and shall be published.

2.   If, having considered any comments made by the public on that application, the Commission decides, in accordance with the procedure referred to in Article 23(2), that the installations will:

(a) as a result of national policies, limit their emissions as much as would be the case if they were subject to the provisions of this Directive;

(b) be subject to monitoring, reporting and verification requirements which are equivalent to those provided for pursuant to Articles 14 and 15; and

(c) be subject to penalties at least equivalent to those referred to in Article 16(1) and (4) in the case of non-fulfilment of national requirements; it shall provide for the temporary exclusion of those installations from the Community scheme.

It must be ensured that there will be no distortion of the internal market.

Article 28

Pooling

1.   Member States may allow operators of installations carrying out one of the activities listed in Annex I to form a pool of installations from the same activity for the period referred to in Article 11(1) and/or the first five-year period referred to in Article 11(2) in accordance with paragraphs 2 to 6 of this Article.

2.   Operators carrying out an activity listed in Annex I who wish to form a pool shall apply to the competent authority, specifying the installations and the period for which they want the pool and supplying evidence that a trustee will be able to fulfil the obligations referred to in paragraphs 3 and 4.

3.   Operators wishing to form a pool shall nominate a trustee:

(a) to be issued with the total quantity of allowances calculated by installation of the operators, by way of derogation from Article 11;

(b) to be responsible for surrendering allowances equal to the total emissions from installations in the pool, by way of derogation from Articles 6(2)(e) and 12(3); and

(c) to be restricted from making further transfers in the event that an operator's report has not been verified as satisfactory in accordance with the second paragraph of Article 15.

4.   The trustee shall be subject to the penalties applicable for breaches of requirements to surrender sufficient allowances to cover the total emissions from installations in the pool, by way of derogation from Article 16(2), (3) and (4).

5.   A Member State that wishes to allow one or more pools to be formed shall submit the application referred to in paragraph 2 to the Commission. Without prejudice to the Treaty, the Commission may within three months of receipt reject an application that does not fulfil the requirements of this Directive. Reasons shall be given for any such decision. In the case of rejection the Member State may only allow the pool to be formed if proposed amendments are accepted by the Commission.

6.   In the event that the trustee fails to comply with penalties referred to in paragraph 4, each operator of an installation in the pool shall be responsible under Articles 12(3) and 16 in respect of emissions from its own installation.

Article 29

Force majeure

1.   During the period referred to in Article 11(1), Member States may apply to the Commission for certain installations to be issued with additional allowances in cases of *force majeure*. The Commission shall determine whether *force majeure* is demonstrated, in which case it shall authorise the issue of additional and non-transferable allowances by that Member State to the operators of those installations.

2.   The Commission shall, without prejudice to the Treaty, develop guidance to describe the circumstances under which *force majeure* is demonstrated, by 31 December 2003 at the latest.

Article 30

Review and further development

1.   On the basis of progress achieved in the monitoring of emissions of greenhouse gases, the Commission may make a proposal to the European Parliament and the Council by 31 December 2004 to amend Annex I to include other activities and emissions of other greenhouse gases listed in Annex II.

2.   On the basis of experience of the application of this Directive and of progress achieved in the monitoring of emissions of greenhouse gases and in the light of developments in the international context, the Commission shall draw up a report on the application of this Directive, considering:

(a) how and whether Annex I should be amended to include other relevant sectors, inter alia the chemicals, aluminium and transport sectors, activities and emissions of other greenhouse gases listed in Annex II, with a view to further improving the economic efficiency of the scheme;

(b) the relationship of Community emission allowance trading with the international emissions trading that will start in 2008;

(c) further harmonisation of the method of allocation (including auctioning for the time after 2012) and of the criteria for national allocation plans referred to in Annex III;

(d) the use of credits from project mechanisms;

(e) the relationship of emissions trading with other policies and measures implemented at Member State and Community level, including taxation, that pursue the same objectives;

(f) whether it is appropriate for there to be a single Community registry;

(g) the level of excess emissions penalties, taking into account, inter alia, inflation;

(h) the functioning of the allowance market, covering in particular any possible market disturbances;

(i) how to adapt the Community scheme to an enlarged European Union;

(j) pooling;

(k) the practicality of developing Community-wide benchmarks as a basis for allocation, taking into account the best available techniques and cost-benefit analysis.

The Commission shall submit this report to the European Parliament and the Council by 30 June 2006, accompanied by proposals as appropriate.

3.   Linking the project-based mechanisms, including Joint Implementation (JI) and the Clean Development Mechanism (CDM), with the Community scheme is desirable and important to achieve the goals of both reducing global greenhouse gas emissions and increasing the cost-effective functioning of the Community scheme. Therefore, the emission credits from the project-based mechanisms will be recognised for their use in this scheme subject to provisions adopted by the European Parliament and the Council on a proposal from the Commission, which should apply in parallel with the Community scheme in 2005. The use of the mechanisms shall be supplemental to domestic action, in accordance with the relevant provisions of the Kyoto Protocol and Marrakesh Accords.

Article 31

Implementation

1.   Member States shall bring into force the laws, regulations and administrative provisions necessary to comply with this Directive by 31 December 2003 at the latest. They shall forthwith inform the Commission thereof. The Commission shall notify the other Member States of these laws, regulations and administrative provisions.

When Member States adopt these measures, they shall contain a reference to this Directive or be accompanied by such a reference on the occasion of their official publication. The methods of making such reference shall be laid down by Member States.

2.   Member States shall communicate to the Commission the text of the provisions of national law which they adopt in the field covered by this Directive. The Commission shall inform the other Member States thereof.

Article 32

Entry into force

This Directive shall enter into force on the day of its publication in the *Official Journal of the European Union.*

Article 33

Addressees

This Directive is addressed to the Member States.

Done at Luxembourg, 13 October 2003.

For the European Parliament	*For the Council*
The President	*The President*
P. COX	G. ALEMANNO

CATEGORIES OF ACTIVITIES REFERRED TO IN ARTICLES 2(1), 3, 4, 14(1), 28 AND 30

1. Installations or parts of installations used for research, development and testing of new products and processes are not covered by this Directive.

2. The threshold values given below generally refer to production capacities or outputs. Where one operator carries out several activities falling under the same subheading in the same installation or on the same site, the capacities of such activities are added together.

Activities	Greenhouse gases
Energy activities	
Combustion installations with a rated thermal input exceeding 20 MW (except hazardous or municipal waste installations)	Carbon dioxide
Mineral oil refineries	Carbon dioxide
Coke ovens	Carbon dioxide
Production and processing of ferrous metals	
Metal ore (including sulphide ore) roasting or sintering installations	Carbon dioxide
Installations for the production of pig iron or steel (primary or secondary fusion) including continuous casting, with a capacity exceeding 2.5 tonnes per hour	Carbon dioxide
Mineral industry	
Installations for the production of cement clinker in rotary kilns with a production capacity exceeding 500 tonnes per day or lime in rotary kilns with a production capacity exceeding 50 tonnes per day or in other furnaces with a production capacity exceeding 50 tonnes per day	Carbon dioxide
Installations for the manufacture of glass including glass fibre with a melting capacity exceeding 20 tonnes per day	Carbon dioxide
Installations for the manufacture of ceramic products by firing, in particular roofing tiles, bricks, refractory bricks, tiles, stoneware or porcelain, with a production capacity exceeding 75 tonnes per day, and/or with a kiln capacity exceeding 4 m^3 and with a setting density per kiln exceeding 300 kg/m^3	Carbon dioxide
Other activities	
Industrial plants for the production of (a) pulp from timber or other fibrous materials	Carbon dioxide
(b) paper and board with a production capacity exceeding 20 tonnes per day	Carbon dioxide

ANNEX II

GREENHOUSE GASES REFERRED TO IN ARTICLES 3 AND 30

Carbon dioxide (CO_2)
Methane (CH_4)
Nitrous Oxide (N_2O)
Hydrofluorocarbons (HFCs)
Perfluorocarbons (PFCs)
Sulphur Hexafluoride (SF_6)

ANNEX III

CRITERIA FOR NATIONAL ALLOCATION PLANS REFERRED TO IN ARTICLES 9, 22 AND 30

1. The total quantity of allowances to be allocated for the relevant period shall be consistent with the Member State's obligation to limit its emissions pursuant to Decision 2002/358/EC and the Kyoto Protocol, taking into account, on the one hand, the proportion of overall emissions that these allowances represent in comparison with emissions from sources not covered by this Directive and, on the other hand, national energy policies, and should be consistent with the national climate change programme. The total quantity of allowances to be allocated shall not be more than is likely to be needed for the strict application of the criteria of this Annex. Prior to 2008, the quantity shall be consistent with a path towards achieving or over-achieving each Member State's target under Decision 2002/358/EC and the Kyoto Protocol.

2. The total quantity of allowances to be allocated shall be consistent with assessments of actual and projected progress towards fulfilling the Member States' contributions to the Community's commitments made pursuant to Decision 93/389/EEC.

3. Quantities of allowances to be allocated shall be consistent with the potential, including the technological potential, of activities covered by this scheme to reduce emissions. Member States may base their distribution of allowances on average emissions of greenhouse gases by product in each activity and achievable progress in each activity.

4. The plan shall be consistent with other Community legislative and policy instruments. Account should be taken of unavoidable increases in emissions resulting from new legislative requirements.

5. The plan shall not discriminate between companies or sectors in such a way as to unduly favour certain undertakings or activities in accordance with the requirements of the Treaty, in particular Articles 87 and 88 thereof.

6. The plan shall contain information on the manner in which new entrants will be able to begin participating in the Community scheme in the Member State concerned.

7. The plan may accommodate early action and shall contain information on the manner in which early action is taken into account. Benchmarks derived from reference documents concerning the best available technologies may be employed by Member States in developing their National Allocation Plans, and these benchmarks can incorporate an element of accommodating early action.

8. The plan shall contain information on the manner in which clean technology, including energy efficient technologies, are taken into account.

9. The plan shall include provisions for comments to be expressed by the public, and contain information on the arrangements by which due account will be taken of these comments before a decision on the allocation of allowances is taken.

10. The plan shall contain a list of the installations covered by this Directive with the quantities of allowances intended to be allocated to each.

11. The plan may contain information on the manner in which the existence of competition from countries or entities outside the Union will be taken into account.

PRINCIPLES FOR MONITORING AND REPORTING REFERRED TO IN ARTICLE 14(1)

Monitoring of carbon dioxide emissions

Emissions shall be monitored either by calculation or on the basis of measurement.

Calculation

Calculations of emissions shall be performed using the formula:

$$\text{Activity data} \times \text{Emission factor} \times \text{Oxidation factor}$$

Activity data (fuel used, production rate etc.) shall be monitored on the basis of supply data or measurement.

Accepted emission factors shall be used. Activity-specific emission factors are acceptable for all fuels. Default factors are acceptable for all fuels except non-commercial ones (waste fuels such as tyres and industrial process gases). Seam-specific defaults for coal, and EU-specific or producer country-specific defaults for natural gas shall be further elaborated. IPCC default values are acceptable for refinery products. The emission factor for biomass shall be zero.

If the emission factor does not take account of the fact that some of the carbon is not oxidised, then an additional oxidation factor shall be used. If activity-specific emission factors have been calculated and already take oxidation into account, then an oxidation factor need not be applied.

Default oxidation factors developed pursuant to Directive 96/61/EC shall be used, unless the operator can demonstrate that activity-specific factors are more accurate.

A separate calculation shall be made for each activity, installation and for each fuel.

Measurement

Measurement of emissions shall use standardised or accepted methods, and shall be corroborated by a supporting calculation of emissions.

Monitoring of emissions of other greenhouse gases

Standardised or accepted methods shall be used, developed by the Commission in collaboration with all relevant stakeholders and adopted in accordance with the procedure referred to in Article 23(2).

Reporting of emissions

Each operator shall include the following information in the report for an installation:

A. Data identifying the installation, including:
  - Name of the installation;
  - Its address, including postcode and country;
  - Type and number of Annex I activities carried out in the installation;
  - Address, telephone, fax and email details for a contact person; and
  - Name of the owner of the installation, and of any parent company.

B. For each Annex I activity carried out on the site for which emissions are calculated:
  - Activity data;
  - Emission factors;
  - Oxidation factors;
  - Total emissions; and
  - Uncertainty.

C. For each Annex I activity carried out on the site for which emissions are measured:
  - Total emissions;
  - Information on the reliability of measurement methods; and
  - Uncertainty.

D. For emissions from combustion, the report shall also include the oxidation factor, unless oxidation has already been taken into account in the development of an activity-specific emission factor.

Member States shall take measures to coordinate reporting requirements with any existing reporting requirements in order to minimise the reporting burden on businesses.

CRITERIA FOR VERIFICATION REFERRED TO IN ARTICLE 15

General Principles

1. Emissions from each activity listed in Annex I shall be subject to verification.

2. The verification process shall include consideration of the report pursuant to Article 14(3) and of monitoring during the preceding year. It shall address the reliability, credibility and accuracy of monitoring systems and the reported data and information relating to emissions, in particular:

    (a) the reported activity data and related measurements and calculations;

    (b) the choice and the employment of emission factors;

    (c) the calculations leading to the determination of the overall emissions; and

    (d) if measurement is used, the appropriateness of the choice and the employment of measuring methods.

3. Reported emissions may only be validated if reliable and credible data and information allow the emissions to be determined with a high degree of certainty. A high degree of certainty requires the operator to show that:

    (a) the reported data is free of inconsistencies;

    (b) the collection of the data has been carried out in accordance with the applicable scientific standards; and

    (c) the relevant records of the installation are complete and consistent.

4. The verifier shall be given access to all sites and information in relation to the subject of the verification.

5. The verifier shall take into account whether the installation is registered under the Community eco-management and audit scheme (EMAS).

Methodology

Strategic analysis

6. The verification shall be based on a strategic analysis of all the activities carried out in the installation. This requires the verifier to have an overview of all the activities and their significance for emissions.

Process analysis

7. The verification of the information submitted shall, where appropriate, be carried out on the site of the installation. The verifier shall use spot-checks to determine the reliability of the reported data and information.

Risk analysis

8. The verifier shall submit all the sources of emissions in the installation to an evaluation with regard to the reliability of the data of each source contributing to the overall emissions of the installation.

9. On the basis of this analysis the verifier shall explicitly identify those sources with a high risk of error and other aspects of the monitoring and reporting procedure which are likely to contribute to errors in the determination of the overall emissions. This especially involves the choice of the emission factors and the calculations necessary to determine the level of the emissions from individual sources. Particular attention shall be given to those sources with a high risk of error and the above-mentioned aspects of the monitoring procedure.

10. The verifier shall take into consideration any effective risk control methods applied by the operator with a view to minimising the degree of uncertainty.

Report

11. The verifier shall prepare a report on the validation process stating whether the report pursuant to Article 14(3) is satisfactory. This report shall specify all issues relevant to the work carried out. A statement that the report pursuant to Article 14(3) is satisfactory may be made if, in the opinion of the verifier, the total emissions are not materially misstated.

Minimum competency requirements for the verifier

12. The verifier shall be independent of the operator, carry out his activities in a sound and objective professional manner, and understand:

   (a) the provisions of this Directive, as well as relevant standards and guidance adopted by the Commission pursuant to Article 14(1);

   (b) the legislative, regulatory, and administrative requirements relevant to the activities being verified; and

   (c) the generation of all information related to each source of emissions in the installation, in particular, relating to the   collection, measurement, calculation and reporting of data.

Appendix 3

EU Directive 2004/101/EC of the
European Parliament and of the Council
amending Directive 2003/87/EC
establishing a scheme for greenhouse gas
emission allowance trading within the
Community in respect of the Kyoto
Protocol's project mechanisms
(adopted 27 October 2004)

DIRECTIVE 2004/101/EC OF THE EUROPEAN PARLIAMENT AND OF THE COUNCIL

of 27 October 2004

amending Directive 2003/87/EC establishing a scheme
for greenhouse gas emission allowance trading within the Community,
in respect of the Kyoto Protocol's project mechanisms

(Text with EEA relevance)

THE EUROPEAN PARLIAMENT AND THE COUNCIL OF THE EUROPEAN UNION,

Having regard to the Treaty establishing the European Community, and in particular Article 175(1) thereof,

Having regard to the proposal from the Commission,

Having regard to the Opinion of the European Economic and Social Committee[1],

After consulting the Committee of the Regions,

Acting in accordance with the procedure laid down in Article 251 of the Treaty[2],

---

1   OJ C 80, 30.3.2004, p. 61.
2   Opinion of the European Parliament of 20 April 2004 (not yet published in the Official Journal) and Council Decision of 13 September 2004 (not yet published in the Official Journal).

Whereas:

(1)   Directive 2003/87/EC[1] establishes a scheme for greenhouse gas emission
allowance trading within the Community ("the Community scheme") in order to
promote reductions of greenhouse gas emissions in a cost-effective and
economically efficient manner, recognising that, in the longer-term, global
emissions of greenhouse gases will need to be reduced by approximately 70%
compared to 1990 levels. It aims at contributing towards fulfilling the
commitments of the Community and its Member States to reduce anthropogenic
greenhouse gas emissions under the Kyoto Protocol which was approved by
Council Decision 2002/358/EC of 25 April 2002 concerning the approval, on
behalf of the European Community, of the Kyoto Protocol to the United Nations
Framework Convention on Climate Change and the joint fulfilment of
commitments thereunder[2].

(2)   Directive 2003/87/EC states that the recognition of credits from project-based
mechanisms for fulfilling obligations as from 2005 will increase the cost-
effectiveness of achieving reductions of global greenhouse gas emissions and shall
be provided for by provisions for linking the Kyoto project-based mechanisms,
including Joint Implementation (JI) and the Clean Development Mechanism
(CDM), with the Community scheme.

(3)   Linking the Kyoto project-based mechanisms to the Community scheme, while
safeguarding the latter's environmental integrity, gives the opportunity to use
emission credits generated through project activities eligible under Articles 6 and
12 of the Kyoto Protocol in order to fulfil Member States' obligations under
Article 12(3) of Directive 2003/87/EC. As a result, this will increase the diversity
of low cost compliance options within the Community scheme leading to a
reduction of the overall costs of compliance with the Kyoto Protocol while
improving the liquidity of the Community market in greenhouse gas emission
allowances. By stimulating demand for JI credits, Community companies will
invest in the development and transfer of advanced environmentally sound
technologies and know-how. The demand for CDM credits will also be
stimulated and thus developing countries hosting CDM projects will be assisted
in achieving their sustainable development goals.

(4)   In addition to the use of the Kyoto project-based mechanisms by the Community
and its Member States, and by companies and individuals outside the Community
scheme, those mechanisms should be linked to the Community scheme in such a
way as to ensure consistency with the United Nations Framework Convention on
Climate Change (UNFCCC) and the Kyoto Protocol and subsequent decisions
adopted thereunder as well as with the objectives and architecture of the
Community scheme and provisions laid down by Directive 2003/87/EC.

---

1   OJ L 275, 25.10.2003, p. 32.
2   OJ L 130, 15.5.2002, p. 1.

(5)   Member States may allow operators to use, in the Community scheme, certified emission reductions (CERs) from 2005 and emission reduction units (ERUs) from 2008. The use of CERs and ERUs by operators from 2008 may be allowed up to a percentage of the allocation to each installation, to be specified by each Member State in its national allocation plan. The use will take place through the issue and immediate surrender of one allowance in exchange for one CER or ERU. An allowance issued in exchange for a CER or ERU will correspond to that CER or ERU.

(6)   The Commission Regulation for a standardised and secured system of registries, to be adopted pursuant to Article 19(3) of Directive 2003/87/EC and Article 6(1) of Decision No 280/2004/EC of the European Parliament and of the Council of 11 February 2004 concerning a mechanism for monitoring Community green-house gas emissions and for implementing the Kyoto Protocol[1], will provide for the relevant processes and procedures in the registries system for the use of CERs during the period 2005–2007 and subsequent periods, and for the use of ERUs during the period 2008–2012 and subsequent periods.

(7)   Each Member State will decide on the limit for the use of CERs and ERUs from project activities, having due regard to the relevant provisions of the Kyoto Protocol and the Marrakesh Accords, to meet the requirements therein that the use of the mechanisms should be supplemental to domestic action. Domestic action will thus constitute a significant element of the effort made.

(8)   In accordance with the UNFCCC and the Kyoto Protocol and subsequent decisions adopted thereunder, Member States are to refrain from using CERs and ERUs generated from nuclear facilities to meet their commitments under Article 3(1) of the Kyoto Protocol and under Decision 2002/358/EC.

(9)   Decisions 15/CP.7 and 19/CP.7 adopted pursuant to the UNFCCC and the Kyoto Protocol emphasise that environmental integrity is to be achieved, inter alia, through sound modalities, rules and guidelines for the mechanisms, and through sound and strong principles and rules governing land use, land-use change and forestry activities, and that the issues of non-permanence, additionality, leakage, uncertainties and socio-economic and environmental impacts, including impacts on biodiversity and natural ecosystems, associated with afforestation and reforestation project activities are to be taken into account. The Commission should consider, in its review of Directive 2003/87/EC in 2006, technical provisions relating to the temporary nature of credits and the limit of 1% for eligibility for land use, land-use change and forestry project activities as established in Decision 17/CP.7, and also provisions relating to the outcome of the evaluation of potential risks associated with the use of genetically modified organisms and potentially invasive alien species in afforestation and reforestation project activities, to allow operators to use CERs and ERUs resulting from land use, land-use change and forestry project activities in the Community scheme from 2008, in accordance with the decisions adopted pursuant to the UNFCCC or the Kyoto Protocol.

---

1   OJ L 49, 19.2.2004, p. 1.

(10)　In order to avoid double counting, CERs and ERUs should not be issued as a result of project activities undertaken within the Community that also lead to a reduction in, or limitation of, emissions from installations covered by Directive 2003/87/EC, unless an equal number of allowances is cancelled from the registry of the Member State of the CERs' or ERUs' origin.

11)　In accordance with the relevant Treaties of Accession, the *acquis communautaire* should be taken into account in the establishment of baselines for project activities undertaken in countries acceding to the Union.

(12)　Any Member State that authorises private or public entities to participate in project activities remains responsible for the fulfilment of its obligations under the UNFCCC and the Kyoto Protocol and should therefore ensure that such participation is consistent with the relevant guidelines, modalities and procedures adopted pursuant to the UNFCCC or the Kyoto Protocol.

(13)　In accordance with the UNFCCC, the Kyoto Protocol and subsequent decisions adopted for their implementation, the Commission and the Member States should support capacity building activities in developing countries and countries with economies in transition in order to help them take full advantage of JI and the CDM in a manner that supports their sustainable development strategies. The Commission should review and report on efforts in this regard.

(14)　Criteria and guidelines that are relevant to considering whether hydro-electric power production projects have negative environmental or social impacts have been identified by the World Commission on Dams in its November 2000 Report "Dams and Development – A New Framework for Decision-Making", by the OECD and by the World Bank.

(15)　Since participation in JI and CDM project activities is voluntary, corporate environmental and social responsibility and accountability should be enhanced in accordance with paragraph 17 of the Plan of Implementation of the World Summit on Sustainable Development. In this connection, companies should be encouraged to improve the social and environmental performance of JI and CDM activities in which they participate.

(16)　Information on project activities in which a Member State participates or authorises private or public entities to participate should be made available to the public in accordance with Directive 2003/4/EC of the European Parliament and of the Council of 28 January 2003 on public access to environmental information[1].

(17)　The Commission may mention impacts on the electricity market in its reports on emission allowance trading and the use of credits from project activities.

1　OJ L 41, 14.2.2003, p. 26.

(18) Following entry into force of the Kyoto Protocol, the Commission should examine whether it could be possible to conclude agreements with countries listed in Annex B to the Kyoto Protocol which have yet to ratify the Protocol, to provide for the recognition of allowances between the Community scheme and mandatory greenhouse gas emissions trading schemes capping absolute emissions established within those countries.

(19) Since the objective of the proposed action, namely the establishment of a link between the Kyoto project-based mechanisms and the Community scheme, cannot be sufficiently achieved by the Member States acting individually, and can therefore by reason of the scale and effects of this action be better achieved at Community level, the Community may adopt measures, in accordance with the principle of subsidiarity as set out in Article 5 of the Treaty. In accordance with the principle of proportionality, as set out in that Article, this Directive does not go beyond what is necessary in order to achieve that objective.

(20) Directive 2003/87/EC should therefore be amended accordingly,

HAVE ADOPTED THIS DIRECTIVE:

<u>Article 1</u>

Amendments to Directive 2003/87/EC

Directive 2003/87/EC is hereby amended as follows:

1) In Article 3, the following points shall be added:

"(k) "Annex I Party" means a Party listed in Annex I to the United Nations Framework Convention on Climate Change (UNFCCC) that has ratified the Kyoto Protocol as specified in Article 1(7) of the Kyoto Protocol;

(1) "project activity" means a project activity approved by one or more Annex I Parties in accordance with Article 6 or Article 12 of the Kyoto Protocol and the decisions adopted pursuant to the UNFCCC or the Kyoto Protocol;

(m) "emission reduction unit" or "ERU" means a unit issued pursuant to Article 6 of the Kyoto Protocol and the decisions adopted pursuant to the UNFCCC or the Kyoto Protocol;

(n) "certified emission reduction" or "CER" means a unit issued pursuant to Article 12 of the Kyoto Protocol and the decisions adopted pursuant to the UNFCCC or the Kyoto Protocol.".

2) The following Articles shall be inserted after Article 11:

"<u>Article 11 a</u>

Use of CERs and ERUs from project activities in the Community scheme

1. Subject to paragraph 3, during each period referred to in Article 11(2), Member States may allow operators to use CERs and ERUs from project activities in the Community scheme up to a percentage of the allocation of allowances to each installation, to be specified by each Member State in its national allocation plan for that period. This shall take place through the issue and immediate surrender of one allowance by the Member State in exchange for one CER or ERU held by the operator in the national registry of its Member State.

2. Subject to paragraph 3, during the period referred to in Article 11(1), Member States may allow operators to use CERs from project activities in the Community scheme. This shall take place through the issue and immediate surrender of one allowance by the Member State in exchange for one CER. Member States shall cancel CERs that have been used by operators during the period referred to in Article 11 (1).

3. All CERs and ERUs that are issued and may be used in accordance with the UNFCCC and the Kyoto Protocol and subsequent decisions adopted thereunder may be used in the Community scheme:

(a) except that, in recognition of the fact that, in accordance with the UNFCCC and the Kyoto Protocol and subsequent decisions adopted thereunder, Member States are to refrain from using CERs and ERUs generated from nuclear facilities to meet their commitments under Article 3(1) of the Kyoto Protocol and under Decision 2002/358/EC, operators are to refrain from using CERs and ERUs generated from such facilities in the Community scheme during the period referred to in Article 11 (1) and the first five-year period referred to in Article 11(2); and

(b) except for CERs and ERUs from land use, land-use change and forestry activities.

Article 11b

Project activities

1. Member States shall take all necessary measures to ensure that baselines for project activities, as defined by subsequent decisions adopted under the UNFCCC or the Kyoto Protocol, undertaken in countries having signed a Treaty of Accession with the Union fully comply with the *acquis communautaire*, including the temporary derogations set out in that Treaty of Accession.

2. Except as provided for in paragraphs 3 and 4, Member States hosting project activities shall ensure that no ERUs or CERs are issued for reductions or limitations of greenhouse gas emissions from installations falling within the scope of this Directive.

3. Until 31 December 2012, for JI and CDM project activities which reduce or limit directly the emissions of an installation falling within the scope of this Directive, ERUs and CERs may be issued only if an equal number of allowances is cancelled by the operator of that installation.

4. Until 31 December 2012, for JI and CDM project activities which reduce or limit indirectly the emission level of installations falling within the scope of this Directive, ERUs and CERs may be issued only if an equal number of allowances is cancelled from the national registry of the Member State of the ERUs' or CERs' origin.

5. A Member State that authorises private or public entities to participate in project activities shall remain responsible for the fulfilment of its obligations under the UNFCCC and the Kyoto Protocol and shall ensure that such participation is consistent with the relevant guidelines, modalities and procedures adopted pursuant to the UNFCCC or the Kyoto Protocol.

6. In the case of hydro-electric power production project activities with a generating capacity exceeding 20MW, Member States shall, when approving such project activities, ensure that relevant international criteria and guidelines, including those contained in the World Commission on Dams November 2000 Report "Dams and Development – A New Framework for Decision-Making", will be respected during the development of such project activities.

7. Provisions for the implementation of paragraphs 3 and 4, particularly in respect of the avoidance of double counting, and any provisions necessary for the implementation of paragraph 5 where the host party meets all eligibility requirements for JI project activities shall be adopted in accordance with Article 23(2).".

3) Article 17 shall be replaced by the following:

"Article 17

Access to information

Decisions relating to the allocation of allowances, information on project activities in which a Member State participates or authorises private or public entities to participate, and the reports of emissions required under the greenhouse gas emissions permit and held by the competent authority, shall be made available to the public in accordance with Directive 2003/4/EC.".

4) In Article 18 the following subparagraph shall be added:

"Member States shall in particular ensure coordination between their designated focal point for approving project activities pursuant to Article 6 (1)(a) of the Kyoto Protocol and their designated national authority for the implementation of Article 12 of the Kyoto Protocol respectively designated in accordance with subsequent decisions adopted under the UNFCCC or the Kyoto Protocol.".

5) In paragraph 3 of Article 19 the following sentence shall be added:

"That Regulation shall also include provisions concerning the use and identification of CERs and ERUs in the Community scheme and the monitoring of the level of such use.".

6) Article 21 shall be amended as follows:

(a) In paragraph 1 the second sentence shall be replaced by the following:

"This report shall pay particular attention to the arrangements for the allocation of allowances, the use of ERUs and CERs in the Community scheme, the operation of registries, the application of the monitoring and reporting guidelines, verification and issues relating to compliance with the Directive and the fiscal treatment of allowances, if any.".

(b) Paragraph 3 shall be replaced by the following:

"3. The Commission shall organise an exchange of information between the competent authorities of the Member States concerning developments relating to issues of allocation, the use of ERUs and CERs in the Community scheme, the operation of registries, monitoring, reporting, verification and compliance with this Directive.".

7) The following Article shall be inserted after Article 21:

"<u>Article 21a</u>

Support of capacity building activities

In accordance with the UNFCCC, the Kyoto Protocol and any subsequent decision adopted for their implementation, the Commission and the Member States shall endeavour to support capacity building activities in developing countries and countries with economies in transition in order to help them take full advantage of JI and the CDM in a manner that supports their sustainable development strategies and to facilitate the engagement of entities in JI and CDM project development and implementation.".

8) Article 30 shall be amended as follows:

(a) In paragraph 2, point (d) shall be replaced by the following:

"(d) the use of credits from project activities, including the need for harmonisation of the allowed use of ERUs and CERs in the Community scheme;";

(b) in paragraph 2 the following points shall be added:

"(l) the impact of project mechanisms on host countries, particularly on their development objectives, whether JI and CDM hydro-electric power production project activities with a generating capacity exceeding 500 MW and having negative environmental or social impacts have been approved, and the future use of CERs or ERUs resulting from any such hydro-electric power production project activities in the Community scheme;

(m) the support for capacity building efforts in developing countries and countries with economies in transition;

(n) the modalities and procedures for Member States' approval of domestic project activities and for the issuing of allowances in respect of emission reductions or limitations resulting from such activities from 2008;

(o) technical provisions relating to the temporary nature of credits and the limit of 1% for eligibility for land use, land-use change and forestry project activities as established in Decision 17/CP.7, and provisions relating to the outcome of the evaluation of potential risks associated with the use of genetically modified organisms and potentially invasive alien species by afforestation and reforestation project activities, to allow operators to use CERs and ERUs resulting from land use, land-use change and forestry project activities in the Community scheme from 2008, in accordance with the decisions adopted pursuant to the UNFCCC or the Kyoto Protocol.";

(c) paragraph 3 shall be replaced by the following:

> "3. In advance of each period referred to in Article 11(2), each Member State shall publish in its national allocation plan its intended use of ERUs and CERs and the percentage of the allocation to each installation up to which operators are allowed to use ERUs and CERs in the Community scheme for that period. The total use of ERUs and CERs shall be consistent with the relevant supplementarity obligations under the Kyoto Protocol and the UNFCCC and the decisions adopted thereunder.

> Member States shall, in accordance with Article 3 of Decision No 280/2004/EC of the European Parliament and of the Council of 11 February 2004 concerning a mechanism for monitoring Community greenhouse gas emissions and for implementing the Kyoto Protocol *, report to the Commission every two years on the extent to which domestic action actually constitutes a significant element of the efforts undertaken at national level, as well as the extent to which use of the project mechanisms is actually supplemental to domestic action, and the ratio between them, in accordance with the relevant provisions of the Kyoto Protocol and the decisions adopted thereunder. The Commission shall report on this in accordance with Article 5 of the said Decision. In the light of this report, the Commission shall, if appropriate, make legislative or other proposals to complement provisions adopted by Member States to ensure that use of the mechanisms is supplemental to domestic action within the Community.

9) In Annex III the following point shall be added:

> "12. The plan shall specify the maximum amount of CERs and ERUs which may be used by operators in the Community scheme as a percentage of the allocation of the allowances to each installation. The percentage shall be consistent with the Member State's supplementarity obligations under the Kyoto Protocol and decisions adopted pursuant to the UNFCCC or the Kyoto Protocol.".

---

*   OJ L 49, 19.2.2004, p. 1.".

Article 2

Implementation

1. Member States shall bring into force the laws, regulations and administrative provisions necessary to comply with this Directive by *. They shall forthwith inform the Commission thereof.

    When Member States adopt these measures, they shall contain a reference to this Directive or be accompanied by such a reference on the occasion of their official publication. The methods of making such reference shall be laid down by the Member States.

2. Member States shall communicate to the Commission the text of the provisions of national law which they adopt in the field covered by this Directive. The Commission shall inform the other Member States thereof.

Article 3

Entry into force

This Directive shall enter into force on the day of its publication in the *Official Journal of the European Union.*

Article 4

Addressees

This Directive is addressed to the Member States.

Done at Strasbourg,

For the European Parliament   For the Council
The President   The President

---

*   12 months after its entry into force.

Appendix 4

EU Guidelines on Allocations of Allowances

Brussels, 7.1.2004
COM(2003)830 final

COMMUNICATION FROM THE COMMISSION

on guidance to assist Member States in the implementation of the criteria listed in
Annex III to Directive 2003/87/EC establishing a scheme for greenhouse gas emission
allowance trading within the Community and amending Council Directive 96/61/EC,
and on the circumstances under which *force majeure* is demonstrated

1.   INTRODUCTION

1.   Directive 2003/87/EC[1] provides for the establishment of a Community-wide greenhouse gas emission allowance trading scheme as of 2005. Pursuant to Article 9 of the Directive, each Member State periodically has to develop a national allocation plan. These plans have to be based on objective and transparent criteria, including those listed in Annex III to the Directive. The first national allocation plans have to be published and notified to the Commission and the other Member States by 31 March 2004. For those Member States joining the Union as of 1 May 2004, the obligation to publish and notify the national allocation plan arises only with the date of accession. The Commission encourages these future Member States to publish and notify national allocation plans also by 31 March 2004.

2.   Article 9 mandates the Commission to develop guidance on the implementation of the criteria listed in Annex III by 31 December 2003. Article 29 mandates the Commission to develop guidance to describe the circumstances under which *force majeure* is demonstrated by the same date. The purpose of this guidance document is three-fold:

   – First, to assist Member States in drawing up their national allocation plans, by indicating the scope of interpretation of the Annex III criteria that the Commission deems acceptable;

   – Second, to support the Commission assessment of notified national allocation plans, pursuant to Article 9(3);

   – Third, to describe the circumstances under which force majeure is demonstrated.

3.   The Directive is a key element of the Community's climate change policy and its objective is to promote reductions of greenhouse gas emissions in a cost-effective and economically efficient manner. It is therefore important to ensure that the emissions trading scheme has a positive environmental outcome. The national allocation plans are the means to achieve this goal. This fact is reflected in the guidance developed in this document.

4.   The Commission will monitor the application of this guidance, and amend it as and when it deems necessary, in particular following any amendments to Annex III pursuant to Articles 22 and/or 30(2)(c) of the Directive.

---

[1] OJ L275, 25.10.2003, p. 32.3

2. GUIDANCE ON THE IMPLEMENTATION OF THE ANNEX III CRITERIA

5. Annex III to Directive 2003/87/EC contains 11 criteria relating to the national allocation plans. The relationships between these criteria can be revealed by categorising them in various ways.

Table 1: Categorisation of the criteria

	Mandatory (M)/ Optional (O)	Total level	Activity/ Sector	Installation level
(1) Kyoto commitments	(M)/(O)	+		
(2) Assessments of emissions development	(M)	+		
(3) Potential to reduce emissions	(M)/(O)	+	+	
(4) Consistency with other legislation	(M)/(O)	+	+	
(5) Non-discrimination between companies or sectors	(M)	+	+	+
(6) New entrants	(O)			+
(7) Early action	(O)			+
(8) Clean technology	(O)			+
(9) Involvement of the public	(M)			
(10) List of installations	(M)			+
(11) Competition from outside the Union	(O)		+	

6. One way of categorising the criteria is on the basis of whether their implementation is mandatory or optional. A Member State has an obligation to apply all elements of criteria (2), (5), (9) and (10), and some elements of the criteria (1), (3) and (4). It can, therefore, choose whether it wants to take specific action with respect to some elements of criteria (1), (3) and (4), and the criteria (6), (7), (8) and (11). The Commission will not reject a plan if all mandatory criteria and mandatory elements of criteria are applied in a correct manner. The Commission will not reject a plan if optional criteria or optional elements of criteria are not applied. However, if these optional criteria or optional elements of criteria or additional transparent and objective criteria are applied, the Commission will assess their application. In all cases, the Commission does require information from a Member State with respect to criteria (7) and (8), even if this is only to state that a criterion has not been applied. In respect of criterion (6) a Member State must state the manner in which new entrants will be able to begin participating in the Community scheme in that Member State.

7.   A second way of categorising the criteria is to distinguish between them depending on whether they are applicable to allowance allocation at the level of all covered installations, at activity or sector level, or at installation level. The Commission's interpretation is presented in Table 1.

8.   The attached common format reflects the fact that criteria apply at different levels, but also that they deal with different aspects, such as technical aspects and Community legislation or policy. For the sake of clarity and in order to facilitate its use by Member States, a recommended common format for establishing and notifying the national allocation plans is attached. The common format will further assist a Member State in drawing up the plan, and, in addition, it will significantly facilitate Member States' scrutiny of each other's plans and increase the accessibility of the plans to stakeholders.

2.1.   Guidance on individual criteria

9.   In the following, the Commission sets out guidance on the implementation of the individual criteria. The criteria are treated individually and in the order in which they are listed in Annex III to the Directive. Cross-references are made in order to highlight relationships between different criteria. The guidance contains an introductory and an analytical section.

2.1.1.   *Criterion (1) – Kyoto commitments*

> *The total quantity of allowances to be allocated for the relevant period shall be consistent with the Member State's obligation to limit its emissions pursuant to Decision 2002/358/EC and the Kyoto Protocol, taking into account, on the one hand, the proportion of overall emissions that these allowances represent in comparison with emissions from sources not covered by this Directive and, on the other hand, national energy policies, and should be consistent with the national climate change programme. The total quantity of allowances to be allocated shall not be more than is likely to be needed for the strict application of the criteria of this Annex. Prior to 2008, the quantity shall be consistent with a path towards achieving or overachieving each Member State's target under Decision 2002/358/EC and the Kyoto Protocol.*

2.1.1.1. Introduction

10.   Criterion (1) makes the link between the total quantity of allowances and the Member State's individual target either under Council Decision 2002/358/EC[2] on the joint fulfilment of commitments under the Kyoto Protocol, or under the Kyoto Protocol itself. For new Member States not referred to in the Decision, their respective targets under the Kyoto Protocol is the reference point under this criterion. While the commitments established for each Member State must be met, the criterion enables a Member State to go *beyond* the 'Kyoto' target. Distributing the effort to meet these targets is a 'zero-sum' exercise, whereby the same result must be achieved however the effort is distributed between covered and non-covered installations and activities, as well as between covered installations.

---

[2] OJ L130, 15.5.2002, p. 1.

11.   Within the scope of the climate change commitment of each Member State, a Member State applying effective policies and measures to sources outside the trading scheme will necessarily be in a position to allocate more allowances to covered installations. National energy policies may also lead to adjustments of the relative contributions to the climate change commitment. If a Member State has committed itself to gradually phase-out nuclear installations on its territory, measures will have to be taken to provide the required levels of electricity. A nuclear phase-out might lead to an increase in greenhouse gas emissions, but would not justify that a Member State does not fulfil its obligations under Decision 2002/358/EC.

12.   The concept of the 'path' reflects the fact that, before the period 2008 to 2012, Member States do not have quantitative targets but are instead required under Article 3(2) of the Kyoto Protocol to make demonstrable progress by 2005 towards meeting their quantitative commitments for 2008 to 2012. Allocations for the period 2005 to 2007 have to be mindful of the targets that will apply from 2008 to 2012. Consequently, it is understood that Member States should be making progress towards their commitments for 2008 to 2012 already in the first trading period of 2005 to 2007. The path is intended to be a trend line, not necessarily a straight one, but one that is leading towards or goes beyond the reductions and limitations called for by the Kyoto Protocol and Decision 2002/358/EC.

2.1.1.2. Analysis

13.   Criterion (1) is largely of a mandatory nature and has to be applied in determining the total quantity of allowances.

14.   While the Directive covers part of a Member State's greenhouse gas emissions, the Kyoto target applies to the total greenhouse gas emissions of a Member State. Hence, a Member State has to decide in the plan what contribution should be made by covered installations to reaching or going beyond the overall commitment in the period 2008 to 2012 and what path it will follow in the period 2005 to 2007.

15.   A Member State has to demonstrate how the chosen total quantity of allowances is consistent with reaching or over-achieving the Kyoto target, taking into account, on the one hand, the proportion of overall emissions that these allowances represent in comparison with emissions from sources not covered by this Directive and, on the other hand, national energy policies. A Member State has to present the chosen path towards reaching or over-achieving the target under Decision 2002/358/EC and the Kyoto Protocol and explain how consistency is ensured between the intended allocation and the path.

16.   In deciding the total quantity, the proportion of overall emissions of covered installations in relation to total emissions is a first element to be taken into account. A Member State should use the most recent data available to determine the proportion. In case a Member State deviates substantially from this proportion, it should give reasons for such deviations. Such reasons may include, *inter alia*, expected structural change in the economy and national energy policy. Consistency with national energy policy may be a reason for an increase or a decrease in the proportion. A Member State phasing-out nuclear

installations over the covered period may increase the proportion, if replacement is not expected to be through carbon-free alternatives. A Member State intending to increase the share of renewable energy or combined heat and power production or other forms of low-carbon or carbon-free power and heat production should decrease the proportion. The Commission recalls that under the provisions of Directive 2001/77/EC[3] on the promotion of electricity produced from renewable energy sources in the internal electricity market all Member States, including the future Member States, have committed themselves to increase the share of electricity from renewable energy sources.

17. The quantity of allowances potentially available for installations covered by the trading scheme needs to be consistent with the forecasted increases or decreases in non-covered activities. Therefore, a Member State should include clear, realistic and substantiated projections of the effectiveness of the policies aimed at non-covered activities in the national allocation plan. Furthermore, a Member State should introduce additional policies and measures to control emissions of non-covered activities in order for all relevant sectors to contribute to the achievement of the target under Decision 2002/358/EC and the Kyoto Protocol.

18. The Commission understands '*likely to be needed*' as forward-looking and linked to the projected emissions of covered installations as a whole, given that this criterion refers to the total quantity of allowances to be allocated. The Commission understands the 'strict application of the criteria in this annex' to comprise the criteria with a mandatory character or containing mandatory elements – i.e. criteria (1), (2), (3), (4) and (5)[4]. In order to satisfy this requirement and fulfil all mandatory criteria and elements, a Member State should not allocate more than is needed, or warranted, by the most constraining of these criteria. It follows that any application of the optional elements of Annex III may not lead to an increase in the total quantity of allowances.

19. The chosen proportion, taking into account criteria (1), (2), (3), (4) and (5), should be multiplied by the annual average emissions allowed under Decision 2002/358/EC and, for new Member States, the Kyoto Protocol in the period 2008 to 2012. This figure could be scaled downwards by an appropriate factor if the Member State intends to go beyond the Kyoto target in 2008 to 2012. In order to determine the total quantity for the period 2005 to 2007 the Member State should scale this amount to the path chosen and multiply the figure by three.

20. As a Party to the Kyoto Protocol, a Member State may use the mechanisms under Articles 6, 12 and 17 (Joint Implementation, Clean Development Mechanism and International Emission Trading) to contribute to compliance with its commitments under the Protocol in the period 2008 to 2012. If a Member State intends to use these mechanisms it may adapt annual average emissions allowed under Decision 2002/358/EC and the Kyoto Protocol in the period 2008 to 2012. In the national allocation plan, a Member State must substantiate any such intentions to use the Kyoto mechanisms. The Commission will base its assessment notably on the state of advancement of relevant legislation or implementing provisions at the national level.

---

[3] OJ L283, 27.10.2001, p. 33.
[4] Criteria (9) and (10) do not relate to the determination of allocated quantities and are therefore not relevant in this context.

A Member State has to determine the total quantity of allowances based on the proportion of overall emissions of covered installations in relation to total emissions. A Member State should use the most recent data available to determine the proportion. In case a Member State deviates substantially from the current proportion, it should give reasons for such deviations.

A Member State must substantiate the intention to use the Kyoto mechanisms.

2.1.2.   *Criterion (2) – Assessments of emissions development*

The total quantity of allowances to be allocated shall be consistent with assessments of actual and projected progress towards fulfilling the Member States' contributions to the Community's commitments made pursuant to Decision 93/389/EEC.

2.1.2.1. Introduction

21.   Pursuant to Decision 93/389/EEC establishing a monitoring mechanism of Community carbon dioxide (CO_2) and other greenhouse gas emissions[5], the Commission undertakes an annual assessment of actual and projected emissions of Member States, in total and by sector and by gas. These assessments are prepared in close cooperation with Member States. Criterion (2) is intended to ensure that the total allocation is consistent with pre-existing, publicly available and objective assessments of actual and projected emissions. The relevant reports that summarise these assessments are COM(2000)749, COM(2001)708, COM (2002)702 and COM(2003)735. The 2000 and 2001 reports cover only the existing Member States, and so are not relevant to new Member States. The 2002 and 2003 reports also cover the new Member States.

22.   Decision 93/389/EEC will be repealed and replaced in early 2004 by Decision 2004/xx/EC on the monitoring of greenhouse gas emissions and the implementation of the Kyoto Protocol[6].

2.1.2.2. Analysis

23.   Criterion (2) is of a mandatory nature and has to be applied in determining the total quantity of allowances.

24.   The Commission conducts assessments under Decision 93/389/EEC, in cooperation with Member States. These assessments cover recent developments of actual emissions of Member States and projected emissions during the period 2008 to 2012, in total and per sector and gas.

25.   Consistency with assessments pursuant to Decision 93/389/EEC will be deemed as ensured if the total quantity of allowances to be allocated to covered installations is not more than would be necessary taking into account actual emissions and projected emissions contained in those assessments. Consistency would not be

---

[5] OJ L167, 9.7.1993, p. 31. Decision as amended by Decision 1999/296/EC (OJ L117, 5.5.1999, p. 35).

[6] This Decision, based on the Commission proposal COM(2003)51, is the subject of a first reading agreement based on the amendments adopted by the European Parliament on 21 October 2003, and is expected to enter into force early in 2004.

ensured if a Member State intended to allocate a total quantity of allowances in excess of actual or projected emissions of covered installations as reported in the assessment for the relevant period.

> Consistency with assessments pursuant to Decision 93/389/EEC will be deemed as ensured if the total quantity of allowances to be allocated to covered installations is not more than actual emissions and projected emissions contained in those assessments.

2.1.3. *Criterion (3) – Potential to reduce emissions*

> *Quantities of allowances to be allocated shall be consistent with the potential, including the technological potential, of activities covered by this scheme to reduce emissions. Member States may base their distribution of allowances on average emissions of greenhouse gases by product in each activity and achievable progress in each activity.*

2.1.3.1. Introduction

26. No definition or further determination of the term 'potential' has been established, and potential should therefore not be limited to technological potential but may include, *inter alia*, economic potential. As the technical options available to reduce emissions by a tonne of carbon dioxide as well as the costs of doing so vary between activities, an allocation may be made to reflect that in some cases a reduction can be achieved at lower cost, and in other cases an equivalent reduction may be more costly. The implication is that more might be asked of activities that can make cheaper reductions, and less might be asked of activities whose reductions are expensive.

27. The second sentence of the criterion makes explicit the possibility for Member States to use benchmarks by product in each activity and achievable progress in each activity. Under a benchmarking approach, an average of emissions per unit of output would be established, and allocations made on the basis of historic, current or expected output quantities. An installation that had lower emissions per unit of output would be given more allowances in relation to current emissions than installations whose emissions were higher per unit of output.

28. Criterion (3) refers to the product in each activity, without defining product. It is implicitly recognised that a given activity could cover several products, so that each activity does not have to be treated as a whole. For example, achievable progress with coal-fired electricity generation is an acceptable basis for the determination of benchmarks. What is achievable by different coal-fired technologies is more limited than what may be achievable in the case of fuel-switching from coal to natural gas. However, the incentive for fuel switching to less carbon intensive fuels would not be affected.

29.   Pursuant to Article 30(2) of the Directive, the Commission shall consider in a future review the *practicality of developing Community wide benchmarks as a basis for allocation*. The Commission notes that the legislators do not consider the application of Community-wide benchmarks to be practicable for the first national allocation plan.

2.1.3.2. Analysis

30.   Criterion (3) is mandatory in part. It has to be applied in determining the total quantity of allowances and it may be applied in determining the quantity per activity.

31.   A Member State should determine the total quantity of allowances resulting from the application of this criterion by comparing the potential of activities covered by the scheme to reduce emissions with the potential of activities not covered. The criterion will be deemed as fulfilled if the allocation reflects the relative differences in the potential between the total covered and total non-covered activities.

32.   A Member State may apply the criterion to determine separate quantities per activity. It should compare the potentials of individual activities covered by the scheme to reduce emissions against each other. If a Member State applies the criterion to determine separate quantities per activity, the criterion will be deemed as fulfilled if the allocation reflects the relative differences in the potential amongst individual covered activities.

33.   A Member State may use the relevant average emissions of covered greenhouse gases by generic product type and achievable progress in each activity to determine the quantity per activity. If a Member State chooses to do so, it should determine the actual average emissions by product using national data, and assess the average emissions by product that could be attained in the relevant period taking into account achievable progress. A Member State should indicate the applied average in the national allocation plan and justify why it considers the chosen average to be an appropriate estimate to incorporate achievable progress. The quantity of allowances per activity should be based on expected output per activity over the relevant period. A Member State should indicate the forecast used and justify why it considers the chosen forecast as the most likely development. In doing so it should also take into account recent output developments in the relevant activities.

34.   In contrast to criterion (7), in which benchmarks may be applied to determine the quantity of allowances by installation, under this criterion the benchmark would be applied to determine the quantity of allowances by activity.

35.   A distinction is made between technological and other potential of activities to reduce emissions. The realisation of the technological potential to reduce emissions within a trading period is limited by factors such as timing, economic viability and legal provisions.

36.   Member States should consider that some measures can be implemented and will have an effect on emissions in the short term, while others may have longer leadtimes and depend on investment cycles. Considering the potential of

measures with a lead-time extending beyond the duration of a trading period will create an incentive for operators to act early.

37. The economic potential of activities to reduce CO_2 emissions should be based on an assessment of abatement costs per tonne of CO_2 equivalent, and not on the economic viability of individual companies or installations belonging to the activity or activities concerned.

38. A Member State may make use of best available techniques reference documents (BREFs) when assessing the potential of activities. A 'best' available technique is defined as a technique, which is 'most effective in achieving a high general level of protection of the environment as a whole'. Therefore, there is not necessarily full coherence between the use of a best available technique and the performance of an installation in terms of covered emissions.

39. In the national allocation plan, a Member State should describe the methodology it has used to assess the potential to reduce emissions. It should preferably base the assessment of the potential on a study made for the purpose of the national allocation plan. In case circumstances and timing do not allow for such a study in the process of elaborating the national allocation plan, recent existing assessments and secondary sources may be used (e.g. peer-reviewed studies). A Member State should indicate sources used and summarise the applied methodology (including major assumptions made) and results.

> A Member State has to apply the criterion to determine the total quantity of allowances. A Member State may apply the criterion to determine the quantities per activity.

2.1.4. Criterion (4) – Consistency with other legislation

> *The plan shall be consistent with other Community legislative and policy instruments. Account should be taken of unavoidable increases in emissions resulting from new legislative requirements.*

2.1.4.1. Introduction

40. Criterion (4) concerns the relationship between allocations under Directive 2003/87/EC and other Community legislative and policy instruments. Consistency between allowance allocations and other legislation is introduced as a requirement in order to ensure that allocation does not contravene the provisions of other legislation. In principle, no allowances should be allocated in cases where other legislation implies that covered emissions had or will have to be reduced even without the introduction of the emission trading scheme. Similarly, consistency implies that if other legislation results in increased emissions or limits the scope for decreasing emissions covered by the Directive account should be taken of this increase.

2.1.4.2. Analysis

41. The first sentence of the criterion is mandatory in nature, while the second one is optional.

42. The first sentence of criterion (4) has to be applied in determining the total quantity, if the Community legislative and policy instruments affect all covered installations, or in determining the quantities for affected installations in other cases.

43. Pursuant to the first sentence in the criterion, consistency with other Community legislative and policy instruments has to be applied in a symmetrical manner. Not only an unavoidable increase in covered greenhouse gas emissions resulting from new Community legislative and policy instruments should be taken into account, but also a decrease in covered emissions resulting from such instruments.

44. A Member State should list all Community legislative and policy instruments it has considered and indicate which ones have been taken into account.

45. 'New' legislative requirements should be understood as legislation and policy instruments that were adopted before the date of submission of the national allocation plan, and will impose relevant obligations on installations covered by the scheme after that date and before the end of the period covered by the national allocation plan. This includes the implementation of relevant parts of the *acquis communautaire* by new Member States following their accession in May 2004.

46. In order to take an unavoidable change in emissions into account, a Member State should consider, firstly, if a change in greenhouse gas emissions from covered installations is in fact due to new requirements, and, secondly, if such a change is unavoidable.

> In order to simplify administrative tasks the Commission recommends a Member State to consider a Community legislative or policy instrument only insofar as it is expected to result, per activity or in total, in a substantial increase or decrease (e.g. 10%) of covered emissions.

2.1.5. Criterion (5) – Non-discrimination between companies or sectors

> *The plan shall not discriminate between companies or sectors in such a way as to unduly favour certain undertakings or activities in accordance with the requirements of the Treaty, in particular Articles 87 and 88 thereof.*

47. Normal state aid rules will apply.

2.1.6. Criterion (6) – New entrants

> *The plan shall contain information on the manner in which new entrants will be able to begin participating in the Community scheme in the Member State concerned.*

2.1.6.1. Introduction

48. The treatment of new entrants, i.e. installations starting operation in the course of a trading period, is one of the important design choices in any emission trad-

ing scheme. The options differ depending on the allocation method chosen for existing installations. If all allowances are sold by a government, no specific decisions are needed for new entrants. If (the majority of) allowances are, however, allocated free of charge, several options are available to integrate new entrants into the scheme.

49.   The definition of new entrants in Article 3 of the Directive[7] puts new installations on an equal footing with existing installations extending capacity. The definition in relation to an updated permit applies only to the extension of an installation, and not to the entire installation, nor to the increased capacity utilisation at an existing installation.

50.   The criterion foresees an informational obligation to state how new entrants will be able to begin participating in the Community scheme. The guidance outlines three options to implement the criterion against the backdrop of relevant Treaty provisions. However, the Commission will also assess any other option notified in a national allocation plan.

2.1.6.2. Analysis

51.   The obligation arising under criterion (6) will be deemed as fulfilled if a Member State explains in the national allocation plan how it intends to ensure access to allowances for new entrants. Thus the criterion will be fulfilled if a Member State indicates that it has decided to have new entrants buy all allowances on the market. There are also other options to treat new entrants. In all cases, the guiding principle is equality of treatment.

52.   The EC Treaty's provisions on the right of establishment in the internal market have to be respected. It is crucial that new entrants have access to allowances, as without such access operators would be prevented from establishing a business in sectors carrying out covered activities. Guaranteeing this freedom is the essence of the second sentence of Article 11(3) of the Directive. Moreover, EU competition law would be applicable in the event that uncompetitive practices with respect to allowances were to be used to erect market entry barriers.

53.   It is important to keep in mind that the new entrants issue is of a temporary nature. In principle, an installation defined as a new entrant in one trading period should no longer fall within this definition when the national allocation plan for the subsequent period is notified.

54.   It follows from the definition that a new entrant is an installation for which no greenhouse gas emission permit has been issued or updated by the date the national allocation plan is notified to the Commission. A Member State may issue or update greenhouse gas emission permits with respect to installations which will start or extend operations with considerable certainty during the relevant trading period. Before issuing or updating a greenhouse gas emission permit, a Member State is recommended to require an operator to demonstrate that it has already obtained the construction permit and any other relevant permits. Once an installation that is expected to start or extend operations during the trading period has obtained a greenhouse gas emission permit or an updated permit, it can be included in the national allocation plan and be allocated

---

[7] See Article 3(h) in Directive 2003/87/EC.

allowances in the same manner as an existing installation. The number of allowances allocated to an installation expected to operate only during part of the trading period should be proportionate to the expected duration of (extended) operations at the installation as a share of the duration of the trading period. The Member State may not withhold foreseen allowances, in case the installation does not, or not at the time intended, start or extend operations, unless it withdraws the greenhouse gas emissions permit.

55.   A Member State has at least three options to enable participation of new entrants: it may have any new entrants buy all allowances on the market, it may make use of the possibility to set aside some allowances for periodic auctioning, or it may foresee a reserve in the national allocation plan to issue allowances to new entrants free of charge.

Having new entrants buy all allowances on the market

56.   A Member State may decide to implement this criterion by having new entrants buy all allowances on the market, as any person (incl. operators) in the Community with or without covered installations may do. The Commission notes that having new entrants buy allowances on the market is in accordance with the principle of equal treatment for the following reasons. Firstly, the Commission notes that the size of the EU-wide allowance market sets the correct conditions for liquidity, which ensures that new entrants will have access to allowances. Secondly, incumbents have made their investments without having been able to take the cost of carbon into account, in contrast to new entrants, who can minimise their carbon costs through investment choices. Thirdly, new installations only fulfil the definition of a new entrant for a limited period of time, i.e. part of a trading period, and the cost of allowances for this limited period (probably less than two years in the first period) can be taken into account in the investment and timing decisions. The Directive guarantees that as of a certain point in time the new entrant will be allocated allowances in the same manner as all other existing installations for the remainder of the lifetime of the installation.

Auctioning

57.   A Member State may enable new entrants to begin participating in the Community scheme and provide access to allowances on the basis of a periodic auction procedure. In accordance with internal market rules a Member State has to allow any person in the Community to participate in such an auction procedure. A Member State has to respect Article 10 of the Directive, pursuant to which a Member State may not auction more than 5% of the total quantity of allowances allocated during the first trading period and 10% in the second trading period.

58.   A Member State should specify what use will be made of any allowances offered in the auction procedure but not purchased. A Member State may cancel remaining allowances and re-issue a corresponding quantity of allowances for auctioning in the subsequent period. The Commission notes that at the end of the first period this option is only available, if a Member State's national legislation provides for such reissue (i.e. banking) of allowances in accordance with Article 13(2) second subparagraph.

59.　The Commission notes that having new entrants buy allowances in an auction is in accordance with the principle of equal treatment for the same reasons as indicated above with respect to having new entrants buy on the market.

Setting aside a reserve

60.　A Member State may provide access to allowances free of charge out of a reserve. If a reserve is set aside a Member State should indicate in the national allocation plan the size of the reserve by stating the absolute quantity of allowances out of the total quantity of allowances. The Member State should justify the size of the reserve with reference to an informed estimate of the expected number of new entrants during the trading period. Up to the quantity in the reserve, new entrants would be issued allowances free of charge according to transparent and objective rules and procedures determined in the national allocation plan. A Member State should describe the methodology by which allowances would be granted to new entrants. If such a method is used, the Commission recommends a Member State to give applicants in possession of a recently granted or updated greenhouse gas emission permit access to allowances on a first-come first-served basis.

61.　In order to respect the principle of equal treatment, the methodology that a Member state uses in order to allocate allowances to new entrants should as far as possible be the same as the one used for comparable incumbents. However, adaptations may be made for justified reasons (cf. guidance criterion (5)). Similarly, all new entrants should be treated in the same way. For instance, the Commission recommends a Member State not to create several reserves dedicated to separate activities, technologies or specific purposes, since they could result in unequal treatment between new entrants.

62.　A Member State should further specify what use will be made of any allowances remaining in the reserve until the end of the period. A Member State may auction any remaining allowances, while respecting Article 10 of the Directive. As in the case of allowances offered for auctioning but not purchased, a Member State may cancel remaining allowances and re-issue a corresponding quantity of allowances into a reserve for the subsequent period. Again, the Commission notes that at the end of the first period this option is only available, if a Member State's national legislation provides for such re-issue (i.e. banking) of allowances in accordance with Article 13(2) second subparagraph.

63.　A Member State should also state in the national allocation plan what transparent procedure it will follow, if new entrants apply for allowances and the reserve set aside for the period is already exhausted.

64.　The Commission notes that the operation of a reserve for new entrants increases the complexity and administrative costs of the emissions trading scheme.

> In case a Member State decides to set aside a reserve from which to grant allowances free of charge, the Commission recommends a Member State not to create dedicated reserves for specific activities, technologies or purposes.

2.1.7. *Criterion (7) – Early action*

> *The plan may accommodate early action and shall contain information on the manner in which early action is taken into account. Benchmarks derived from reference documents concerning the best available technologies may be employed by Member States in developing their National Allocation Plans, and these benchmarks can incorporate an element of accommodating early action.*

2.1.7.1. Introduction

65.   The accommodation of early action is considered as desirable from a fairness point of view. Those installations that have already reduced greenhouse gas emissions in the absence of or beyond legal mandates should not be disadvantaged *vis-à-vis* other installations that have not undertaken such efforts. The application of this criterion necessarily implies fewer allowances available for installations that have not undertaken early action.

66.   Neither the criterion nor the Directive contains a definition of early action and in which way it may be accommodated. Therefore, a Member State has a degree of freedom how to define, and whether and how to accommodate early action. This freedom is only limited by other Annex III criteria and provisions derived from the Treaty. The guidance on this criterion outlines limitations imposed by these other criteria and provisions, and contains options how to accommodate early action, if a Member State decides to do so.

67.   The second sentence of the criterion builds on the reference to benchmarks in criterion (3). It repeats the possibility for Member States to use benchmarks and points out that such benchmarks are one possible option to accommodate early action. Furthermore, the reference documents developed under Directive 96/61/EC concerning integrated pollution prevention and control[8] are inferred as a potential source for developing benchmarks.

2.1.7.2. Analysis

68.   Criterion (7) is optional and should, if applied, be used to determine the quantity of allowances allocated to individual installations.

69.   'Early action' is to be understood as actions undertaken in covered installations to reduce covered emissions before the national allocation plan is published and notified to the Commission. In line with criterion (4), only measures that operators undertook beyond requirements arising from Community legislation can qualify as early action. More stringent national legislation, applying to all covered installations in total or carrying out an activity, will be reflected in the potential to reduce emissions (cf. criterion (3)). Thus, early action is limited to reductions of covered emissions beyond reductions made pursuant to Community or national legislation, or to actions undertaken in the absence of any such legislation. A parallel can also be drawn to the Community guidelines on State aid for environmental protection, which prohibit public investment aid

---

[8] OJ L257, 10.10.1996, p. 26.

with respect to investments that merely bring companies into line with Community standards already adopted but not yet in force.

70.   Member States have several options to accommodate early action that operators have undertaken in existing installations. Three possible methods are elaborated below, but the Commission will also assess other methods.

Choosing an early base period

71.   The first option to accommodate early action is to base the allocation on historical emissions by applying a relatively early base period. If operators are allocated allowances corresponding to a proportion of the historical emissions from installations, those operators who have invested to reduce emissions since the base period will receive an allocation which covers a larger share of current emissions than an operator who has not made such investments. A Member State using this approach would have to verify that the difference in emission levels over time was not due to installations having implemented legal requirements.

72.   The drawback of this approach is that reliable and comparable data for emissions in an early base period may not be available, and the number of operator changes since the base period will increase with time, making it more difficult to establish reliable and complete records.

73.   An alternative is to use a recent multi-year base period and then allow an operator to choose one early year when it had higher emissions. Emissions data from one of the years in the recent period would then be replaced by data from the early year. This would increase the average annual emissions on which the allocation was based. In line with the restrictions described above, a Member State wanting to substitute data in this way would have to verify that the difference in emission levels over time was not due to installations implementing legal requirements.

Making a two-round allocation at installation level

74.   After determining the total quantity of allowances, a share of the available allowances is set aside. The allowances set aside would be used in a second round, after an initial distribution to all installations, to give a bonus to those installations in which operators have undertaken early action. Operators would have to apply to be taken into account in the second round and would have to demonstrate that the measures they propose to be accepted as early action comply with a pre-established definition of early action. A Member State should indicate in the national allocation plan the list of measures recognised as early action, and should specify for the relevant installations which measures have been accommodated as early action and the corresponding number of allowances allocated.

Using benchmarks

75.   A Member State may accommodate early action by using benchmarks derived from reference documents concerning the best available techniques. Benchmarks accommodate early action because they imply that a more carbon efficient installation receives more allowances than a less carbon efficient one, which is not necessarily the case with a base period driven allocation formula.

76.   In contrast to criterion (3), in which benchmarks (average emissions by product incorporating achievable progress) may be applied to determine the quantity of allowances by activity, under this criterion the benchmark would be applied to determine the quantity of allowances by installation.

77.   In order to apply the benchmarking approach, a Member State should first group homogenous installations and then apply a benchmark to each of these groups. Installations in a group should be sufficiently homogenous with respect to their input or output characteristics, such that it is feasible to apply the same type of benchmark to them. If benchmarking is used to determine allocations per installation carrying out energy activities, the Commission recommends to group installations by input fuels and to apply separate input-derived benchmarks. It should be described in the national allocation plan on what basis the grouping of installations was carried out and the respective benchmarks chosen (cf. criterion. (3)).

78.   In order to determine the allocation to an installation, the benchmark needs to be multiplied by an output value. A Member State should indicate in the national allocation plan the output values applied and justify why it considers them appropriate. A Member State may use the most recent actual output data or a forecast for the trading period, which should be clearly substantiated in the national allocation plan.

79.   Due to the ex-ante nature of the allocation decision pursuant to Article 11(1) a Member State may not base the allocation to an installation on actual output data in the trading period, i.e. data unknown at the time the national allocation plan is established but known during the course of the trading period.

80.   A benchmarking approach should not result in an allocation to installations in an activity of more allowances than determined per activity in accordance with criterion (3). A Member State would also have to verify that the installations whose emissions are lower than the benchmark have not arrived at their particular level of emissions, as a result of the implementation of legal requirements.

81.   Alternatively, a Member State may use benchmarks in a simplified way to accommodate early action. If a Member State determines allocations at installation level with a base period approach, it may use benchmarks in order to determine and apply an installation-specific correction factor to a base period driven formula. In this way the allocation to installations performing better than average is increased, while the allocation to installations performing below average is decreased. Such corrections should result in a net balance of zero across all installations concerned.

If a Member State applies this criterion, it should be used in determining the quantity of allowances allocated to individual installations. A Member State should not include measures as early action if they were taken in order to comply with legislative requirements.

If benchmarks are used to determine allocations per installation for energy activities, the Commission recommends to group installations by input fuels and to apply separate input-derived benchmarks.

2.1.8. *Criterion (8) – Clean technology*

> The plan shall contain information on the manner in which clean technology, including energy efficient technologies, are taken into account.

2.1.8.1. Introduction

82. This criterion enables a Member State to take account of clean technologies in setting allocations, but does not define what constitutes such clean technology.

83. While emission trading will promote and reward the application of low-carbon technologies, this criterion is related to the criteria on potential and early action. This guidance outlines these links.

2.1.8.2. Analysis

84. Criterion (8) is optional and should, if applied, be used to determine the quantity of allowances allocated at installation level.

85. Information is required from a Member State on the application of criterion (8). Thus, the criterion will be deemed as fulfilled if a Member State states that it makes no specific provisions to take clean technologies, including energy efficient technologies, into account.

86. Criterion (8) can be seen as the extension of criterion (3) to the installation level. An installation using a clean or energy efficient technology has a lower technological reduction potential than a comparable installation not using such a technology. It follows that the use of a clean or energy efficient technology should not be rewarded under this criterion with respect to an installation belonging to an activity which has a relatively low technological reduction potential. The reduced technological reduction potential of such an installation would already have been covered in the implementation of criterion (3).

87. Moreover, there is a link between criterion (7) on early action and criterion (8), since an early action will typically have been an investment in a clean or energy efficient technology. The Commission recommends a Member State not to apply both criteria (7) and (8) to the same installation, unless it can be shown that the early action did not consist of an investment in a clean or energy efficient technology.

88. In addition, the use of clean technologies, including energy efficient technologies, should only be taken into account under this criterion with respect to installations using such technologies before the national allocation plan is published and notified to the Commission. The Commission notes that this criterion should not be applied to clean technologies that do not result in emissions covered by the Directive.

89. By clean or energy efficient technologies, the Commission understands technologies that have resulted in lower direct emissions of covered greenhouse gases than the alternative technologies that could realistically have been deployed by the installation concerned would have led to. In determining the difference in emission levels between direct emissions from combined heat and power and an alternative technology, the latter may consist of on-site separate power and heat production.

90. As regards energy production, the Commission will accept as clean or energy efficient technologies those technologies for which it has approved State aid under the Community guidelines on State aid for environmental protection. The following list is not exhaustive:
    - High efficiency combined heat and power production. Member States may apply national definitions of 'high efficiency' cogeneration production, unless such a definition has been adopted in Community law.
    - District heating, other than high efficiency combined heat and power.

91. As regards other industrial technologies than energy production, a Member State should justify why a particular technology is to be considered as clean or energy efficient. A minimum requirement is that the technology constitutes a 'best available technique' as defined in Council Directive 96/61/EC of 24 September 1996 concerning integrated pollution prevention and control and that it was being used by the installation at the date of submission of the national allocation plan. However, since a 'best' available technique is defined as one which is 'most effective in achieving a high general level of protection of the environment as a whole', it will, in addition, have to be shown that the technique is particularly performing in limiting emissions of covered greenhouse gases.

92. Where a waste gas from a production process is used as a fuel by another operator, the distribution of allowances between the two installations is a matter for Member States to decide. For that purpose, a Member State may choose to allocate allowances to the operator of the installation transferring the waste gas, provided this is done on the basis of a pre-established criterion, compatible with the existing criteria of Annex III and the Treaty. This paragraph applies independently of whether a Member State chooses to apply criterion (7) or criterion (8) in accordance with paragraph 108.

> If a Member State takes clean technology, including energy efficient technologies, into account, it should do so by applying either criterion (7) or criterion (8) but not both.

2.1.9. Criterion (9) – Involvement of the public

> *The plan shall include provisions for comments to be expressed by the public, and contain information on the arrangements by which due account will be taken of these comments before a decision on the allocation of allowances is taken.*

2.1.9.1. Analysis

93. This criterion is of a mandatory nature.

94. A Member State will be considered to have implemented criterion (9), if it describes in the national allocation plan how it makes the plan available to the public for comments, and how it provides for due account to be taken of any comments received. The plan should be made available in a manner which enables the public to comment on it effectively and at an early stage. This means that the

public is informed, whether by public notices or other appropriate means such as electronic media, about the plan, including its text, and that also other relevant information is made available, including *inter alia* information about the competent authority to which comments or questions may be submitted.

95.   A Member State should provide for a reasonable time frame for submitting comments, and coordinate the deadline for comments to be submitted by the public with the national decision-making procedure, so that due account can be taken of comments before the decision on the national allocation plan. 'Due account' is to be understood as meaning that comments are to be taken into account if appropriate with reference to the criteria in Annex III or to any other objective and transparent criteria applied by the Member State in the national allocation plan. A Member State should inform the Commission of any intended modifications following public participation subsequent to the publication and notification of the national allocation plan and before taking its final decision pursuant to Article 11. Feedback is to be provided, in a general form, to the public about the decision taken and the main considerations upon which it is based.

96.   It should be noted that the possibility for the public to comment on the national allocation plan provided for under this criterion constitutes a second round of public consultation. Pursuant to Article 9(1) of the Directive, the comments resulting from a first round of consultation of the public on the basis of the draft plan should, where pertinent, already have been integrated into the national allocation plan prior to notification of the plan to the Commission and to the other Member States. For the overall public participation (consultation and taking account of comments) to be effective, the first round of public consultation is of particular importance. The rules described under this criterion should also be applied to the first round of consultation.

> A Member State should inform the Commission of any intended modifications subsequent to the publication and notification of the national allocation plan before taking its final decision pursuant to Article 11.

2.1.10. *Criterion (10) – List of installations*

> *The plan shall contain a list of the installations covered by this Directive with the quantities of allowances intended to be allocated to each.*

2.1.10.1. Introduction

97.   This criterion provides for the transparency of the national allocation plan. It implies that the quantities of allowances per installation are indicated, and therefore visible to the general public, when the plan is submitted to the Commission and other Member States.

2.1.10.2. Analysis

98.   This criterion will be deemed as fulfilled if a Member State has respected its obligation to list all the installations covered by the Directive. This includes installations to be temporarily excluded in the first period pursuant to Article 27 and installations to be unilaterally included in any period pursuant to Article 24.

99.   As mentioned under criterion (5), combustion installations with a rated thermal input of more than 20 MW can be found in several sectors. A Member State should therefore indicate the main activity carried out at the site where the combustion installation is located, e.g. 'paper' for a combustion installation which is part of the paper production process. A Member State should list installations by main activity, and provide subtotals of all data at activity level.

100.   A Member State has to indicate the total quantity of allowances intended to be allocated to each installation and should indicate the quantity issued in each year to each installation following Article 11(4).

101.   Article 11(4) constitutes an obligation to issue a share of the total quantity to each installation each year. Hence a Member State could issue a large proportion of the allowances in the first year(s) of a period and issue only a small share in the remaining year(s) of a period. Alternatively, a Member State could issue a small proportion of the allowances in the first year(s) of a period and issue a large share in the remaining year(s) of a period. Such approaches, in particular if taken by several Member States, may result in low market liquidity in the initial years, so that the market may fail to provide a sufficiently robust price signal. Such a signal is vital for the allowance market to provide operators of covered installations with an orientation whether to implement measures on-site or rather acquire allowances. Therefore, the Commission makes a recommendation on the proportion to be issued in each year.

102.   Furthermore, a Member State should in principle issue to all operators included in the plan equivalent, but not necessarily equal, annual shares in order to avoid undue discrimination (cf. criterion (5)).

> The Commission recommends a Member State to issue each year a share that does not deviate substantially from equal proportions over the period.

2.1.11. Criterion (11) – Competition from outside the Union

> *The plan may contain information on the manner in which the existence of competition from countries or entities outside the Union will be taken into account.*

2.1.11.1. Introduction

103.   The European Union has repeatedly affirmed its commitment to respect the Kyoto target. At the same time, the European Union, at the Lisbon European Council in March 2000, set itself the strategic goal to become the most competitive and dynamic knowledge-based economy in the world, capable of sustainable economic growth with more and better jobs and greater social cohesion. Emission allowance trading is a cost-effective instrument, which allows industrial activities covered by the Directive to keep the costs of contributing to the Community's climate change commitments low. The implementation of the Kyoto Protocol will give companies in the European Union the chance for a head-start in the gradual transition to a carbon-constrained global economy, since carbon efficiency may be an important source of competitive advantage in the future, as much as labour or capital productivity today.

In the short-term these commitments may imply increased costs for some companies and sectors.

2.1.11.2. Analysis

104. Criterion (11) is optional and should only be used, if applied, in determining the quantity of allowances per activity, since any effect from competition from countries or entities outside the Union would be one affecting all installations carrying out a certain activity.

105. A Member State should not use the mere existence of competition from outside the Union as a reason to apply this criterion. The Commission considers this criterion to be applicable exclusively to cases where covered installations belonging to a specific activity would be rendered significantly less competitive directly and predominantly as a result of a major difference in climate policies between the EU and countries outside the EU. In assessing any such differences in climate policy, a Member State should take into account any relevant measures that competitors outside the EU are subject to, including voluntary initiatives, technical regulation, taxes and emissions trading, and not judge solely on the basis of whether or not the country concerned has a quantified emissions commitment and has ratified the Kyoto Protocol.

106. A Member State should not take the existence of competition from outside the Union into account in such a manner as to improve the competitive position of installations carrying out an activity *vis-à-vis* competitors outside the Union as compared to their competitive position in the absence of the EU emissions trading scheme. It should be noted that incorrect application of this criterion may constitute export aid, which is incompatible with the EC Treaty.

107. If a Member State deems it necessary to take account of competition from outside the Union, it should also consider applying other options outside the national allocation plan.

108. A Member State should keep in mind, when applying this criterion to individual activities, that where the mandatory criterion (3) is applied at activity level, installations carrying out activities with a relatively large potential to reduce emissions should still receive a smaller share of allowances in relation to emissions, compared to installations carrying out activities with a relatively small potential to reduce emissions.

109. The existence of competition should only be taken into account in the national allocation plan by a modification of the quantity of allowances per activity, without a change in the total quantity of allowances determined in accordance with the criteria (1) to (5).

> If competition from outside the Union is taken into account in the national allocation plan, the criterion should only be applied in determining the quantity of allowances allocated at activity level, without a change in the total quantity of allowances.
>
> If a Member State deems it necessary to take account of competition from outside the Union, it should also consider applying other options outside the national allocation plan.

3.   GUIDANCE ON CIRCUMSTANCES UNDER WHICH *FORCE MAJEURE* IS
      DEMONSTRATED

Article 29

1.   During the period referred to in Article 11(1), Member States may apply to the Commission for certain installations to be issued with additional allowances in cases of force majeure. The Commission shall determine whether force majeure is demonstrated, in which case it shall authorise the issue of additional and non-transferable allowances by that Member State to the operators of those installations.

2.   The Commission shall, without prejudice to the Treaty, develop guidance to describe the circumstances under which force majeure is demonstrated, by 31 December 2003 at the latest.

---

110.   In principle, allocation decisions are made by Member States before the beginning of the relevant trading period, thereby avoiding uncertainty in the allowance market. A limited provision allows the issuance of additional non-transferable allowances in exceptional and unforeseeable circumstances in the first period of the Community scheme.

111.   Article 29 derogates from the general principle of the Community scheme, under which Member States allocate allowances before the beginning of the relevant trading period. Applications for *force majeure* allowances may therefore cause uncertainty in the allowance market, and, if granted, give an advantage to certain companies, which affects trade between Member States. Article 29 is therefore explicitly without prejudice to the Treaty, and the Commission will carefully consider the justification and potential effects of any application for such allowances.

112.   Companies may seek insurance against various risks that could result in increased emissions, but insurance policies normally do not cover circumstances of *force majeure*. The Commission will not consider circumstances that could have been insured to constitute *force majeure*.

113.   Circumstances of *force majeure* are, by their nature, difficult to anticipate. The Commission considers these to be exceptional and unforeseeable circumstances, which cause a substantial increase in annual direct emissions of greenhouse gases covered by Directive 2003/87/EC at an installation, which could not have been avoided even if all due care had been exercised. The circumstance must have been beyond the control of the operator of the installation concerned and of the Member State submitting an application to the Commission under Article 29 of the Directive with respect to the installation of that operator.

114.   Circumstances that the Commission may consider to be *force majeure* include notably natural disasters, war, threats of war, terrorist acts, revolution, riot, sabotage or acts of vandalism.

115.   The presence of *force majeure* has to be demonstrated at installation level and on a case-by-case basis.

116. An application under Article 29 of the Directive should include, in respect of each installation, the Member State's best estimate of the increase in emissions resulting from the circumstance for which *force majeure* is pleaded and a substantiation of that estimate.

117. A Member State should submit an application under Article 29 to the Commission by 31 January the year following the year of the trading period during which the circumstance occurred for which *force majeure* is pleaded.

<u>ANNEX</u>

COMMON FORMAT FOR THE NATIONAL ALLOCATION PLAN 2005 TO 2007

1.   DETERMINATION OF THE TOTAL QUANTITY OF ALLOWANCES

What is the Member State's emission limitation or reduction obligation under Decision 2002/358/EC or under the Kyoto Protocol (as applicable)?

What principles, assumptions and data have been applied to determine the contribution of the installations covered by the emissions trading Directive to the Member State's emission limitation or reduction obligation (total and sectoral historical emissions, total and sectoral forecast emissions, least-cost approach)? If forecast emissions were used, please describe the methodology and assumptions used to develop the forecasts.

What is the total quantity of allowances to be allocated (for free and by auctioning), and what is the proportion of overall emissions that these allowances represent in comparison with emissions from sources not covered by the emissions trading Directive? Does this proportion deviate from the current proportion of emissions from covered installations? If so, please give reasons for this deviation with reference to one or more criteria in Annex III to the Directive and/or to one or more other objective and transparent criteria.

What policies and measures will be applied to the sources not covered by the emissions trading Directive? Will use be made of the flexible mechanisms of the Kyoto Protocol? If so, to what extent and what steps have been taken so far (e.g. advancement of relevant legislation, budgetary resources foreseen)?

How has national energy policy been taken into account when establishing the total quantity of allowances to be allocated? How is it ensured that the total quantity of allowances intended to be allocated is consistent with a path towards achieving or overachieving the Member State's target under Decision 2002/358/EC or under the Kyoto Protocol (as applicable)?

How is it ensured that the total quantity of allowances to be allocated is not more than is likely to be needed for the strict application of the criteria of Annex III? How is consistency with the assessment of actual and projected emissions pursuant to Decision 93/389/EEC ensured?

Please explain in Section 4.1 below how the potential, including the technological potential, of activities to reduce emissions was taken into account in determining the total quantity of allowances.

Please list in Section 5.3 below the Community legislative and policy instruments that were considered in determining the total quantity of allowances and state which ones have been taken into account and how.

If the Member State intends to auction allowances, please state the percentage of the total quantity of allowances that will be auctioned, and how the auction will be implemented.

2. **DETERMINATION OF THE QUANTITY OF ALLOWANCES AT ACTIVITY LEVEL (IF APPLICABLE)**

By what methodology has the allocation been determined at activity level? Has the same methodology been used for all activities? If not, explain why a differentiation depending on activity was considered necessary, how the differentiation was done, in detail, and why this is considered not to unduly favour certain undertakings or activities within the Member State.

If the potential, including the technological potential, of activities to reduce emissions was taken into account at this level, please state so here and give details in Section 4.1 below.

If Community legislative and policy instruments have been considered in determining separate quantities per activity, please list the instruments considered in Section 5.3 and state which ones have been taken into account and how.

If the existence of competition from countries or entities outside the Union has been taken into account, please explain how.

3. **DETERMINATION OF THE QUANTITY OF ALLOWANCES AT INSTALLATION LEVEL (+ ANNEX I)**

By what methodology has the allocation been determined at installation level? Has the same methodology been used for all installations? If not, please explain why a differentiation between installations belonging to the same activity was considered necessary, how the differentiation by installation was done, in detail, and why this is considered not to unduly favour certain undertakings within the Member State.

If historical emissions data were used, please state whether they have been determined in accordance with the Commission's monitoring and reporting guidelines pursuant to Article 14 of the Directive or any other set of established guidelines, and/or whether they have been subject to independent verification.

If early action or clean technology were taken into account at this level, please state so here and give details in Sections 4.2 and/or 4.3 below.

If the Member State intends to include unilaterally installations carrying out activities listed in Annex I below the capacity limits referred to in that Annex, please explain why, and address, in particular, the effects on the internal market, potential distortions of competition and the environmental integrity of the scheme.

If the Member State intends temporarily to exclude certain installations from the scheme until 31 December 2007 at the latest, please explain in detail how the requirements set out in Article 27(2)(a)–(c) of Directive 2003/87/EC are fulfilled.

4.   TECHNICAL ASPECTS

4.1.   Potential, including technological potential

> Has criterion (3) been used to determine only the total quantity of allowances, or also the distribution of allowances between activities covered by the scheme?
>
> Please describe the methodology (including major assumptions made) and any sources used to assess the potential of activities to reduce emissions. What are the results obtained? How is it ensured that the total quantity of allowances allocated is consistent with the potential?
>
> Please explain the method or formula(e) used to determine the quantity of allowances to allocate at the total level and/or activity level taking the potential of activities to reduce emissions into account.
>
> If benchmarking was used as a basis for determining the intended allocation to individual installations, please explain the type of benchmark used, and the formula(e) used to arrive at the intended allocation in relation to the benchmark. What benchmark was chosen, and why is it considered to be the best estimate to incorporate achievable progress? Why is the output forecast used considered to be the most likely development? Please substantiate the answers.

4.2.   Early action (if applicable)

> If early action has been taken into account in the allocation to individual installations, please describe in which manner it is accommodated. Please list and explain in some detail the measures that were accepted as early action and what the criteria for accepting them were. Please demonstrate that the investments/actions to be accommodated led to a reduction of covered emissions beyond what followed from any Community or national legislation in force at the time the action was taken.
>
> If benchmarks are used, please describe on what basis the grouping of installations to which the benchmarks are applied was made and why the respective benchmarks were chosen. Please also indicate the output values applied and justify why they are considered appropriate.

4.3.   Clean technology (if applicable)

> How has clean technology, including energy efficient technologies, been taken into account in the allocation process?
>
> If at all, which clean technology has been taken into account, and on what basis does it qualify as such? Have any energy production technologies intended to be taken into account been in receipt of approved State aid for environmental protection in any Member State? Please state whether any other industrial technologies intended to be taken into account constitute 'best available techniques' as defined in Council Directive 96/61EC, and explain in what way it is particularly performing in limiting emissions of covered greenhouse gases.

5.   COMMUNITY LEGISLATION AND POLICY

5.1.   Competition policy (Articles 81–82 and 87–88 of the Treaty)

> If the competent authority has received an application from operators wishing to form a pool and if it is intended to allow it, please attach a copy of that application to the national allocation plan. What percentage of the total allocation will the pool represent? What percentage of the relevant sector's allocation will the pool represent?

5.2.   Internal market policy – new entrants (Article 43 of the Treaty)

> How will new entrants be able to begin participating in the EU emissions trading scheme?
>
> In the case that there will be a reserve for new entrants, how has the total quantity of allowances to set aside been determined and on what basis will the quantity of allowances be determined for each new entrant? How does the formula to be applied to new entrants compare to the formula applied to incumbents of the relevant activity? Please also explain what will happen to any allowances remaining in the reserve at the end of the trading period. What will apply in case the demand for allowances from the reserve exceeds the available quantity of allowances?
>
> Is information already available on the number of new entrants to expect (through applications for purchase of land, construction permits, other environmental permits etc.)? Have new or updated greenhouse gas emission permits been granted to operators whose installations are still under construction, but whose intention it is to start a relevant activity during the period 2005 to 2007?

5.3.   Other legislation or policy instruments

> Please list other Community legislation or policy instruments that were considered in the establishment of the national allocation plan and explain how each one has influenced the intended allocation and for which activities.
>
> Has any particular new Community legislation been considered to lead to an unavoidable decrease or increase in emissions? If yes, please explain why the change in emissions is considered to be *unavoidable*, and how this has been taken into account.

6.   PUBLIC CONSULTATION

> How is this national allocation plan made available to the public for comments?
>
> How does the Member State provide for due account to be taken of any comments received before a decision on the allocation of allowances is taken?
>
> If any comments from the public received during the first round of consultation have had significant influence on the national allocation plan, the Member State should summarise those comments and explain how they have been taken into account.

7.   CRITERIA OTHER THAN THOSE IN ANNEX III TO THE DIRECTIVE

> Have any criteria other than those listed in Annex III to the Directive been applied for the establishment of the notified national allocation plan? If yes, please specify which ones and how they have been implemented.
>
> Please also justify why any such criteria are not considered to be discriminatory.

8.   ANNEX I – LIST OF INSTALLATIONS

> Please submit a matrix containing the following information:
> – Identification (e.g. name, address) of each installation
> – The name of the operator of each installation
> – The number of the greenhouse gas emissions permit
> – The unique (EPER) identifier of the installation
> – The main activity, and, if applicable, other activities carried out at the installation
> – Total quantity of allowances to be allocated for the period, and the annual breakdown, for each installation
> – Whether the installation has been unilaterally included or temporarily excluded and whether it is part of a pool
> – Annual data per installation, including emission factors if emissions data are used, which have been used in the allocation formula(e)
> – A subtotal per activity of data used and number of allowances allocated

Index

Added to the page number 'f' denotes a figure, 't' denotes a table and 'n' denotes a footnote.

A

A6SC (Article 6 Supervisory Committee) xliii, 53, 55–7
Aarhus Convention 267
AAUs xlii, 10
  fungibility 17
  potentially available for EITs during the commitment period 202t
  recycling into energy projects 210
  sales 209
  use in the EU trading regime 131, 332
Abatement Certificates 234
absolute emission limits 237
absolute trading regimes
  linking relative trading regimes with 88
  v.relative trading regimes 86–8
  see also 'cap-and-trade' schemes
accession countries
  anticipated 2010 emissions as a percentage of Kyoto target 203f
  establishment of baselines for JI and CDM projects 134, 221–2
  industries 226
  JI projects 138–9, 215
  NAPs see NAPs, in accession countries
accreditation
  DOEs
    procedures 41
    registration fee 52–4
    withdrawal 40, 50
  for IEs 57
accreditation standards for DOEs 39–40
'ACEA Agreement' 84
acid rain management in Japan 242–3
acquis communautaire and baselines 134, 215, 221–2, 228–9
Activities Implemented Jointly see AIJ
Ad Hoc Group on the Berlin Mandate (AGBM) 4
adaptation
  costs and benefits xxxix
  share of proceeds for 29, 50, 52
Adaptation Fund 52
ADB (Asia Development Bank) CDM projects in China 249
additionality xlii, 44, 221, 269, 284, 289–303
  of AIJ projects 11, 13

Gold Standard approach 274
  and JI 54, 221
additionality determination 302
  difference between baseline determination and 296–302
  need for 298–9
    environmental credibility 298
    financial and technological flows to host countries 298
    incentives for host countries to take up PAMs and targets 299
    no crowding out of real projects 298
  problems 299–302
    additionality testing in the CDM rules 300–2
    gaming of numbers or exaggeration of barriers 299
    'Grubb's paradox' and rising CER prices 299–300
    higher transaction costs 299
additionality testing 44, 133, 300–2, 303
  stricter 280
adjustments, application of 22–5
administrative costs
  of the CDM 52–4
  of the JI 56–8
AEs 30
  accreditation as DOEs see accreditation, as DOEs
afforestation/reforestation projects xli, 46–9
  see also monoculture plantations
Africa and CDM 258, 266
AGBM (Ad Hoc Group on the Berlin Mandate) 4
Agenda 20 267
AIJ 11–15
  in Central and Eastern European (CEE) countries 210–14
  pilot phase 11–13, 20
  substantive criteria 12–14
AIJ projects 6
  additionality 11
  in Central and Eastern European (CEE) countries 210–11, 212
  and the Kyoto mechanisms 13–15
  number 13
  uniform reporting format (URF) 12

AIPs *see* Annex I Parties
Alberta 240
ALGAS (Asia Least-Cost Greenhouse Gas
    Abatement) Project 250, 255
allocation of emission allowances 88–90
    in the ET Directive 110–18, 180
        allocation check 116–17
        compliance issues 119–20
        division of competence between the
            Member States and the Community
            111–12
        guidance 115–16
        implementation at the national level
            117–18
        methodology 113–14
        relation between state aid rules and
            112–13
    implications for linkage 333
    methods 89–90
    short-term 162
    see also EU Guidelines on Allocations of
        Allowances; NAPs
allowances, emission *see* emission allowances
Annex B Parties
    and IET 26
    and linkage of emissions trading
        programmes 331–2, 334–5
Annex I Parties ix
    and AIJ 11
    and CDM xli, 20
    and the Compliance Committee 62–3
    and the CPR 27–8
    and DNAs 20
    and emissions trading 232–41
    and equity issues 15–17
    and IET xli, xlii, 28
    initial assigned amount 17
    and JI xli, 53
    legally binding targets xl
    mitigation commitments xxxviii, xxxix–xl
    and the Protocol participation
        requirements 22
    reporting commitments 5
    representation on the EB/CDM 35
    and supplementarity 16
    supplementary reporting and review
        requirements xl
Annex II Parties xxxviii, 22
    and AIJ 11
annual inventories 21, 22–4
Applicant Entities *see* AEs
Argentina and CDM 255–6
Article 6 Supervisory Committee (A6SC) xl,
    53, 55–8
Asia and the CDM 248–55
Asia Development Bank (ADB) CDM projects
    in China 249

Asia Least-Cost Greenhouse Gas Abatement
    (ALGAS) Project 250, 255
ASPEN software 166–7
assigned amount units *see* AAUs
assigned amounts
    establishment 21
    and fungibility 17–18
    submission of supplementary information
        on 24
AT Biopower project methodology 295
auctioning of emission allowances 89–90
    and the ET Directive 114, 162
    in the ETS 187
Australia
    and emissions trading 232–4
    ratification of the Kyoto Protocol 166,
        232–3
Austria
    emission reduction target 80t
    JI projects 217
awareness-building programmes for the CDM
    312–3

B
'backstop and covenant' system 323–4
banking
    in the ETS 195–6
    into the commitment period 333–4
    Kyoto units 17
barrier assessment for additionality 300, 302
baseline and credit regimes 4
    and the allocation of allowances 88–9
    see also international baseline and credit
        schemes; relative trading regimes
baseline determination 290–6
    data needs 296, 302, 317
    difference between additionality
        determination and 296–302
    international procedures 292–6
    for small-scale projects 291–2
baselines 44–5, 87, 289–303
    and the *acquis communautaire* 134, 215,
        221–2, 228–9
    Gold Standard approach 274
BAT criterion for the IPPC Directive 82, 123,
    222
    discretion in application 124
BAT reference documents (BREFS) 123, 124
BAU scenario *see* 'business-as-usual' scenario
Belgium
    CDM and JI tenders 218
    emission reduction target 80t
benchmarking 195
'best available techniques' *see* BAT
bilateral CDM projects 30
BioCarbon Fund 260
BM case

characteristics 170
market features 174–7
Bonn Agreements 1, 7, 99
and the CPR 27
and JI 53
and sinks 47
Bonn Convention 341
Bonn-Marrakesh Accords case *see* BM case
Brazil
and the CDM 6, 256–8, 280
and emissions trading 243
BREFS (BAT reference documents) 123, 124
British thermal unit (BTU) tax 4
Bujagali hydroelectric plant 280–1
Bulgaria
AIJ projects 210t, 211t
and the decommissioning of nuclear
reactors 222
and the ERU Procurement Tender
(ERUPT) 216
and JI
memoranda of understanding 219t
ranking of conditions on three scales
218t
scope for low-cost 217t
status administratively 214t
potentially available AAUs during the
commitment period 202t, 203f
burden-sharing agreement *see* EU
burden-sharing agreement
Bureau of the Compliance Committee 62
Bush Administration
and non-engagement with the climate
problem 8
and ratification of the Kyoto Protocol 7
'business', pollution reduction as 92
'business-as-usual' scenario xxxi, xxxvi,
302, 306
and the EU 75
buyer criteria for CDM projects 278

C
Cambodia 318
Canada
and CDM 258–9
and emissions trading 238–9
programme for 323–6
and the 'Green Investment Scheme' (GIS)
209
'cap-and-trade' schemes 4, 86, 93
US proposals 237–8
see also absolute trading regimes; IET
capacity building in the CDM *see* CDM
capacity building
caps xl
and the ET Directive 110–11, 128, 193
lax 284

and the Linking Directive proposal 137
stringency 94
car manufacturers and negotiated agreements
84
carbon credits
Central and Eastern European (CEE)
countries as exporters 200
value 282–3
carbon markets 153
analysis of impact through five cases
167–81
characteristics of cases 169–71
market features of cases 171–80
methodology and assumptions 167–9
interaction between xxx
potential participants 166–7
purchasers 259–60
carbon pools 46–9
'carbon professionals' xviii
Carbon Ring Consortium 252
carbon sinks *see* sinks
carbon taxes 157
US 4
CC case
characteristics 170–1
market features 177, 178t
CCX (Chicago Climate Exchange) 238–9
CD4CDM guide 318
CDCF (Community Development Carbon
Fund) 260–1, 278
CDET programme 323–6
CDM xxxi, xl, xli, 6, 9–11t, 29–54
additionality *see* additionality
administrative costs 52–4
and afforestation/reforestation projects 46–9
and Africa 258
and Asia 248–53
baselines *see* baselines
and Canada 258–9
and DCs 7
DNAs *see* DNAs
funding issues 50–4
implementation 245–9
institution-building 312
programmes 314–5
institutions and procedures 32–43
interest in 9
in Latin America 255–8
linkage with the EU trading regime
127–30
monitoring and verification and
certification requirements 49
'prompt start' 33, 46, 53
share of proceeds provision 29, 50, 52–4
and sink projects 1–2, 7, 9, 46
structural weaknesses 285
and sustainable development *see*
sustainable development and CDM

CDM *continued*
    validation and registration requirements
        42–9
    vs. JI 220
CDM awareness-building programmes
        312–13
CDM capacity building 305–20
    challenges 317–8
        donor competition 318
        printed CDM manuals 318–9
        sourcing cheap CERs 318
        weaning host countries from foreign aid
            318
        workshop tourism 318
    donor activities 310–17
        future directions 316–7
        spreading the message 310–14
        supporting institutions 314–15
        supporting project proposals 315–16
    provision of funding 259
    requirements for implementation 306–10
CDM credits *see* CERs
CDM institution-building programmes
        314–15
CDM Investment Guide for Brazil 257
CDM manuals 318–19
CDM market 278–83
    in 2004 279–80
    link with the ETS xxx
    sustainable development contribution
        280–3
CDM modalities 15, 29
    and baselines 44
    and the CDM project cycle 41–2
    and DOEs 39–41
    and ratification of the Kyoto Protocol 20
CDM project cycle 29–33, 306
    and afforestation/reforestation 47
    design 31
    and DOEs 39
    information and review provisions and
        stakeholders and observers 41–3
    issuance of CERs 32
    monitoring 32
    tasks and actors 306–10
    validation and registration 31–3
    verification and certification 33
CDM project development programmes
        315–16
CDM projects 227, 285
    in 2004 279, 285
    in China 249–50
    conversion of AIJ projects into 13, 15
    crowding out of real 298
    development of proposals 315–16
    funding 50–2
        assistance 52

geographic imbalance 51–3, 266
    in India 252
    in Japan 245–6
    in Malaysia 250–51
    models 29–31
    oversight 20
    potential benefits 268–9
    public participation 18–20, 267
    smaller community-based and designed
        285
    and stakeholder involvement 18–20
    tools to assess eligibility and sustainability
        269–78
    transaction costs 299, 306
    validation and registration requirements 2,
        42–49
CDM Registry 22, 29, 30, 50
CDM rules 37–9, 246, 305
    and additionality testing 300–2
CDM sink projects 166
CDM workshop tourism 318
CDM-AP 40
Central and Eastern European (CEE) countries
        200–29
    AIJ in 210–14
    European emissions trading 222–9
        JI, the *acquis communautaire* and
            double counting 227
        JI and the ETS 226
        keeping 'hot air' out of the system
            223–4
        more emission reductions without 'hot
            air' 225
        NAPs *see* NAPs, in accession countries
        other climate policy and JI baselines
            220–1
    future potential of JI 218–22
        effect of Russian non-ratification 220
        JI, baselines and the *acquis
            communautaire* 221–2
        JI vs. CDM 220
    IET and 'hot air' 203–5
    JI projects 215–9
    mechanism participation requirements
        213–16
CERs xviii, xli, 9, 10t, 248
    and double counting 137–8
    from nuclear facilities 60
    from sink projects 47
    fungibility 17
    generation 29, 44, 289, 290, 305–6
        methodological and institutional
            procedures xli
    issuance 29, 32, 42, 49–51
    low-cost 281–83, 284
    sourcing 318
    use in the EU trading regime 131, 135–7,
        332

utilization from non-Annex I country
  CDM projects 246
see also ICERS; tCERS
'CERT model' 311
certification of the CDM project cycle 32
certification reports for CDM projects 49
Certified Emission Reduction Unit
    Procurement Tender *see* CERUPT
certified emission reductions *see* CERs
certifiers, role in the CDM process 309
CERUPT 8, 112, 249, 252, 260, 279, 280,
    315
  screening processes 278
CH₄ emissions, measurability 91
Chicago Climate Exchange (CCX) 238–9
China
  and CDM 248–50
    governance structure 248–9
  and emissions trading 241–3
Clean Air Act (US) 15
Clean Development Mechanism *see* CDM
Climate Change Committee 108, 115
climate change mitigation 5
  costs and benefits xxxvii
  see also mitigation commitments
Climate Change Plan for Canada 238, 240,
    323
climate change problem xxxvi–xxxvii, 206
  non-engagement of the Bush
    Administration 8
climate change regime, future of *see* future
    climate change regime
Climate Stewardship Act 332
CMS 341
CO₂ emissions 336
  measurability 91
  reduction xxxvii
    by cars 84
CO₂ law 328–9
co-decision procedure of the EU 100
cohesion countries 78
comitology procedure of the EU 107, 126
command and control regulation xvii, 4
  and the EU 81–2
  replacement with market-based
    mechanisms 85–6
  see also IPPC Directive; 'negotiated
    command and control instruments'
commitment period, banking into 303–5
commitment period reserve (CPR) 24, 27–9
commitments xxxi
  differentiation xxxviii, xxix, xl, 336–8
  future 336–8
    absolute emission limits 337
    emission intensity limits 337–8
    types 338
  overseeing implementation xxxviii

see also mitigation commitments;
    pre-commitment period reports;
    reporting commitments
Community Development Carbon Fund
    (CDCF) 260–1, 278
compliance
  cost 343
  in the international context 5, 8
  see also non-compliance
Compliance Committee xl, xlii, 62–5
compliance procedures
  acceptance 25
  in the ET Directive 118–21
  under the Kyoto Protocol 61–7
    mechanism-relevant 65–7
Conference of the Parties *see* COPs
consultants role in the CDM process 309
  training courses 317
Convention on the Conservation of Migratory
    Species of Wild Animals 341
COP-1 xxxii, 4, 211
COP-2 5
COP-3 xxxii, 6, 12
COP-4 61
COP-5 12, 255
COP-6 1, 7, 61, 209
COP-7 1, 7, 8, 12, 13, 14, 16, 46, 52, 166,
    306
  and compliance procedures and
    mechanisms xli, 61
COP-8 12, 14, 36, 46, 52, 66
COP-9 xli, 2, 7, 13, 23, 36, 47, 52, 53, 134,
    166
COP-10 24
COP/MOP xl, 1, 2, 33–5
  and the CDM 33–5, 51
  and designation of DOE status 39, 40
  endorsement of COP decisions 14
  and IET 26
  and JI 53, 54–6
  and non-compliance with the Kyoto
    Protocol 62
COPs xxiv, xxxii, 1, 2, 167
  and AIJ projects 12
  and CDM 14, 33
  and the definition of trading rules 26
  and JI 55
  main decisions related to the Kyoto
    mechanisms 3t
  transition of mechanism modalities to the
    COP/MOP from 14, 33
coverage of GHGs by trading regimes 90–2
  in the ET Directive 105–109
    overlap with the IPPC Directive 122–5
CPR (commitment period reserve) 24, 27–9
crediting period for CDM projects 44
credits xli, 4

credits *continued*
  in the AIJ pilot phase 11
  creation at international level 4
  and linkage between the Kyoto project-
    based mechanisms and the EU
    trading regime 128, 129–30
    influx into the EU trading regime 131,
      136–8
  non-Kyoto market 264
  see also carbon credits; emission
    allowances; permits
Croatia
  AIJ projects 210t, 211t
  and JI
    memoranda of understanding 219t
    ranking of conditions on three different
      scales 218t
    scope for low-cost 217t
    status administratively 218t
  potentially available AAUs during the
    commitment period 206t
cross-cutting mechanism issues 14–20
  adoption and review of mechanism
    modalities 14–15
  equity issues 15–16, 93–4
  fungibility 17–18
  stakeholder involvement *see* stakeholder
    involvement
  supplementarity *see* supplementarity
Czech Republic
  AIJ projects 210t, 211t, 212
  and JI
    memoranda of understanding 219t
    ranking of conditions on three different
      scales 218t
    scope for low-cost 217t
    status administratively 214t
  NAP 226
  potentially available AAUs during the
    commitment period 202t, 203f

D
DANIDA 314t, 315, 316t
data
  for baselines 296, 302, 317
  and determination of allocation
    methodologies 194
  on sinks 24
DCs
  and CDM 6
  and CDM projects 39
  commitments xli
    future 336–8
  and DNAs 20
  DOEs in 51–3
  and the EB/CDM 33
    membership 35

economic impacts of response measures
    xl–xli
  emission levels xxxvi
  and emissions trading 241–4
  and the evolution of the Kyoto mechanisms
    4–5, 6, 9
  and JI 54
  and prioritization of domestic action 206
deforestation 243
Denmark
  emission reduction target 80t
  see also DANIDA
'descending clock auction' 90
design of the CDM project cycle 31
Designated National Authorities *see* DNAs
designated operational entities *see* DOEs
developing countries *see* DCs
differentiated responsibilities, principle of xl
differentiation in commitments xxxix–xl,
    xli–xlii, 336–8, 342
dispatch modelling 295
DNAs 20, 21, 30, 231, 246, 308
  functions 38–9
  national project approval criteria 270
  setting up 314–5
DOEs 30
  accreditation *see* accreditation
  geographic and regional distribution 51–2
  procedural requirements 45–6
  responsibilities and functions
    39–43
  verification and certification by 32
domestic action 205–9
  emissions trading as a disincentive 94
  limiting hot air 207–9
  moral reasons for 205–6
  political reasons for 206
  prioritizing 16
  and technology forcing 206–7
domestic GHG trading programmes 322–9
  application in an international context 5
  linkage 329–30
    issues raised by design 332–4
    methods of linkage 330–2
    rationale 329–30
Domestic Kyoto Mechanism Support Centre
    245
donor programmes for CDM 309–16
  competition 317
double counting and JI in accession countries
    137–9, 228
downstream trading regimes 91, 92, 333
draft PDDs 32
'Dutch situation' 112

E
EB/CDM xli, 2, 8, 29–31, 34–8

baseline and monitoring methodologies 45
board meetings 36
and the CDM Registry 50
formal reviews 41–2
functions 34–5
funding 52
and issuance of CERs 32
membership 35–6
observers and stakeholders 37
panels and liaison with SBSTA 38
relationship with the COP/MOP 33–4
voting 37–8
EBRD (European Bank for Reconstruction and Development) 216
ECCP 96–7, 99
ECCP Working Group 1 on Flexible Mechanisms (WG 1) 97, 98–9, 114
linking project-based mechanisms with the trading regime 130–1
Economies in Transition *see* EITs
EEA (European Economic Area) states, application of the ET Directive to 108
efficiency in climate change agreements 342
EIA component of the Gold Standard 275
EIAs 266–7
EITs
and application of adjustments 23
and 'hot air' 5
and IET 200–3
and implementation of national systems 20–21
and JI 6, 7, 54, 200–3
and prioritization of domestic action 206
El Gallo hydro project 296
electricity production, renewable energy *see* renewable energy
electricity sector
approaches for baselines 294–6
in the EU 193
see also hydropower projects; nuclear projects
eligibility requirements 19
and the Enforcement Branch of the Compliance Committee 66–7
for JI in Central and Eastern European (CEE) countries 214–15
see also participation/eligibility requirements
emission allowances 4
allocation *see* allocation of emission allowances
and the ET Directive 109–10
emission caps *see* caps
emission intensity limits 337–8
emission limit values (ELVs) 81–2
emission limitation commitments *see* commitments

Emission Reduction Purchase Agreements (ERPAs) 284, 309–10
emission reduction units *see* ERUs
emission rights 88
emissions
causes xxxvi
contribution of non-trading sectors 158
impacts xxxvi–xxxvii
measurability 90–1
as morally wrong 93
reductions by the EU 75–6
regulation
point of 91–2, 333
under the IPPC Directive 122–3
see also national emissions
emissions budgets 338
emissions intensity targets 337
emissions trading 76, 83, 85–94, 232–45
absolute v. relative regimes 86–8
allocation of allowances *see* allocation of emissions allowances
and Annex I Parties 232–41
comparison with environmental taxation 92–3
concept 86–92
coverage 90–2
in DCs 241–4
ethical dimension 92–4
arguments against 93–4
arguments in favour 94–5
and the EU xviii–xix
from the early notion of joint implementation to 153–4
'safety valve' 342
under the Kyoto Protocol 152–63
see also IET; international GHG emissions trading
Emissions Trading Framework Directive *see* ET Directive
Energy Efficiency Benchmarking Covenant 84
energy project types, baseline rules 291–2
Energy Savings Law 240
energy taxes
effect on emissions trading 155
US 4
Enforcement Branch of the Compliance Committee 23, 25, 64–5
and mechanism-relevant procedures 65–7
enforcement measures in the ET Directive 118–21
environmental credibility and the need for additionality determination 298
environmental impact assessments *see* EIAs
'environmental integrity' and selection of projects for the EU trading regime 132

environmental NGOs
  and auctioning 114
  and emissions trading 92
  and the 'Green Investment Scheme' (GIS)
      209
  opposition to the Linking Directive
      proposal 130–1
environmental policy
  EU 81–6
  relationship between technology
      development and 206–7
environmental taxation 84
  comparison with emissions trading 92–3
EPER (European Pollutant Emission Register)
      123
equity in climate change agreements 342
equity issues for emissions trading 15–16,
      93–4
  and the distribution of CDM projects 51
ERPAs (Emission Reduction Purchase
      Agreements) 284, 309–10
ERTs 21, 22
  and the application of adjustments 23–4
  and questions of implementation 64, 65
ERU Procurement Tender (ERUPT) 8, 112,
      215–17, 236
  financing 221
ERUs xli, 10t, 53, 55, 221
  and double counting 137–8
  from nuclear facilities 60
  fungibility 17
  transfers and acquisitions 57–8, 213
  use in the EU trading regime 131, 136,
      137, 332
Esti dam 281
Estonia
  AIJ projects 210t, 211t
  and JI
    memoranda of understanding 219t
    ranking of conditions on three different
        scales 218t
    scope for low-cost 217t
    status administratively 214t
  potentially available AAUs during the
      commitment period 202t, 203f
ET Directive xviii, 76, 139, 322, 355–70
  and the achievement of targets under the
      burden-sharing agreement and
      Kyoto Protocol 116
  adoption 100–101
  core elements 101–26
    allocation of allowances *see* allocation
        of emission allowances, ET Directive
    compulsory participation 103–5
    coverage of sectors, gases and Member
        States 105–8
    enforcement 118–21

  permits and allowances 108–10
  relation with the IPPC Directive 121–5
  documents related to 354–5
  key dates and documents relating to
      development 95t
  key dates for implementation 102–3t
  links to the Kyoto project-based
      mechanisms 126–39
    provisions 127–8
    reasons for exclusion of a direct link
        126–7
  and state aid rules 113
  trading regime 86
ET proposal
  adoption 99
  and links with the Kyoto project-based
      mechanisms 126
  relationship between the IPPC Directive
      and 122–5
ethical dimension of emissions trading 92–4
ETS xvii–xix, xxx, 7, 75–140, 160–2, 163,
      180, 231, 322
  adoption 139
  background to development 97–101
  effect of Russian non-ratification 220
  ET Directive *see* ET Directive
  and the IET 160–2
  implementation 140, 183–98
    monitoring and verification 119, 125,
        197
    NAPs *see* NAPs
    permitting procedures *see* permits, ET
        Directive procedure
    registries 197–8
  and JI 226–7
  and linkage xviii, xxx, 331, 332, 334
    desirability 335
    to the Norwegian emissions trading
        programme 328
  Linking Directive proposal *see* Linking
      Directive proposal
  openness xx
  reasons for xx
  short-term allocations and industry
      investment cycles 162
  start date 135
ETS installations, allocation of emission
      allowances 186
EU
  co-decision procedure 100
  comitology procedure 107, 126
  development of emissions allowance
      trading 95–101
  and emissions trading xx–xxi
  environmental policy 81–6
  and the evolution of the Kyoto mechanisms
      4, 5, 6–9

and the 'Green Investment Scheme' (GIS) 209
Kyoto targets 80, 185–6
negotiated agreements 84
ratification of the Kyoto Protocol 75–6, 99
see also accession countries
EU allowances *see* EUAs
EU burden-sharing agreement 77–81, 111–12
  role of the ET Directive and NAPs in achieving targets 116
EU case
  characteristics 170t, 171
  market features 177, 179t, 180
EU emissions allowance trading scheme *see* ETS
EU Guidelines on Allocations of Allowances 395–424
EU state aid rules, relation between the allocation of allowances and 112–13
EU-wide emissions trading 76
EUAs 131, 197, 198
  restrictions on the transfer 133
European Band for Reconstruction and Development (EBRD) 216
European Climate Change Programme *see* ECCP
European Economic Area (EEA) states, application of the ET Directive to 108
European Pollutant Emission Register (EPER) 123
Exchange Allowances (XAs) 238
Exchange Offsets (XOs) 238
Executive Board of the CDM *see* EB/CDM
expert review teams *see* ERTs

F
Facilitative Branch of the Compliance Committee 63–4
fee, moral difference between a fine and 93
financial sanctions 159
financial sector
  and the ETS xx
  role in the CDM process 309
financing JI projects 221
fine, moral difference between a fee and 93
Finland
  emission reduction target 80t
  JI projects 218
flexibility mechanisms *see* Kyoto mechanisms
foreign aid, weaning host countries from 318
formal reviews by the EB/CDM 41–2
France, emission reduction target 80t
free of charge allocation 89
  and the ET Directive 114

in the ETS 187
to new entrants 195
funding issues and the CDM 50–3
fungibility 17–18
  of units traded under the EU trading regime 131
future climate change regime 335–43
  international emissions trading 341–2
  participation 338–41
  stringency of commitments 336–8

G
G-77 xl
geographic imbalance in CDM projects 51–2, 266
Germany
  climate change policy 104, 105, 111
  emission reduction target 80t, 186
  ETS allocation process 188–90
GERT (Greenhouse Gas Emission Reduction Trading project) 240
GHGs
  coverage by trading regimes 90–2
  emissions *see* emissions
  types xxxvi
GIS ('Green Investment Scheme') 209, 221
Gold Standard 272–6
Gold Standard PDDs 276
Gold Standard technologies 273t
government capacity and the ETS 196
government-level participation requirements for the CDM 42–3
governments
  and IET 157–9, 163
  role in the CDM process 308
grandfathering 89
  in the ETS 187
  see also free of charge allocation
Greece, emission reduction target 80t
Green Aid Plan 243
'Green Investment Scheme' (GIS) 209, 221
Green Paper on emissions trading 96–8
Greenhouse Gas Abatement Scheme 233–4
greenhouse gas allowance trading regime *see* UK greenhouse gas allowance trading regime
Greenhouse Gas Emission Reduction Trading project (GERT) 240
greenhouse gases *see* GHGs
grid emission factors, calculation 292
Grubb's paradox 299–300, 302

H
Hague Missed Compromise case *see* MC case
Helio International's Criteria and Indicators for Appraising CDM Projects 270, 272t
HFC-23 projects 281–2

host Annex I Parties and JI 53, 54
host countries xli
   actions for JI to be successful 213
   approval of CDM projects 266
      criteria 270
   financial and technological flows to 298
   and foreign aid 318
   foundations for participation in the CDM
      305–19
   incentives to take up PAMs and targets
      299
host country governments, role in the CDM
      process 308
'hot air' 168, 180, 298, 302, 306
   as an incentive for broader participation
      339
   and the Central and Eastern European
      (CEE) countries 203–5
      keeping it out of the system 223–4
      potential for emission reductions
         without 225
   creation 94
   detrimental effect 5
   and Grubb's paradox 300
   limiting 207–9
hot spots 83, 94
Hungary
   AIJ projects 210t, 211t
   and JI
      memoranda of understanding 219t
      ranking of conditions on three different
         scales 218t
      scope for low-cost 217t
      status administratively 214t
   potentially available AAUs during the
      commitment period 202t, 203f
hydropower projects 280–1
   inclusion in the EU trading regime 134
   see also El Gallo hydro project

I
ICERs xli, 17, 48
   inclusion in the EU trading regime 134
ID case 180
   characteristics 169, 170t
   market features 171–3
IEs 53–4, 57
   and validation of Track 2 JI projects 59–60
IET xxix, xl, 9–10t, 25–8, 86, 337,
      341–2
   and Central and Eastern European (CEE)
      countries 200–29
      'hot air' 203–5
   distinctions between industry-based
      allowances trading regimes and
      157–9
   and EITs 200–3

   implementation 232–45
   and JUSCANNZ 6
   linkages 28, 76, 331–2, 344–5
   participation requirements 27
   principles and supplementarity 26–7
   relation between the ETS and 160–2
   restraints 28
   and stakeholder involvement 18–19
   suspension of eligibility 65–6
IET modalities 26
   adoption and review 14–15
IETA (International Emissions Trading
      Association) 309
IGOs (intergovernmental organizations) 34
impacts of GHG emissions xxxvi–xxxviii
'independent entities' *see* IEs
India
   and CDM 250–2, 280
   CDM awareness-building programme 313
   and emissions trading 243–4
indirect links between domestic GHG trading
      programmes 330
Indonesia
   and CDM 252–4, 280
   NSS 311–2, 317
   ratification of the Kyoto Protocol 308
industrial sectors
   allocation of emission allowances 186–7,
      193, 198
   coverage by the ET Directive 105–8
   in New Zealand 245
   see also electricity sector
industrialized country emission cuts 205–6
industry
   in accession countries 236
   and auctioning 89–90, 114
industry investment cycles, short-term
      allocations and 162
industry-based allowances trading regimes
      163
   distinctions between emissions trading
      under the Kyoto Protocol and
      157–9
ineligibility *see* non-eligibility
installations
   and allocation of emission allowances 187
      effect of closures 195
      new entrants 195
   coverage by the ET Directive 105–8
      overlap with the IPPC Directive 105,
         122
   identification 194
institution-building programmes for the CDM
      314–15
'institutional parallelism' and JI 54
integrated permitting 85
Integrated Pollution Prevention and Control

Directive (1996) *see* IPPC Directive
intergovernmental organizations (IGOs) 34
Intergovernmental Panel on Climate Change
    see IPCC
Interim Measures for CDM Project Activities
    248
international baseline and credit schemes 29
international CDM rules *see* CDM rules
international development agencies 259
international emissions trading agreements
    links with domestic 331–2
    see also ETS
International Emissions Trading Association
    (IETA) 309
international GHG emissions trading
    making the case 154–7
    see also IET
international procedures to derive baseline
    rules 292–6
international relations, IET as an instrument
    159
international rules on Kyoto mechanisms xxx,
    1–67
IPCC xxxvi–xxxvii, xxxix, 335–6
IPCC Revised Guidelines for National
    Greenhouse Gas Inventories (1996)
    21
IPPC Directive 81–2, 85, 196
    BAT criterion 123, 124
    regulation of GHG emissions through
        122–3
    relation of the ET Directive with 121–5
IPPC permits 81–2, 123, 196
    and ET permits 125
Ireland, emission reduction target 80t
issuance of CERs 29, 32, 42, 49–50
Italian Carbon Fund 260
Italy
    emission reduction target 80t
    JI projects 218

J
Japan
    and the Commission's approach to double
        counting 138–9
    and emissions trading 240–1
        programme for 326–7
    and the 'Green Investment Scheme' (GIS)
        209
    and JI 245
    support for China 243
    see also NEDO
JDET programme 326–7
JI xxix, xl, xli, 9t–10t, 53–60
    adoption and review of modality 14–15
    and EITs 6, 7, 200–3
    history and characteristics 54

identification of host country actions for
    success 213
implementation 244–5
institutions 54–7
linkage with the EU trading regime
    127–30
participation of Annex I Parties xl, 53
participation of Central and European
    (CEE) countries *see* Central and
    Eastern European (CEE) countries
participation/eligibility 57–8
UNFCCC reference to 4, 153
JI credits *see* ERUs
JI market, linkage with the ETS xxx
JI modalities 15, 54
JI projects 55
    in accession countries 138–9
        effect of the ETS 226–7
    in Central and Eastern European (CEE)
        countries 215–8
    conversion of AIJ projects into 13–14
    establishment of baselines 134, 221–2,
        227, 228–9
    oversight 20
    stakeholder involvement 18–19
JI Track 1 xli, 53, 58–9, 213
    eligibility criteria 58t
JI Track 2 xli, 53, 59–60, 213–4
    eligibility criteria 58t
'joint fulfilment' 79
joint implementation *see* JI
JUSCANNZ 4
    and IET 6
    and JI 54
    and mechanism eligibility 25, 66
        reinstatement of 67
    and the 'right to trade' 26

K
Kyoto Compliance Procedures 61–7
Kyoto Mechanism Fund 245
Kyoto mechanism modalities
    adoption and review 14–15
    see also CDM modalities; IET modalities;
        JI modalities
Kyoto mechanisms xxix–xxxi, xli–xlii,
    9–10t
    and AIJ projects 13–14
    compliance procedures and mechanisms
        65–7
    cross-cutting features *see* cross-cutting
        mechanisms issues
    evolution 4–10
    future 341–2
    implementation 166, 231–60
    international rules xxx, 1–67

Kyoto mechanisms *continued*
    linkage of the ETS with xxx, 8–9
    links between the ET Directive and the
        project-based mechanisms 126–39
    negotiations 1–2
        context and historical background
            4–10
        related COP decisions 3t
        trading through 166–80
        use by Argentina 256
        use by EU Member States 116
    see also CDM; IET; JI
Kyoto Protocol xxix, xli 152, 336
    compliance procedures and mechanisms
        61–7
    emission limitation commitments 336
    entry into force xxx, 286
    moral stance on GHG emissions 93
    ratification *see* ratification of the Kyoto
        Protocol
    role of the ET Directive and NAPs in
        achieving targets 116
    and sharing research results 339–40
Kyoto Protocol Initial Deal case *see* ID case
Kyoto targets, EU 80, 185–6
Kyoto units xli, xlii, 15
    and fungibility 17–18
    use in the EU trading regime 128, 129,
        131
    see also ERUs, CERs, AAUs, RMUs

L
land-use, land-use change and forestry *see*
    LULUCF
Latin America, CDM in 255–8
Latvia
    AIJ projects 210t, 211t
    and JI
        memoranda of understanding 219t
        ranking of conditions on three different
            scales 218t
        scope for low-cost 217t
        status administratively 214t
    potentially available AAUs during the
        commitment period 202t, 203f
lawyers, role in the CDM process 309–10
LDCs (least developed countries) 52
leakage issues for standard CDM projects 44
limitation commitments *see* commitments
Linking Directive proposal 8–9, 76, 127,
    130–37, 141, 231, 331
    double counting and JI in accession
        countries 137–9
    exclusion of sink projects 9
    how much to link 136–7
    how to link 131

    what projects to link with 131–5
    when to link 135–6
Lithuania
    AIJ projects 210t, 211t
    and the decommissioning of nuclear
        reactors 222
    and JI
        memoranda of understanding 219t
        ranking of conditions on three scales
            218t
        scope for low-cost 217t
        status administratively 214t
    potentially available AAUs during the
        commitment period 202t
long term policy choices 158
long-term CERs *see* ICERs
LULUCF 166
    constraints on additions to assigned
        amounts 16
    Good Practice Guidance 24
    inclusion in the EU trading regime 134
    sequestration achieved by xl, xlii
    see also afforestation/reforestation projects
LULUCF projects 60
Luxembourg, emission reduction target 80t

M
McCain Bill 238
Malaysia and CDM 250
mandatory emission trading schemes 259
Mandatory Renewable Energy Target (MRET)
    233
market-based mechanisms 4
    introduction in the EU xvii, 82–6
    replacement of command and control
        regulation with 85–6
    see also emissions trading
Marrakesh Accords xli, 1, 7
    adoption 8
    and allocation of tasks within the CDM
        project cycle 306, 307f
    and the Article 6 Supervisory Committee
        (A6SC) 55
    baseline approaches 290–1
    and the commitment period reserve (CPR)
        27
    and compliance procedures 25, 62, 159
    and the EB/CDM 34, 35
    and equity issues 15
    and fungibility 17
    and JI 54
    and supplementarity 16
    and sustainable development 265–6
MAUT (multi-attribute utility theory) 276
MC case
    characteristics 169, 170t
    market features 173–5

MEAs (multilateral environmental
    agreements) and trade measures
    against non-parties 340–1
measurability of GHG emissions 90–1
mechanism modalities *see* Kyoto mechanism
    modalities
mechanism-relevant compliance procedures
    25, 65–7
Methodology Panel 292–3, 294, 296
mitigation commitments xl, 9, 163
    and the Kyoto Protocol xxxii–xxxvii
    see also commitments
monitoring
    in an international context 5
    of the CDM project cycle 32
    and the ET Directive 119, 125, 197
    methodologies for CDM projects 45
monitoring plans for CDM projects 49
monoculture plantations 280, 281
Montreal Protocol 340
morally wrong, emission trading as 93
Morocco and CDM project approval criteria
    270
MRET (Mandatory Renewable Energy Target)
    233
multi-attribute utility theory (MAUT) 276
multi-attributive assessment of CDM 276–8
multi-project baselines 44
multilateral CDM projects 30
multilateral environmental agreements
    (MEAs) and trade measures against
    non-parties 340–1

N
N$_2$O emissions, measurability 91
NAPs (national allocation plans) xx, 113,
    114, 116–17, 118, 160, 184–96
    in accession countries 184, 222
        process concerns 223
        public participation 225–6
    achieving targets under the burden-sharing
        agreement and Kyoto Protocol 116
    allocation methods 184–7
    assessment 114–5
    drafting 115
    major challenges 192–6
        banking 195–6
        data constraints 194
        government capacity 196
        new entry and closure 195
        policy equivalence 194–5
        sectoral approaches 193
        total cap 193
    and state aid 113
National Committee on Climate Change
    (NCCC) 252–3
national communications xl

national emission caps *see* caps
national emissions, and the allocation of
    allowances 185–6
national emissions trading schemes 28
    approaches 232–45
    see also domestic GHG trading
        programmes
national ETS registries 197–8
national registries 22
    New Zealand as a source of expertise 245
    and stakeholder involvement 18–19
National Strategy Studies (NSS) programme
    310–12
national systems 21–2
NCCC (National Committee on Climate
    Change) 252–3
near term policy choices 158
NEDO 315, 316t
negotiated agreements in the EU 84
'negotiated command and control
    instruments' 84
Negotiated Greenhouse Agreements (NGAs)
    235
negotiations on the Kyoto mechanisms 1–2
    context and historical background 4–10
The Netherlands
    domestic measures 215
    emission reduction target 80t, 112, 186
    JI projects 216–7
    see also CERUPT; Energy Efficiency
        Benchmarking Covenant; ERU
        Procurement Tender (ERUPT); NO$_X$
        trading regime
The Netherlands Clean Development Facility
    260
new entrants to the ETS
    and allocation of allowances 195
    and permitting procedures 196
New South Wales *see* NSW
New Zealand
    and emissions trading 234–6
    and JI 244–5
        memoranda of understanding 219t
NGAs (Negotiated Greenhouse Agreements)
    235
NGOs 18
    and the EB/CDM 34, 41
    role in the CDM process 306, 310
    see also environmental NGOs
Nicaragua and CDM project approval criteria
    271
nitrous oxides trading regime *see* NO$_X$ trading
    regime
non-Annex I Parties xl
    and CDM xli, 20, 29
    and the Compliance Committee 62, 63
    and mitigation commitments xxxviii, 29

non-Annex I Parties *continued*
  and national registries 50
  representation on the EB/CDM 35
non-compliance
  and the COP/MOP 61
  ET Directive's penalty regime 120
  and financial sanctions 159
  penalties 333
  sanctions in the ET Directive 119–20
  see also compliance
non-eligibility with participation/eligibility
  conditions 65–6
non-governmental organizations *see* NGOs
non-Kyoto market for credits 264
non-parties to the Kyoto Protocol 20
  agreements with xxx, 341
  trade measures against 340–1
non-trading sectors contribution to GHG
  emissions 158, 186
Norway, emissions trading programme 327–8
NO$_X$ trading regime 88, 124
  and state-aid 113
NSS (National Strategy Studies) programme
  310–12
NSW, emissions trading 233–4
NSW Abatement Certificates 234
NSW Greenhouse Gas Abatement Scheme
  (NSW Scheme) 233–4
nuclear projects
  and the CDM 42
  inclusion in the EU trading regime 131,
  133
  as JI projects 60
nuclear reactors, decommissioning 222

O
OAMDL (*Oficina Argentina del Mecanismo
  para un Desarollo Limpio*) 255–6
observers at the EB/CDM 37, 41
ODA (Official Development Assistance) xxxviii
  Gold Standard approach 274
  use to fund CDM projects 44, 50–1,
  267–8
OEs, training of domestic 317
oil-producing DCs xl–xli
  and the EB/CDM 33, 35
Operational Entities (OEs), training of
  domestic 317
opt-out choice in the ETS 194–5
optional emissions budgets 338

P
Pacific Islands, provision of overseas
  development aid 245
PAMs xl, 11
  and the EU 5
  future direction 338
Panels established by the EB/CDM 38

participation, broadening 338–41
participation/eligibility requirements 19–25
  and acceptance of compliance procedures
  25
  annual inventories & adjustments 21,
  22–4
  commitment period reserve (CPR) 27–8
  commitment period reserves (CPRs) 24
  DNAs *see* DNAs
  eligibility assessment, consequences and
  reinstatement 25
  establishing assigned amounts and
  pre-commitment period reports 21
  for IET 27
  for JI 57–8
  national registries *see* national registries
  national systems 21–2
  non-eligibility 65–7
  for PPs in the CDM project cycle 43
  Protocol ratification *see* ratification of the
  Kyoto Protocol
  supplemental information 24
  see also eligibility requirements; public
  participation
Parties to the Kyoto Protocol xl
  see also Annex B Parties; Annex I Parties;
  Annex II parties; host Annex I
  Parties; non-Annex I Parties
PCF 8, 216, 217, 260, 279
  screening processes 278
PDDs
  for CDM projects 31
  validation process 31–2, 41
  for Track 2 JI projects 59–60
  see also draft PDDs; Gold Standard PDDs
penalties for non-compliance 333
  in the ET Directive 120
  see also sanctions
permits 4
  creation at the international level 4
  ET Directive procedure 108–9, 196
    relation with IPPC Directive procedure
    125
  use in command and control regulation 81
  see also integrated permitting; IPPC
  permits
The Philippines and CDM 254–5
Pilot Emission Reductions Trading project
  (PERT) 240
Plantar 281
Plenary of the Compliance Committee 62
Poland
  AIJ projects 210t, 211t
  and the ERU Procurement tender (ERUPT)
  216
  and JI
    memoranda of understanding 219t

ranking of conditions on three scales
218t
scope for low-cost 217t
status administratively 214t
PCF projects 217
potentially available AAUs during the
commitment period 202t, 203f
POLES model 167
policies and measures *see* PAMs
pollution
IPPC Directive definition 123
as morally wrong 93
see also GHG emissions
pooling 104
portfolio CDM projects 30
Portugal, emission reduction target 80t
PPMs (production processes or methods),
trade restrictions based on 340
PPs
for CDM projects 30-2
participation requirements 43
for JI projects 59-60
pre-commitment period reports 21
'pre-validation' of the CDM project cycle
31-2, 44, 49
principles of IET 26-7
printed CDM manuals 318-9
private sector
and CDM projects 8, 29
and the ETS xx
production cost model 295
production processes or methods (PPMs),
trade restrictions based on 340
Project Design Documents *see* PDDs
project developers for the CDM process
308-9
training courses 317
Project Participants *see* PPs
project-based mechanisms *see* CDM; JI
project-specific baselines 44, 290-91
Projects Mechanism, New Zealand 235-6
'prompt start', CDM 33, 46, 52
Prototype Carbon Fund *see* PCF
public funding for CDM projects 50-1
public participation
in accession country NAPs 225-6
in CDM projects 18-19, 41, 267, 285
and the ET Directive 125

R
ratification of the Kyoto Protocol 19-20,
308, 345n
by Australia 166, 232-3
by the EU 76-7, 99
by Indonesia 308
by Russia 166, 180, 204-5, 209, 220,
226-7, 264

by the US 7, 99, 264, 286
RECs ('Renewable Energy Certificates') 233
reforestation/afforestation projects xli, 47-8
'Regional Greenhouse Gas Initiative' (RGGI)
237
registered monitoring plans for CDM projects
49
registration
of the CDM project cycle 31-2, 41-2
of CDM projects 42-8
registries *see* national ETS registries; national
registries
'regulatory additionality' 221-2
reinstatement of mechanism eligibility 67
relative trading regimes
linking absolute trading regimes with 88
v. absolute trading regimes 86-8
see also baseline and credit regimes
removal units *see* RMUs
renewable energy 228-9
in Australia 233
baseline methodologies for projects 295-6
in Japan 240
'Renewable Energy Certificates' (RECs) 233
Renewable Energy (Electricity) Act (2002)
(REEA) 233
reporting commitments 5
and the ET Directive 119, 125-6
research and development, cooperation on
339-40
Revised Guidelines for National Greenhouse
Gas Inventories (IPCC) (1996) 21
RGGI ('Regional Greenhouse Gas Initiative')
237
'right to pollute' 90
Rio Declaration xxxvii, 267
RMUs xlii, 24
fungibility 17
use in the EU trading regime 137, 332
Romania
AIJ projects 210t, 211t
and the ERU Procurement Tender
(ERUPT) 216
and JI
memoranda of understanding 218t
ranking of conditions on three scales 218t
scope for low-cost 217t
status administratively 214t
potentially available AAUs during the
commitment period 202t
Russia
AIJ projects in 210t, 211t, 212
and the ERU Procurement Tender
(ERUPT) 216
and JI
memoranda of understanding 219t
ranking of conditions on three scales
218t

Russia *continued*
    scope for low-cost 217t
    status administratively 214t
    potentially available AAUs during the
        commitment period 201, 202t
    and ratification of the Kyoto Protocol
        166, 180, 204–5, 209, 220, 226–7,
        264
    use of emissions trading revenue 208–9

S
'safety valve' for emissions trading 342
sanctions
    in the ET Directive 119–20
    see also financial sanctions; penalties for
        non-compliance; trade sanctions
SBSTA (Subsidiary Body for Scientific and
        Technological Advice) 38, 50
sector-based CDM 286
sequestration achieved by LULUCF xl, xli
share of proceeds provision in CDM 29, 50,
        52–3
short-term allocations and industry
        investment cycles 162
sinks 166, 180
    availability 7
    inclusion
        in the CDM 1–2, 7, 9, 47
        in the EU trading regime 9, 131,
            133–4
    provision of accurate data on 24
    see also afforestation/reforestation projects;
        LULUCF projects
Slovak Republic
    AIJ projects 210t, 211t
    and the decommissioning of nuclear
        reactors 222
    and JI
        memoranda of understanding 219t
        ranking of conditions on three different
            scales 218t
        scope for low-cost 217t
        status administratively 214t
    potentially available AAUs during the
        commitment period 202t, 203f
Slovenia
    AIJ projects 211t
    and JI
        memoranda of understanding 219t
        ranking of conditions on three different
            scales 218t
        scope for low-cost 217t
        status administratively 214t
    potentially available AAUs during the
        commitment period 202t, 203f
small-scale afforestation/reforestation projects
        48

small-scale CDM projects 46
small-scale JI projects 60
small-scale projects
    barrier assessment for additionality 300
    baseline rules 291–2
SO_2 allowances trading programmes
    China 242–3
    US 4, 153–4, 162
South Africa and CDM project approval
        criteria 270
South Korea and CDM 280
SouthSouthNorth Network 271
    modular programme for training courses
        317
Spain, emission reduction target 80t
stakeholder involvement 18–19, 41, 267, 285
    in the CDM project cycle 41, 43
    Gold Standard approach 275–6
    see also public participation
standardised baselines 290
state aid rules in the EU, relation between the
        allocation of allowances and
        112–3
Subsidiary Body for Scientific and
        Technological Advice (SBSTA) 38,
        50
sulphur dioxide trading programmes *see* SO_2
        allowances trading programmes
supplemental information on assigned
        amounts 24
supplementarity 16, 136
    and IET 26–7, 204
supplementarity restrictions on technology
        and innovation 207
suspension, cases of 67
sustainability matrix 275
sustainable development and CDM 263–86,
        305
    contribution 265–9
    contribution of the CDM market 280–3
    definition of criteria 316–7
    future issues and options 284–6
    Gold Standard approach 274–6
    tools to assess eligibility and sustainability
        269–78
Sutter's Sustainability Check-Up for CDM
        Projects 276–8, 317
Sweden
    emission reduction target 80t, 186
    and the ETS 198
Switzerland
    emissions trading programme 328–9
    participation in the ETS 108

T
TAR (Third Assessment Report) (2001)
    xxxvi

taxation measures 248
  see also energy taxes; environmental
    taxation
tCERs xli, 17, 48
  inclusion in the EU trading regime 134
TEC (Total Emissions Control) policies
    241–2, 243
technology development
  cooperation on 339–40
  emissions trading
    as an incentive 92
    as a disincentive 94
  relationship between environmental policy
    and 206–7
  see also Gold Standard technologies
temporary CERs *see* tCERs
Thailand and CDM 250
Third Assessment report (TAR) (2001) xxxvi
Total Emissions Control (TEC) policies
    241–2, 243
Track 1 *see* JI Track 1
Track 2 *see* JI Track 2
trade sanctions 340–1
Trading among Committed Countries case *see*
    CC case
Trading restricted to Europe case *see* EU case
trading restrictions 334
transaction costs of CDM projects 299, 306
transaction logs 22, 28
Triptique Approach 78–9
Triptique variants 78–9
'tropical hot air' *see* 'hot air'

U
UK
  emission reduction target 80t, 186
  ETS allocation process 190–2
UK greenhouse gas allowance trading regime
    88, 124
  coverage 91, 92
  establishment of targets and coverage 90
  state aid aspects 113
Ukraine
  AIJ projects 210t, 211t
  and JI
    memoranda of understanding 219t
    ranking of conditions on three scales
      218t
    scope for low-cost 217t
    status administratively 214t
  potentially available AAUs during the
    commitment period 202t
Umbrella group 5
UN Conference on Environment and
    Development (UNCED) xxxvii

UN Millennium Development Goals xxxi
UNCTAD, CDM awareness-building
    programme 312t
UNDP
  CDM awareness-building programme
    312t
  CDM institution-building programme
    314t
UNEP, CDM institution-building programme
    314t
UNFCCC xxix, xxxvii–xxxix, 335, 336
  and JI 4, 11, 153
  main institutions xxxix
UNIDO, CDM awareness-building
    programmes 312–13t, 314
unilateral CDM projects 30
unilateral links between domestic GHG
    trading programmes 330, 332
United Kingdom *see* UK
United Nations Framework Convention on
    Climate Change *see* UNFCCC
upstream trading regimes 91, 333
US 180
  agreements with 341
  CDM awareness-building programme in
    India 313
  and emissions trading 236–8
  and the evolution of the Kyoto mechanisms
    3–4, 171
  proposals for cap-and-trade scheme 237–8
  ratification of the Kyoto Protocol 7, 99,
    264, 286
  support for China 242–3
  see also Bush Administration; NO$_X$
    trading regime; SO$_2$ allowances
    trading programme, US

V
Vale de Rosario bagasse co-generation
    methodology 296
validation
  of the CDM project cycle 31–2, 41–2
  of CDM projects 42–8
  of Track 2 JI projects 59–60
validators, role in the CDM process 309
verification
  in an international context 5
  of the CDM project cycle 32, 49
  and the Et Directive 119, 125, 197
verification reports for CDM projects 49
voluntary agreements (VAs) 103
  in Canada 103, 240
  in the EU 84
  in Germany 104, 105
  in Japan 240–1

voluntary agreements (VAs) *continued*
    policy equivalence between ETS allocations
        and 194–5
    in the UK 103
    in the US 238–9

W
women
    membership of the Article 6 Supervisory
        Committee (A6SC) 56
    membership of the EB/CDM 35
Working Group 1 on Flexible Mechanisms
        (WG 1) *see* ECCP Working Group
        1 on Flexible Mechanisms (WG 1)

workshop tourism 318
World Band NSS (National Strategy Studies)
        programme 310–12
World Bank CDM institution-building
        programme 314t
World Bank Prototype Carbon Fund *see* PCF
World Summit on Sustainable Development
        (WSSD) agreement 267
World Trade Organization (WTO) and trade
        sanctions 340–1

X
XAs (Exchange Allowances) 238
XOs (Exchange Offsets) 238